D1006778

NEW PATHWAYS IN
INORGANIC CHEMISTRY

HARRY JULIUS EMELÉUS

NEW PATHWAYS IN
INORGANIC CHEMISTRY

EDITED BY

E. A. V. EBSWORTH
A. G. MADDOCK
A. G. SHARPE

CAMBRIDGE
AT THE UNIVERSITY PRESS
1968

Published by the Syndics of the Cambridge University Press
Bentley House, 200 Euston Road, London N.W. 1
American Branch: 32 East 57th Street, New York, N.Y. 10022

© Cambridge University Press 1968

Library of Congress Catalogue Card Number: 68–26984
Standard Book Number: 521 07254 9

Printed in Great Britain
at the University Printing House, Cambridge
(Brooke Crutchley, University Printer)

546
N532

11. 70

Richard Abel & Co.

25 Nov, 1969

Dedicated to

HARRY JULIUS EMELÉUS

to commemorate his sixty-fifth birthday and his
contributions to inorganic chemistry

CONTENTS

List of authors   *page* xv

Preface by E. A. V. EBSWORTH, A. G. MADDOCK and
A. G. SHARPE   xvii

CHAPTER I

Organometallic cations
by Howard C. Clark

1 Organometallic ions in solution   2

2 Organometallic ions in the solid state   3

3 Organometallic cations of Group III metals   4

4 Organometallic cations of Group IV metals   6

5 Organometallic cations of Group V metals   10

6 Nature of the cation–anion interactions   10

CHAPTER 2

Unusual coordination numbers
by A. J. Downs

1 Introduction   15

2 Experimental methods   18

3 Conditions affecting coordination numbers   24

   3.1 Size and steric effects   24

   3.2 Nature of the central atom   26

   3.3 Electrostatic effects   30

   3.4 Nature of ligand   33

CHAPTER 3

Gallium hydride and its derivatives
by N. N. Greenwood

1   Introduction   *page* 37

2   Tetrahydridogallates, $MGaH_4$ 38

3   Trimethylamine adducts of GaH_3 40

4   Other amine derivatives of GaH_3 43

5   Tertiary phosphine derivatives of GaH_3 47

6   Other phosphine derivatives of GaH_3 49

7   Adducts of GaH_3 with arsines   53

8   Adducts with oxygen and sulphur donors   53

9   Trimethylamine adducts of halogenogallanes, $GaH_{3-n}X_n$ 55

10 Uncoordinated gallium hydride and halogenogallanes   59

11 Mixed hydrides of boron and gallium   61

12 Conclusion   62

CHAPTER 4

Properties of donor solvents and coordination chemistry in their solutions
by V. Gutmann

1   Introduction   65

2   The donor number   66

3   The donor number of the solvent and the formation of
    anionic complexes   70

4   The donor number of the solvent and ionization of halides   73

5   The donor number and autocomplex-formation   73

6   Other factors influencing coordination chemistry in solution   74

7 The coordination forms of cobalt(II) towards different X⁻ ligands in different solvents   *page* 77

   7.1 In nitromethane ($DN_{SbCl_5} = 2·7$)   77

   7.2 In acetonitrile ($DN_{SbCl_5} = 14·3$)   78

   7.3 In propane-1,2-diolcarbonate ($DN_{SbCl_5} = 15·1$)   78

   7.4 In ethylene sulphite ($DN_{SbCl_5} = 15·3$)   79

   7.5 In acetone ($DN_{SbCl_5} = 17$)   80

   7.6 In water ($DN_{SbCl_5} = 18$) and ethyl alcohol   80

   7.7 In trimethylphosphate ($DN_{SbCl_5} = 23$)   81

   7.8 In *NN*-dimethylacetamide (DMA) ($DN_{SbCl_5} = 27·8$)   82

   7.9 In dimethylsulphoxide ($DN_{SbCl_5} = 29·7$)   83

8 Conclusion   84

CHAPTER 5

Perfluoropseudohalides and the chemistry of chlorofluoromethylsulphenyl compounds

by Alois Haas

1 Introduction   87

2 Preparation and properties of compounds having the general formulae   90

   2.1 FX   90

   2.2 $F_nY^mX_{m-n}$ 91

   2.3 $F_n(YZ)^{m-2}X_{m-(n+2)}$ 94

   2.4 $(CF_3)_n(YZ)^{m-2}X_{m-(n+2)}$ 96

   2.5 $(CF_3)_nY^mX_{m-n}$ 96

3 Preparation and properties of $Cl_{3-n}F_nCS$-pseudohalides   102

4 Contributions to CF_3S-chemistry   104

5 Infrared and ${}^{19}F$ n.m.r. spectroscopic investigations of $Cl_{3-n}F_nCS$-compounds   105

6 Summary   110

CHAPTER 6
Polyfluoroalkyl silicon compounds
by R. N. Haszeldine

1 Introduction *page* 115

2 Synthesis 116

3 Nucleophilic attack on silicon in polyfluoroalkylsilicon
compounds 120

   3.1 Amine complexes 120

   3.2 Hydrolysis 121

4 Polyfluoroalkyl silicon polymers 121

5 Thermal breakdown of polyfluoroalkyl silicon compounds 123

   5.1 γ-Fluoro-compounds 123

   5.2 β-Fluoro-compounds 124

   5.3 α-Fluoro-compounds 125

6 Carbenes from polyfluoroalkyl silicon compounds 132

7 General comment 134

CHAPTER 7
Fluoroalkylmercurials
by J. J. Lagowski

1 Introduction 137

2 Coordination chemistry 139

3 Fluoroalkylmercurials as synthetic intermediates 143

   3.1 Aluminum 143

   3.2 Boron 144

   3.3 Zinc 145

   3.4 Phosphorus 145

CHAPTER 8

Catenation in inorganic silicon compounds
by A. G. MacDiarmid

1 Introduction   *page* 149

2 Synthesis of the silicon–silicon bond   150

  2.1 Primary syntheses   150

  2.2 Secondary syntheses   153

3 Parent polysilanes and their derivatives   155

4 Physical properties associated with the silicon–silicon bond   163

5 Cleavage of the silicon–silicon bond   167

CHAPTER 9

Metal–metal interaction in paramagnetic clusters
by R. L. Martin

1 Introduction   175

2 The exchange effect and spin–spin coupling   177

3 HDVV model of spin–spin interaction   181

  3.1 Binuclear clusters   182

  3.2 Trinuclear clusters   186

  3.3 Tetranuclear clusters   190

4 Mechanisms of spin–spin coupling   191

  4.1 Historical speculations   192

  4.2 Origin of superexchange   193

  4.3 Metal–metal interactions   200

  4.4 Magnetic dipolar coupling   210

5 Experimental results   211

  5.1 Dimers   211

  5.2 Trimers   218

  5.3 Tetramers   227

CHAPTER 10

Amides as non-aqueous solvents
by R. C. Paul

1 Solubility of substances in amides   *page* 234

2 Formation of solvates in amides   238

3 Infrared spectral study of the complexes of amides with Lewis
   acids   240

   3.1 Dimethylformamide–Lewis acid systems   241

   3.2 Formamide–Lewis acid systems   242

   3.3 Acetamide–Lewis acid systems   243

4 Conductance studies in amides   244

   4.1 Lewis acids   244

   4.2 Nitrogen bases   248

   4.3 Protonic acids   252

   4.4 Quaternary ammonium salts   256

5 Solvolytic reactions in amides   257

6 Acid-base neutralization reactions in amides   258

CHAPTER 11

Defect aggregation in solid state chemistry
by A. L. G. Rees

1 The interaction of point defects   263

2 Statistics of defect populations   266

3 Premonitory effects   270

4 A premonitory model of melting   271

5 Conclusion   281

CHAPTER 12

Transition metal derivatives of silicon, germanium, tin and lead

by F. G. A. Stone

1	Introduction	*page* 283
2	Synthesis	284
3	Structure and bonding	290
4	Chemical reactivity	296
5	Conclusion	299

CHAPTER 13

Chemistry of coordination compounds of Schiff bases

by B. O. West

1	Introduction	303
2	The synthesis of complexes	304
3	The mechanism of the formation of Schiff base complexes	305
4	Stereochemical properties of Schiff base complexes	307
5	Complexes with bidentate ligands	308
6	Complexes with tridentate ligands	314
7	Complexes with tetradentate ligands	315
8	Complexes with pentadentate and hexadentate ligands	318
9	The stability of Schiff base complexes	319
	9.1 Dissociative stability	319
	9.2 Metal exchange	320
	9.3 Stability to hydrogenation of the C=N group	321
10	Future developments	322

CHAPTER 14

Fluorosulphates

by A. A. Woolf

1   Comparison of fluorosulphates with parent fluorides   *page* 327

2   Preparation of fluorosulphates   337

    2.1   Direct insertion of sulphur trioxide   337

    2.2   Reactions in fluorosulphuric acid   337

    2.3   Reactions in other solvents   339

    2.4   Radical reactions   340

    2.5   Other methods   341

3   The HF–SO$_3$ systems   342

4   Carbonium fluorosulphates   348

5   Halogen fluorosulphates   351

6   Hydrolytic stability of fluorosulphates   354

7   Thermal stability of fluorosulphates   355

8   Metal fluorosulphates   357

9   Conclusions   359

Author Index   363

Subject Index   376

LIST OF AUTHORS

Professor H. C. Clark. *Department of Chemistry, University of Western Ontario.*

Dr A. J. Downs. *Department of Inorganic Chemistry, University of Oxford*

Professor N. N. Greenwood. *Department of Inorganic Chemistry, University of Newcastle-upon-Tyne*

Professor V. Gutmann. *Institut für Anorganische Chemie, Technische Hochschule Wien*

Dr A. Haas. *Anorganisch-Chemisches Institut der Universität, Gottingen*

Professor R. N. Haszeldine. *Department of Chemistry, The University of Manchester Institute of Science and Technology*

Professor J. J. Lagowski. *Department of Chemistry, University of Texas*

Professor A. G. MacDiarmid. *The John Harrison Laboratory of Chemistry, University of Pennsylvania*

Professor R. L. Martin. *Department of Inorganic Chemistry, University of Melbourne*

Professor R. C. Paul. *Department of Chemistry, University of Panjab, Chandigarh*

Dr A. L. G. Rees. *Division of Chemical Physics, C.S.I.R.O., Melbourne*

Professor F. G. A. Stone. *Department of Inorganic Chemistry, University of Bristol*

Professor B. O. West, *Department of Chemistry, Monash University, Melbourne*

Dr A. A. Woolf. *Department of Chemistry, Bath University of Technology*

PREFACE

Harry Julius Emeléus went to Imperial College from Hastings Grammar School, and immediately after graduation began research in the Laboratory of H. B. Baker. His first published work was a paper with W. E. Downey[1] on the luminescent oxidation of phosphorus, in which it was shown that the glow of phosphorus and the flame of the element burning in air have the same ultraviolet spectrum. Subsequently it was shown that the combustion of phosphorus(III) oxide and phosphine also produced the same spectrum, and the glow was attributed to slow oxidation of phosphorus(III) oxide formed in a preliminary non-luminous oxidation.[2, 3] Studies of the inhibition of the glow by organic vapours were also made.[4]

Further spectroscopic studies were made on the phosphorescent flames of carbon disulphide,[5] ether,[5] arsenic,[6] sulphur,[7] and (with R. H. Purcell) the phosphorescence of phosphorus(V) oxide illuminated with ultraviolet light.[8] A paper on light emission from phosphorescent flames of ether, acetaldehyde, propionaldehyde and hexane also includes vapour pressure data for these substances at low temperatures.[9]

The visit to Karlsruhe to work with Stock was in many ways the turning-point in Emeléus's career, introducing him as it did to the great German exponents of preparative inorganic chemistry. Nothing was published in his name from Karlsruhe, however, and his next papers came from Princeton, where he held a Commonwealth Fellowship and collaborated with H. S. Taylor, first on the photosensitization by ammonia of the polymerization of ethylene,[10] and later on the photochemical decomposition of amines,[11] the photochemical interaction of amines and ethylene,[11] and the photochemical reaction of carbon monoxide with ammonia and with amines.[12]

Sir Hugh Taylor has contributed the following recollections of Emeléus as a visitor at Princeton.

'In the years 1929–31, Harry Julius Emeléus arrived at Princeton University from England as a Commonwealth Fund Fellow in the Department of Chemistry. He came rich in research experience gained in London University and in Germany as an 1851 Exhibition Scholar with Professor Stock. He joined a distinguished group of research students who had gathered from all parts of the world to enjoy the new research facilities available in the Frick Chemical Laboratory of the University. The opening of the research laboratories in September 1929 coincided with the arrival of Eméleus. The Auditorium had already been in operation during the last term of the 1928–9 academic year, a fire in the John C. Green School of Science having destroyed the lecture halls for Freshman Chemistry in the old building at the beginning of 1929, thus necessitating the early use of a part of the new quarters being erected. The Inaugural Lectures at the official opening of the new Laboratory were addressed to the theme: The Rates of Chemical Reactions. A distinguished group of European and American chemists assembled for the occasion. Professor Max Bodenstein, Dr Michael Polanyi, and Dr Karl Bonhoeffer were among the speakers from Germany. Chain reactions, the parahydrogen conversion and atom–molecule reactions were the themes of the German addresses. Nobel Laureate Jean B. Perrin, the delegate from France, was unable to be present but was represented by his son, Francis Perrin, who addressed himself to the question of unimolecular reactions. Professor F. G. Donnan, from University College, London, gave the general Inaugural Lecture, and Professor Sir James Irvine of St Andrews described his work on the sugars. Irving Langmuir was the representative of the United States and gave a résumé of his researches on reactions at solid surfaces and the spreading of films on water surfaces. Professors Bodenstein, Donnan, Irvine and Perrin, and Dr Langmuir were the recipients of honorary degrees. It was a gala occasion with which to welcome the new Commonwealth Fellow from England whose interests lay in inorganic chemistry and chemical kinetics.

'There were research students from many countries in the "golden age" of physico-chemical research at Princeton that spanned the decade from the opening of the Frick Laboratory to the outbreak of World War II. Robert Spence, who spent three years from 1928–1931 as Commonwealth Fellow, was a Durham University organic chemist, who metamorphosed into a physical chemist during his stay, and returned to Britain to academic pur-

suits and ultimate high place in the United Kingdom Atomic
Energy Authority at Harwell; Montefiore Barak, also a Commonwealth Fund Fellow and a Ph.D. from England, accompanied
Eméleus in the years 1929–31 and returned to a professional
scientific career in British industry; M. G. Evans succeeded these
in 1933–4 as Rockefeller Foundation Fellow. He had worked in
Manchester with Professor Michael Polanyi and subsequently
occupied a Chair of Physical Chemistry in Leeds until his early
death terminated his brilliant career; Hans L. J. Bäckström from
Sweden worked in the old laboratories and in Frick from 1927 to
1930; George Kistiakowsky, a refugee from Russia and coming
from Bodenstein's laboratory in Berlin, was also in residence from
the same period until his departure to Harvard University in 1930;
James Kenneth Dixon was a National Research Council, U.S.A.,
Fellow during the opening year of the laboratory. They formed a
goodly company to welcome Henry Eyring to Princeton for
lectures in the spring of 1931 and to his continuous service from
late 1931 to his professorial years from 1936 to 1946. Hetero-
geneous catalysis, adsorption, photochemical and photosensitized
reactions, and reaction rate theory were the principal topics of
research in those beginning years.

'Three communications of Eméleus to the *Journal of the
American Chemical Society* in 1930 and 1931 indicate the range of
his researches in Princeton, the earliest one on 'Photosensitization
by Ammonia', the second on 'The Photochemical Interaction of
Ethylene and Ammonia', and a final communication on 'The
Photochemical Decomposition of Amines and the Photochemical
Interaction of Amines and Ethylene'. The topics of research in the
Physical Chemistry Section of the Laboratory during the period of
Eméleus's stay show the problems under active investigation. In
those he shared by reason of the intimacy of contact between
graduate research workers and faculty. These topics included the
reaction of atomic hydrogen with hydrocarbons; the photosensitized
and photochemical decomposition of hydrazine; the reactions of
hydrocarbons under the influence of excited mercury; the thermal
decomposition of metal alkyls yielding radicals which initiated
polymerization of alkenes; the surface reactions of atoms and
radicals; the chain characteristics of the ethylene–oxygen reaction,

and of the photo-polymerization of styrene and vinyl acetate; catalytic reactions of sulphur compounds present in petroleum. In the theoretical field there was a study of photodecomposition of molecules having diffuse band spectra, and the initial phases of the concept of adsorption processes requiring activation energies. In 1930–1 the measurements of the velocity of adsorption processes on oxide catalysts were in full swing. The parahydrogen conversion at surfaces was being measured to distinguish between physical, or van der Waals, and chemical sorption. Kistiakowsky showed, in the initial research coming out from the new laboratory, that the reaction between oxygen atoms and carbon monoxide required a measurable activation energy in contrast to the very low energies of activation reported by Polanyi in the Inaugural Lectures for the reactions of sodium atoms with alkyl chlorides, bromides, and iodides. The reactions between sodium atoms and fluorides were later shown by Polanyi's students to require considerable activation energies. It was the eve also of Eyring's advent to Princeton during which the whole area of the theory of rate processes in chemistry was fashioned. This was the setting which the Frick Laboratory provided in the early 'thirties in which Eméleus could enrich his background and prepare for the great career in chemistry which lay ahead.

'One other aspect of his life in Princeton must be recorded, for it held a fateful significance for his future life. For exercise and entertainment, relaxation and leisure, Harry Eméleus joined a folk-dancing group in which one of the chemistry professors, a Scot, Alan W. C. Menzies, was an active member. Early in the first year of Eméleus's residence in Princeton he met Mary Catherine Horton, a member of the Princeton University community who shared his enthusiasm for the folk-dancing. From dancing partners they quickly became an engaged couple looking forward to matrimony. The harsh regulations of the Commonwealth Fellowship officials stood in their way, since it was a rule of the fellowship that fellows should not marry during the term of the award; and all fellows must return to the country from which they had come to the United States. There was therefore a long period of waiting for the Lawrenceville girl and her English gallant. The rules, however, were observed meticulously and, if memory serves me right, the two happy people were married in New York on their way to the boat which would take them, man and wife, back to England. If memory still serves, an officer of the Commonwealth Foundation in New York City sponsored the happy event.'

After his return to Imperial College Emeléus continued to work on subjects similar to those which had engaged his earlier attention, but the influence of his visits to Karlsruhe and Princeton was quickly apparent. With G. H. Cheeseman, he investigated the ultra-violet spectra of phosphine, arsine, and stibine,[13] and with H. L. Riley he studied the luminous reduction of selenium dioxide by organic compounds.[14] Emeléus and R. G. H. Damerall[15] examined the oxidation of arsenic in nitrous oxide and in oxygen, and Emeléus and L. J. Jolley[16, 17] the kinetics of the thermal and photochemical decomposition of methylamine. In the first of many publications on the chemistry of the silicon hydrides and their derivatives, Emeléus and K. Stewart described the kinetics of the oxidation of mono-silane,[18] and this investigation was later extended to the oxidation of disilane and trisilane.[19]

Quite different work carried out at the same time involved the investigation of the isotopic composition of water from different operations and sources. In collaboration with F. W. James, A. King, T. G. Pearson, R. H. Purcell, and H. V. A. Briscoe it was found that distillation, freezing, and adsorption were all capable of effecting some degree of separation of isotopes, and the very accurate density determination method developed for this work was then used to show that naturally occurring water exhibits a small variation in deuterium content.[20] In an important paper with J. S. Anderson and the foregoing collaborators, it was shown that whereas N-hydrogen atoms in methylamine undergo rapid exchange with heavy water, hydrogen in ammine complexes of cobalt(III) exchanges only very slowly.[21] In related work with E. R. Roberts and H. V. A. Briscoe, a number of deuterated amines were prepared and their physical properties examined.[22, 23]

Except for a single communication on the decomposition of tris-trifluoromethyl arsenic, the last of Emeléus's papers on kinetics were published just before, and during, the War. With L. A. Bashford and H. V. A. Briscoe he investigated the oxidation of chloroacetylene and bromoacetylene,[24] with E. R. Gardner he examined the oxidation of monogermane and digermane,[25] and with H. H. G. Jellinek he studied the kinetics of pyrolysis of digermane.[26] Thereafter he became increasingly involved in the field of preparative inorganic chemistry, in which an unerring instinct reinforced his scientific ability, and in which, by virtue of his chapters in Emeléus and Anderson's *Modern Aspects of Inorganic Chemistry*, he at once became an influential contributor.

In a series of investigations on derivatives of silane, A. G. Maddock, C. Reid, and Eмеléus prepared mono- and di-iodo derivatives by the interaction of silane, hydrogen iodide, and aluminium iodide, and made the fluorides SiH_3F and SiH_2F_2 by the action of antimony trifluoride on the corresponding chlorides.[27–30] Eméléus and Maddock later isolated the hydride Si_4H_{10} from the products of the reaction of magnesium silicide with dilute hydrochloric acid and examined its properties.[31] Related work carried out in this period included studies of the reactions of the chloride SiH_3Cl with amines (with N. Miller),[32] and of the thermal and photochemical oxidation of the chlorosilanes (with A. J. E. Welch).[33] Several investigations of the preparation of fluoro-derivatives of non-metals by halogen exchange were also made, notably of alkyl and aryl silicon fluorides (with C. J. Wilkins),[34] of similar derivatives of sulphur, selenium, and tellurium (with H. G. Heal),[35] of alkyl difluoroarsines (with L. H. Long),[36] and of carbonyl and sulphuryl chlorofluorides (with J. F. Wood).[37] Many alkyl and aryl silicon compounds were made by the action of Grignard reagents on hexachlorodisiloxane (Eméléus and D. S. Payne)[38] or trichlorosilane (Eméléus and S. R. Robinson).[39] Most of this work was done during the War at Imperial College, but some of it was continued after Eméléus moved to Cambridge, first as Reader but shortly afterwards as Professor of Inorganic Chemistry.

During the remainder of his period at Imperial College Eméléus was predominantly concerned with studies bearing directly on the War. He was assisted in this work by D. F. Rushman, P. J. Stewart, R. F. Hudson, J. R. Knowles, B. A. J. Lister, R. A. Cox and B. Atkinson. Although little of this work was published many observations were made that bore fruit in more favourable times. The tragic death of Knowles, during experiments on incendiary materials, marred the pleasure of the achievement of this time.

Emeléus's growing interest in fluorine chemistry was made apparent by his Tilden Lecture in 1942,[40] and it was considerably increased by his visit to the United States on behalf of the British government during the latter part of the war. In the United States he was attached to the Manhattan Project at Oakridge, Tennessee, and he was especially concerned with chemical problems related to the diffusion plant using UF_6 for the separation of ^{235}U.

Emeléus's first work at Cambridge was on the halogen fluorides. With A. A. Banks, R. N. Haszeldine, and V. Kerrigan[41] he showed that iodine pentafluoride reacts with carbon tetraiodide to form the much sought-after trifluoroiodomethane, the key to a large number of synthetic processes. Emeléus, Maddock, G. L. Miles, and A. G. Sharpe studied the action of bromine trifluoride on compounds of the heavy elements,[42] and Emeléus and Sharpe subsequently prepared the polyhalides $KBrF_4$ and KIF_6.[43] The discovery by Banks, Emeléus, and A. A. Woolf[44] that bromine trifluoride is a fairly good conductor of electricity was soon followed by the preparation of the compound BrF_3, SbF_5 and the demonstration that it could be titrated conductimetrically in bromine trifluoride against $AgBrF_4$, leading to the formulation of the reaction as:[45]

$$BrF_2^+SbF_6^- + Ag^+BrF_4^- \rightarrow AgSbF_6 + 2BrF_3$$

The direct formation of complex fluorides by the action of bromine trifluoride on, for instance, borates and phosphates, and the preparation of nitrosonium or nitronium complex fluorides by the action of the reagent on mixtures of nitrosyl chloride or dinitrogen tetroxide and non-metallic oxides were discovered shortly afterwards.[46] Complexes of vanadium, niobium, and tantalum pentafluorides were made by Emeléus and V. Gutmann by similar methods.[47]

The formation of the mercury derivatives CF_3HgI and $(CF_3)_2Hg$ by irradiating CF_3I with ultraviolet light in the presence of mercury and by the action of silver, copper, or cadmium amalgam on trifluoromethylmercuric iodide was the first report of the existence of a perfluoroalkyl organometallic compound.[48] J. Banus, Emeléus, and Haszeldine showed that the formation of CF_3 and C_2F_5 radicals on irradiation of the iodides opened up many new possibilities; nucleophilic substitution of the iodine atom was, however, found to be impossible.[49] Trifluoromethyl derivatives of phosphorus and arsenic were made by F. W. Bennett, G. R. A. Brandt, Eméleus and Haszeldine by heating the elements with trifluoroiodomethane,[50] and the study of these compounds and their oxidation products was continued by these workers and by E. G. Walaschewski, and R. C. Paul.[51-54] Later, Eméleus and J. H. Moss, and J. W. Dale, Eméleus, Haszeldine, and Moss extended the work to derivatives of antimony[55,56] and selenium.[57]

Several other lines of work were pursued during this period. The electrical conductivities and electrolyses of the iodine chlorides and of the ether and acetic acid adducts of boron trifluoride were examined in detail in collaboration with N. N. Greenwood and R. L. Martin.[58,59] Eméleus and J. Thompson prepared the nitrogen derivative $N(CF_3)_2F$ by the action of trimethylamine on cobalt trifluoride.[60] Eméleus and F. G. A. Stone examined the polymerization of alkene oxides and vinyl compounds in the presence of diborane and the complex reaction of the latter compound with hydrazine.[61] The electrochemical fluorination of dimethyl sulphide in anhydrous hydrogen fluoride, studied by A. F. Clifford, K. H. El-Shamy, Eméleus, and Haszeldine, led to the isolation of the sulphur derivatives $CF_3 . SF_5$ and $(CF_3)_2SF_4$.[62] In a brief return to earlier interests, P. B. Ayscough and Eméleus examined the kinetics of pyrolysis of the compounds $(CF_3)_3As$ and $(CH_3)_3As$.[63] Finally, the reactions of iodosilane with white phosphorus, arsenic, and sulphur, by providing routes to the compounds $P(SiH_3)_3$, $As(SiH_3)_3$, and $(SiH_3)_2S$ opened up a new branch of silicon chemistry; this work, by B. J. Aylett, Eméleus, and Maddock,[64] was extended in collaboration with A. G. MacDiarmid.[65]

The preparation of methyliodosilane and methyldiiodosilane from methylsilane, hydrogen iodide, and aluminium iodide, by Eméleus, M. Onyszchuk, and W. Kuchen[66] led to the synthesis of many derivatives of the methylsilyl group by Eméleus and Onyszchuk[67] and Eméleus and E. A. V. Ebsworth,[68] who also studied the interaction of boron halides and methylsilyl derivatives of oxygen, sulphur, and nitrogen.

The chemistry of dimethyliodosilane was investigated at the same time by Eméleus and L. E. Smythe.[69] Alkali metal derivatives of stannane, germane, phosphine, arsine, and stibine were shown to be formed, in collaboration with S. F. A. Kettle and K. Mackay, by the action of sodium in liquid ammonia on the hydrides; some of these were isolated and their reactions studied.[70, 71]

In the fluorine field, Eeméleus and H. C. Clark re-examined the properties of vanadium, niobium, and tantalum pentafluorides and tungsten hexafluoride,[72] and Eeméleus and J. D. Smith showed that tristrifluoromethyl phosphine will displace one or two molecules of carbon monoxide from nickel carbonyl.[73] Studies were also made of the conductance of $P(CF_3)_3Cl_2$ in acetonitrile (with G. S. Harris[74]), of adducts of the titanium tetrahalides (with G. S. Rao[75]), and of the interaction of bis(trifluoromethyl)mercury and potassium halides, leading to the formation of the complexes $KHg(CF_3)_2X$ and $K_2Hg(CF_3)_2X_2$ (with J. J. Lagowski[76]).

In collaboration with K. Wade, it was shown that diborane reacts with alkyl cyanides to give adducts which decompose to form substituted borazoles,[77] and with G. J. Videla it was found that a reduced cobalt catalyst is effective in the preparation of substituted borazoles from boron trichloride and ammonium or alkyl-ammonium chlorides.[78]

Much of the work on fluoroalkyl derivatives of metals and non-metals carried out in the 1960s was concerned with the CF_3S group. Eeméleus and S. N. Nabi condensed trifluoromethane sulphenyl chloride with ammonia, amines, and phosphine,[79] and Eeméleus and Pugh studied the reactions of the mercury derivative $(CF_3S)_2Hg$ with chlorides of phosphorus and arsenic.[80] The silver salt $AgSCF_3$ was obtained (with D. E. MacDuffie) by interaction of silver(I) fluoride and carbon disulphide at 140 °C.[81] A. J. Downs, Ebsworth, and Eeméleus[82] studied the interaction of $Hg(SCF_3)_2$ and mercury(II) halides, acetate, trifluoroacetate, and nitrate by means of Raman spectroscopy; and Eeméleus and A. Haas prepared trifluoromethanesulphenyl cyanide and other pseudohalides by the action of the appropriate silver salt on the chloride.[83] With K. J. Packer, N. Welcman, and R. G. Cavell, phosphorus and arsenic derivatives containing both CF_3 and CF_3S groups were prepared,[84, 85] and with Cavell and J. M. Miller the addition of chlorine to the compounds $(CF_3)_3P$ and $(C_6H_5)_3P$, and the conversion of the resulting dichlorides to sulphides, were studied.[86, 87]

In an exploratory examination of the synthetic utility of the microwave discharge, with G. H. Cady and B. Tittle, the electrolysis of liquid sulphur dioxide containing water was studied,[88] and Eméleus and Tittle then used the method to prepare sulphur pentafluoride chloride from SF_6, S_2F_{10}, or SF_4 and chlorine.[89] The chemistry of sulphur pentafluoride chloride was studied, along with that of disulphur decafluoride and the peroxide $(SF_5)_2O_2$, with Packer.[90] Mixed silyl and perfluoroalkyl derivatives of selenium were made from iodosilane and the mercury derivative

$$Hg(SeCF_3)_2.^{90}$$

The compound $(CF_3)_2PI$ was shown by Eméleus and J. Grobe to react with manganese carbonyl and iron pentacarbonyl, forming $Mn_2(CO)_8P(CF_3)_2I$ and $[Fe(CO)_3P(CF_3)_2I]_2$ respectively,[92] and several amino-derivatives were made by W. R. Cullen and Eméleus.[93] In a study (with G. L. Hurst) of the action of cyanogen on silver(II) fluoride at 105 °C, the remarkable heterocylic compound

$$\begin{array}{c} F_2C\!\!-\!\!CF_2 \\ |\qquad| \\ N\!\!=\!\!N \end{array}$$

was prepared.[94] Mercury(II) fluoride, on the other hand, yielded the compound $[(CF_3)_2N]_2Hg$,[95] the starting-point for further work described later. The last group of miscellaneous researches to be mentioned here included an infrared study of the structures of perthiocyanic and isoperthiocyanic acids (with Haas and N. Sheppard),[96] the preparation of triphenyllead cyanide and an examination of the action of hydrogen chloride on derivatives of diplumbane (with P. R. Evans),[97] the preparation of arsenic trifluoroacetate and perfluoroalkyl derivatives of selenium (with M. J. Dunn),[98] the formation of cyclic esters of Group IV elements from dihydric phenols and dialkyl silicon and tin dichlorides (with J. J. Zuckerman),[99] and the reaction of hexachlorodisilane with bases (with R. Tufail).[100]

Among reactions of the compound $[(CF_3)_2N]_2Hg$ which have been exploited for synthetic purposes are those with disulphur dichloride and sulphur dichloride, giving $(CF_3)_2NSCl$ and $[(CF_3)_2N]_2S$ (Eméleus and B. W. Tattershall[101]); with selenium tetrachloride, giving $[(CF_3)_2N]Se$ (R. C. Dobbie and Emeléus[102]), and with bromine giving $(CF_3)_2NBr$. Ultraviolet irradiation of the last compound gives $(CF_3)_2NN(CF_3)_2$, and addition of it across ethylene, tetrafluoroethylene, or acetylene takes place readily.[103] The corresponding chloride $(CF_3)_2NCl$ in turn reacts with mercurials R_2Hg, giving $(CF_3)_2NHgR$;[104] with cyanogen chloride, giving $(CF_3)_2N.N{=}CCl_2$ (which is converted by mercuric fluoride into $[(CF_3)_2NN(CF_3)]_2Hg$);[105] with tristrifluoromethylphiophine, giving $(CF_3)_2NP(CF_3)_2$ (H. G. Ang and Emeléus[106]); and with phosphorus trifluoride, giving $(CF_3)_2NPF_3Cl$ (Emeléus and T. P. Onak[107]). Many other lines remain to be explored.

The researches described here represent only a part of Emeléus's influence on inorganic chemistry. Himself essentially an originator, he has seen many areas of chemistry which he initiated developed extensively by others, and his kindness and generosity towards his colleagues have been the cause of much successful work by his former students after they have left his group. Honours bestowed on him have included the C.B.E., the Fellowship of the Royal Society, the Presidency of the Chemical Society and of the Royal Institute of Chemistry, the Lavoisier Medal of the French Chemical Society, the Stock Medal of the German Chemical Society, the Davy Medal of the Royal Society and the George Fisher Baker Lectureship at Cornell University. Always a modest man, he has never claimed that the results of his work were of outstanding fundamental significance; but his position in the development of experimental inorganic chemistry is secure, his influence on the expansion of interest in the subject since 1945 has been enormous, and he has shown, more than anyone else during the past two decades, that although inorganic chemistry has become a science it has also remained an art.

E. A. V. EBSWORTH

A. G. MADDOCK

A. G. SHARPE

University Chemical Laboratory
Cambridge

REFERENCES

1   H. J. Emeléus and W. E. Downey, *J. chem. Soc.* 1924, **125**, 2491.
2   H. J. Emeléus, *J. chem. Soc.* 1925, **127**, 1362.
3   H. J. Emeléus, *Nature, Lond.* 1925, **115**, 460.
4   H. J. Emeléus, *J. chem. Soc.* 1926, p. 1336.
5   H. J. Emeléus, *J. chem. Soc.* 1926, p. 2948.
6   H. J. Emeléus, *J. chem. Soc.* 1927, p. 783; 1929, p. 1846.
7   H. J. Emeléus, *J. chem. Soc.* 1928, p. 1942.
8   H. J. Emeléus and R. H. Purcell, *J. chem. Soc.* 1927, p. 788.
9   H. J. Emeléus, *J. chem. Soc.* 1929, p. 1733.
10 H. S. Taylor and H. J. Emeléus, *J. Am. chem. Soc.* 1930, **52**, 2150.
11 H. J. Emeléus and H. S. Taylor, *J. Am. chem. Soc.* 1931, **53**, 3370.
12 H. J. Emeléus, *Trans. Faraday Soc.* 1932, **28**, 89.
13 G. H. Cheesman and H. J. Emeléus, *J. chem. Soc.* 1932, p. 2847.
14 H. J. Emeléus and H. L. Riley, *Proc. R. Soc.* A, 1933, **140**, 378.
15 A. G. H. Damerell and H. J. Emeléus, *J. chem. Soc.* 1934, p. 974.
16 H. J. Emeléus and L. J. Jolley, *J. chem. Soc.* 1935, p. 929.
17 H. J. Emeléus and L. J. Jolley, *J. chem. Soc.* 1935, p. 1612.
18 H. J. Emeléus and K. Stewart, *J. chem. Soc.* 1935, p. 1182.
19 H. J. Emeléus and K. Stewart, *J. chem. Soc.* 1936, p. 677.
20 H. J. Emeléus, F. W. James, A. King, T. G. Pearson, R. H. Purcell and H. V. A. Briscoe, *J. chem. Soc.* 1934, pp. 1207, 1948.
21 J. S. Anderson, R. H. Purcell, T. G. Pearson, A. King, F. W. James, H. J. Emeléus and H. V. A. Briscoe, *J. chem. Soc.* 1937, p. 1492.
22 H. J. Emeléus and H. V. A. Briscoe, *J. chem. Soc.* 1937, p. 127.
23 E. R. Roberts, H. J. Emeléus and H. V. A. Briscoe, *J. chem. Soc.* 1939, p. 41.
24 L. A. Bashford, H. J. Emeléus and H. V. A. Briscoe, *J. chem. Soc.* 1938, p. 1358.
25 H. J. Emeléus and E. R. Gardner, *J. chem. Soc.* 1938, p. 1900.
26 H. J. Emeléus and H. H. G. Jellinek, *Trans. Faraday Soc.* 1944, **40**, 93.
27 A. G. Maddock, C. Reid and H. J. Emeléus, *Nature, Lond.* 1939, **144**, 328.
28 H. J. Emeléus and C. Reid, *J. chem. Soc.* 1939, p. 1021.
29 H. J. Emeléus, A. G. Maddock and C. Reid, *J. chem. Soc.* 1941, p. 353.
30 H. J. Emeléus and A. G. Maddock, *J. chem. Soc.* 1944, p. 293.
31 H. J. Emeléus and A. G. Maddock, *J. chem. Soc.* 1946, p. 1131.
32 H. J. Emeléus and N. Miller, *J. chem. Soc.* 1939, p. 819.
33 H. J. Emeléus and A. J. E. Welch, *J. chem. Soc.* 1939, p. 1928.
34 H. J. Emeléus and C. J. Wilkins, *J. chem. Soc.* 1944, p. 454.
35 H. J. Emeléus and H. G. Heal, *J. chem. Soc.* 1946, p. 1126.
36 L. H. Long, H. J. Emeléus and H. V. A. Briscoe, *J. chem. Soc.* 1946, p. 1123.
37 H. J. Emeléus and J. F. Wood, *J. chem. Soc.* 1948, p. 2183.
38 H. J. Emeléus and D. S. Payne, *J. chem. Soc.* 1947, p. 1590.
39 H. J. Emeléus and S. R. Robinson, *J. chem. Soc.* 1947, p. 1592.

40 H. J. Emeléus, *J. chem. Soc.* 1942, p. 441.
41 A. A. Banks, H. J. Emeléus, R. N. Haszeldine and V. Kerrigan, *J. chem. Soc.* 1948, p. 2188.
42 H. J. Emeléus, A. G. Maddock, G. L. Miles and A. G. Sharpe, *J. chem. Soc.* 1948, p. 1991.
43 A. G. Sharpe and H. J. Emeléus, *J. chem. Soc.* 1948, p. 2135; 1949, p. 2206.
44 A. A. Banks, H. J. Emeléus and A. A. Woolf, *J. chem. Soc.* 1949, p. 2816.
45 H. J. Emeléus and A. A. Woolf, *J. chem. Soc.* 1949, p. 2865.
46 H. J. Emeléus and A. A. Woolf, *J. chem. Soc.* 1950, pp. 164, 1050.
47 H. J. Emeléus and V. Gutmann, *J. chem. Soc.* 1949, p. 2979; 1950, p. 2115.
48 H. J. Emeléus, and R. N. Haszeldine, *J. chem. Soc.* 1949, p. 2948.
49 J. Banus, H. J. Emeléus and R. N. Haszeldine, *J. chem. Soc.* 1950, p. 3041; 1951, p. 60.
50 F. W. Bennett, G. R. A. Brandt, H. J. Emeléus and R. N. Haszeldine, *Nature, Lond.* 1950, **166**, 225.
51 G. R. A. Brandt, H. J. Emeléus and R. N. Haszeldine, *J. chem. Soc.* 1952, pp. 2198, 2549, 2552.
52 F. W. Bennett, H. J. Emeléus and R. N. Haszeldine, *J. chem. Soc.* 1953, p. 1565; 1954, p. 3598.
53 H. J. Emeléus, R. N. Haszeldine and E. G. Walaschewski, *J. chem. Soc.* 1953, p. 1552.
54 H. J. Emeléus, R. N. Haszeldine and R. C. Paul, *J. chem. Soc.* 1954, pp. 881, 3896; 1955, p. 563.
55 H. J. Emeléus and J. H. Moss, *Z. anorg. Chem.* 1956, **282**, 24.
56 J. W. Dale, H. J. Emeléus, R. N. Haszeldine and J. H. Moss, *J. chem. Soc.* 1957, p. 3708.
57 J. W. Dale, H. J. Emeléus and R. N. Haszeldine, *J. chem. Soc.* 1958, p. 2939.
58 N. N. Greenwood and H. J. Emeléus, *J. chem. Soc.* 1950, p. 987.
59 N. N. Greenwood, R. L. Martin and H. J. Emeléus, *J. chem. Soc.* 1950, p. 3030; 1951, p. 1328.
60 J. Thompson and H. J. Emeléus, *J. chem. Soc.* 1949, p. 3080.
61 H. J. Emeléus and F. G. A. Stone, *J. chem. Soc.* 1950, p. 2755; 1951, p. 840.
62 A. F. Clifford, K. H. El-Shamy, H. J. Emeléus and R. N. Haszeldine, *J. chem. Soc.* 1953, p. 2372.
63 P. B. Ayscough and H. J. Emeléus, *J. chem. Soc.* 1954, p. 3381.
64 B. J. Aylett, H. J. Emeléus and A. G. Maddock, *J. inorg. nucl. Chem.* 1955, **1**, 187.
65 H. J. Emeléus, A. G. MacDiarmid and A. G. Maddock, *J. inorg. nucl. Chem.* 1955, **1**, 194.
66 H. J. Emeléus, M. Onyszchuk and W. Kuchen, *Z. anorg. Chem.* 1956, **283**, 74.
67 H. J. Emeléus and M. Onyszchuk, *J. chem. Soc.* 1958, p. 604.
68 E. A. V. Ebsworth and H. J. Emeléus, *J. chem. Soc.* 1958, p. 2150.
69 H. J. Emeléus and L. E. Smythe, *J. chem. Soc.* 1958, p. 609.

70 H. J. Emeléus and S. F. A. Kettle, *J. chem. Soc.* 1958, p. 2444.
71 H. J. Emeléus and K. M. Mackay, *J. chem. Soc.* 1961, p. 2676.
72 H. C. Clark and H. J. Emeléus, *J. chem. Soc.* 1957, p. 2119, 4778; 1958, p. 190.
73 H. J. Emeléus and J. D. Smith, *J. chem. Soc.* 1958, p. 527.
74 H. J. Emeléus and G. S. Harris, *J. chem. Soc.* 1959, p. 1494.
75 H. J. Emeléus and G. S. Rao, *J. chem. Soc.* 1958, p. 4245.
76 H. J. Emeléus and J. J. Lagowski, *Proc. chem. Soc.* 1958, p. 231.
77 H. J. Emeléus and K. Wade, *J. chem. Soc.* 1960, p. 2614.
78 H. J. Emeléus and G. J. Videla, *Proc. chem. Soc.* 1957, p. 288; *J. chem. Soc.* 1959, p. 1306.
79 H. J. Emeléus and S. N. Nabi, *J. chem. Soc.* 1960, p. 1103.
80 H. J. Emeléus and H. Pugh, *J. chem. Soc.* 1960, p. 1108.
81 H. J. Emeléus and D. E. MacDuffie, *J. chem. Soc.* 1961, p. 2597.
82 A. J. Downs, E. A. V. Ebsworth and H. J. Emeléus. *J. chem. Soc.* 1961, p. 3187; 1962, p. 1254.
83 H. J. Emeléus and A. Haas, *J. chem. Soc.* 1963, p. 1272.
84 H. J. Emeléus, K. J. Packer and N. Welcman, *J. chem. Soc.* 1962, p. 2529.
85 R. G. Cavell and H. J. Emeléus, *J. chem. Soc.* 1964, p. 5825.
86 R. G. Cavell and H. J. Emeléus, *J. chem. Soc.* 1964, p. 5896.
87 H. J. Emeléus and J. M. Miller, *J. inorg. nucl. Chem.* 1966, **28**, 662.
88 G. H. Cady, H. J. Emeléus and B. Tittle, *J. chem. Soc.* 1960, p. 4138.
89 H. J. Emeléus and B. Tittle, *J. chem. Soc.* 1963, p. 1644.
90 H. J. Emeléus and K. J. Packer, *J. chem. Soc.* 1962, p. 771.
91 E. A. V. Ebsworth, H. J. Emeléus and N. Welcman, *J. chem. Soc.* 1962, p. 2290.
92 H. J. Emeléus and J. Grobe, *Angew. Chem.* 1962, **74**, 467.
93 W. R. Cullen and H. J. Emeléus, *J. chem. Soc.* 1959, p. 372.
94 H. J. Emeléus and G. L. Hurst, *J. chem. Soc.* 1962, p. 3276.
95 H. J. Emeléus and G. L. Hurst, *J. chem. Soc.* 1964, p. 396.
96 H. J. Emeléus, A. Haas and N. Sheppard, *J. chem. Soc.* 1963, pp. 3165, 3168.
97 H. J. Emeléus and P. R. Evans, *J. chem. Soc.* 1964, pp. 510, 511.
98 H. J. Emeléus and M. J. Dunn, *J. inorg. nucl. Chem.* 1965, **27**, 269, 752.
99 H. J. Emeléus and J. J. Zuckerman, *J. organomet. Chem.* 1964, **1**, 328.
100 H. J. Emeléus and R. Tufail, *J. inorg. nucl. Chem.* 1967, **29**, 2081.
101 H. J. Emeléus and B. W. Tattershall, *J. chem. Soc.* 1964, p. 5892.
102 R. C. Dobbie and H. J. Emeléus, *J. chem. Soc.* 1964, p. 5894.
103 H. J. Emeléus and B. W. Tattershall, *Z. anorg. Chem.* 1964, **327**, 147.
104 R. C. Dobbie and H. J. Emeléus, *J. chem. Soc.* (*A*), 1966, p. 367.
105 R. C. Dobbie and H. J. Emeléus, *J. chem. Soc.* (*A*), 1966, p. 933.
106 H. G. Ang and H. J. Emeléus, *Chem. Commun.* 1966, p. 460.
107 H. J. Emeléus and T. P. Onak, *J. chem. Soc.* (*A*), 1966, p. 1291.

1

ORGANOMETALLIC CATIONS

HOWARD C. CLARK

The proposition that many organic derivatives of metals belonging to main groups of the Periodic Table are capable of ionization has long been accepted, and in general terms is expressed by the equation

$$R_aMX_b \rightarrow R_aM^{b+} + bX^-$$

This, of course, assumes that X is a singly charged anion, such as halide, although doubly charged anions such as SO_4^{2-} may also be involved. Specific examples of such processes are

$$(CH_3)_3SnBr \rightarrow (CH_3)_3Sn^+ + Br^-$$

and $$(C_6H_5)_2TlCl \rightarrow (C_6H_5)_2Tl^+ + Cl^-$$

For the purposes of this article, discussion will be restricted to derivatives of indium, thallium, tin, lead, and antimony, for most of which a number of studies have been made. While equations such as those above refer to the formation of singly charged cations, ions of higher charge may also be proposed, for example,

$$(CH_3)_3SbCl_2 \rightarrow (CH_3)_3Sb^{2+} + 2Cl^-$$

and $$CH_3SnCl_3 \rightarrow CH_3Sn^{3+} + 3Cl^-$$

As the ionic charges increase, it becomes less likely that such ions can exist as discrete species. In solution, extensive solvation should occur readily, so that even for species R_nN^+, a more accurate representation must include solvent molecules bound to the central metal atom. For more highly charged species (e.g. R_2Sn^{2+} or RSn^{3+}) in aqueous solution, hydrolysis and condensation reactions occur extensively, giving rise to ions such as $R_2Sn(OH)(OH_2)_x^+$. Hence, in any solvent of sufficient solvating power to bring about ionization, the coordination of the central metal atom is far greater than that expected for the simple discrete ion, and 'free' (i.e. non-solvated) cations are not present. In the solid state, structural studies may indicate a non-ionized, molecular structure, or an aggregate of free, discrete ions R_aM^{b+}, bX^-. In the discussion that follows, we will be particularly concerned with the question of whether or not discrete organometallic cations exist in the solid state.

This question has particular interest since, in a species R_aM^{b+}, a is always less than the usual or maximum coordination number, and is frequently an odd, and less common, number (e.g. $a = 3$ or 5). These ions therefore might afford an opportunity to study metallic derivatives of the less common geometries. Moreover, the metal–carbon bonds in these ions are highly covalent and in general are inert kinetically, so that the over-all stability is considerable. An additional point of interest lies in the relationship which these ions provide between the inorganic chemistry of these metallic elements and their organometallic chemistry.

1. Organometallic ions in solution

It has long been known that organic compounds of many main group metals produce conducting solutions when dissolved in water and other polar solvents. At an early stage it was claimed[1,2] that triorganotin derivatives, R_3SnX, gave tetrahedral $R_3Sn(OH_2)^+$ ions in aqueous solution, and not planar R_3Sn^+ ions, this being based on the resolution of the optically active methylethyl-n-propyltin camphor sulphonate. However, numerous workers have found it impossible to repeat this resolution. Somewhat later, Kraus[3] described the electrolytic behaviour of triorganotin derivatives in non-aqueous solvents, particularly in the presence of Lewis bases such as pyridine. He also attributed the acidity of aqueous solutions of di- and tri-organotin di- and mono-halides to the following type of equilibria:

$$(CH_3)_3SnCl + H_2O \rightleftharpoons [(CH_3)_3SnOH_2]^+ + Cl^-$$
$$[(CH_3)_3Sn(OH_2)]^+ + H_2O \rightleftharpoons (CH_3)_3SnOH + H_3O^+$$

Likewise, the conductance of aqueous solutions of $(CH_3)_2TlOH$ has been attributed[4] to the formation of $(CH_3)_2Tl(H_2O)_x^+$ and OH^- ions. In these and in other cases, however, the early evidence gave no information as to the number of solvent molecules coordinated per metal atom and hence of the geometry of the organometallic species. More recently, extensive e.m.f. studies of the hydrolytic equilibria involved in aqueous solutions of these organometallic derivatives, together with the application of spectroscopic methods, have produced much useful information. This has recently been summarized in an excellent review by Tobias.[5] In brief, the ions $(CH_3)_2Tl^+$, $(CH_3)_2Sn^{2+}$, and $(CH_3)_2Pb^{2+}$ have been shown to be linear while the ions $(CH_3)_3Sn^+$ and $(CH_3)_3Pb^+$ are planar. These conclusions, of course, refer only to the shape of the metal–carbon skeleton, and it seems probable that the linear ions also have four solvent molecules in the first solvation sheath, giving the metal an over-all octahedral configuration. Probably, also, the planar ions should best be regarded as being trigonal bipyramidal in shape owing to solvation by two solvent molecules. Clearly, the structure and stability of these organometallic ions in solution are derived from the stabilizing effect of the coordinated solvent molecules; 'free' organometallic ions cannot be expected and the effect of solvation is to stabilize the metal with a high coordination number. Although brief reference will be made later to solutions in some non-aqueous solvents, the article by Tobias[5] should be read for a fuller discussion of aqueous solutions.

2. Organometallic ions in the solid state

While electrolytic properties can provide unambiguous evidence of the existence of organometallic cations in solution, it is often less easy to decide whether crystalline structures are to be regarded as ionic aggregates. Since the organometallic group in a compound R_aMX_b is generally large and contains covalent metal–carbon bonds, crystallization in typically ionic lattices does not usually occur. The crystals that are obtained are generally rather soft with low melting points, both of which properties are more character-istic of molecular rather than ionic crystals. It is therefore neces-sary to determine the detailed crystal structure in order to observe any 'interaction' between the anionic and cationic groupings. This can ideally be done by diffraction methods, as in a single crystal structure determination by X-ray diffraction. Alternatively, much useful information can be obtained by the use of infrared spectro-scopic investigations. Even with such studies as these, there remains the problem of interpreting any observed interaction in terms of the conventional, but only qualitative, chemical terms, 'covalent' and 'ionic' bonds.

3. Organometallic cations of Group III metals

Since the metal must be reasonably electropositive to allow the formation of an organometallic cation, we will in most cases confine our discussion to the heavier and more electropositive metals of each group—in this case indium and thallium.

The organometallic chemistry of indium has not been extensively studied. Although triorganoindium compounds are well known, little investigation has been made of diorganoindium (III) derivatives, R_2InX, until very recently. It has now been concluded from infrared studies[6] that like their dimethylgallium analogues, dimethylindium fluoride, chloride and iodide, are dimeric, with bent dimethylindium groups and halogen bridges. The reactions of these halides with ligands such as pyridine, and triphenylphosphine, give the neutral and apparently tetrahedral mono-adducts, $(CH_3)_2InX.L$. Preliminary studies[7] suggest that compounds such as $[(CH_3)_2In]_2SO_4$ can be prepared, for which the infrared spectra indicate a considerable interaction between the $(CH_3)_2In$ and anionic groups. So far, then, there is no evidence for the existence of discrete R_2In^+ ions in the solid state.

Many more studies have been made of organothallium cations, and this is not surprising since the R_2TlX compounds are by far the most stable class of organothallium derivatives. The ion R_2Tl^+ is isoelectronic with R_2Hg and the derivatives R_2TlX are stable, salt-like compounds giving highly conducting aqueous solutions. The halides $(CH_3)_2TlX$ were investigated structurally in 1934;[8] the thallium atoms are six coordinated, with linear $(CH_3)_2Tl$ groups. These structures have frequently been described as ionic lattices, but there seems no reason to suppose that the halogens cannot equally well be described as bridging halogen atoms and that the thallium–halogen bond has a reasonable degree of co-valency. Certainly, at present, it has not been possible to distinguish between these two models. When the anion is a weak donor group, such as ClO_4^-, BF_4^- or NO_3^-, infrared studies[9] of the dimethylthallium derivatives are consistent with structures containing free, linear $(CH_3)_2Tl^+$ ions and free anions. However, for the related perchlorates, tetrafluoroborates, etc., of triorganotin groups, it is known that strictly anhydrous conditions are necessary for their preparation and manipulation, if authentic spectroscopic results are to be obtained, and that traces of moisture are sufficient to give spectra consistent with ionic structures. A comparison of Shier and Drago's results with those obtained[7] for dimethylthallium compounds prepared under anhydrous conditions show no major differences. Hence, although X-ray structural studies are highly desirable for at least one of this class of compounds, the present indications remain that dimethylthallium derivatives may have, but not that they must have, purely ionic structures.

To a considerable extent, however, the structure is dependent on the nature of the organic groups bonded to thallium. Although the dimethylthallium halides contain linear $(CH_3)_2Tl$ groups, the bis-(pentafluorophenyl)thallium halides are apparently dimeric, with bent $(C_6F_5)_2Tl$ groups, each thallium atom being tetrahedrally co-ordinated.[10] Moreover, while dimethylthallium sulphate can be considered ionic, $2(CH_3)_2Tl^+SO_4{}^{2+}$, the analogous $[(C_6F_5)_2Tl]_2SO_4$ is thought to have the structure

$$
\begin{array}{ccccc}
R_f & & O & & R_f \\
\diagdown & & \| & & \diagup \\
Tl & \!\!-O-\!\! & S & \!\!-O-\!\! & Tl \\
\diagup & & | & & \diagdown \\
R_f & & O & & R_f \\
\end{array}
$$

These conclusions are based entirely on infrared spectral results, namely the observation of two Tl–C stretching vibrations which can only occur if $(R_f)_2Tl$ is non-linear, and of the C_{2v} symmetry of the SO_4 group. While it is reasonable to conclude that these features can only arise from some strong interaction between $(R_f)_2Tl$ and SO_4, such as is shown above, there is no basic information as to the nature of the interaction. Certainly it is not observed for dimethylthallium derivatives, and in view of the many four, five, and six coordinate $(C_6F_5)_2Tl$ derivatives described,[10, 11] it cannot be a consequence of the steric requirements of the bulky C_6F_5 groups. Further studies of a wider range of diorganothallium derivatives would be very worth while.

It is also of some interest that, while the $(CH_3)_2Tl$ group is linear in the presence of ligands such as Cl, ClO_4, and BF_4, a strong donor ligand can cause some distortion. Thus, dimethylthallium perchlorate on reaction with pyridine gives $(CH_3)_2TlClO_4 \cdot py$, which has been formulated[12] as $(CH_3)_2Tlpy^+ClO_4^-$. Arguing again only on infrared evidence, it has been concluded[12] that the cation has a slightly distorted T-shape, the $(CH_3)_2Tl$ portion being not quite linear. At first sight, then, the suggestion[9] that a pyridine solution of $(CH_3)_2TlClO_4$ should contain a non-linear $(CH_3)_2Tl$ group and a free ClO_4^- ion seems reasonable. However, the solvation of $(CH_3)_2Tl$ would surely proceed beyond the formation of $(CH_3)_2Tlpy^+$ and should certainly produce a symmetrical solvating effect about a linear $(CH_3)_2Tl^+$ ion.

4. Organometallic cations of Group IV metals

The usual sharp distinction between silicon and germanium on the one hand and tin and lead on the other is again apparent as far as the formation of organometallic ions are concerned. Arguments have been advanced for many years that many triorgano- and diorgano-tin and lead derivatives are essentially ionic, whereas all the analogous silicon and germanium compounds are considered covalent. Thus, the conclusion that triorganotin fluorides were ionic, in contrast to the covalent character of the related silicon and germanium fluorides, was based on the much higher melting points and lower solubilities of the former. However, these properties are now best attributed to the polymeric nature of the fluorides rather than their ionicity. Thus, the crystal structure of trimethyltin fluoride[13] is composed of planar trimethyltin groups, linked through non-linear, unsymmetrical Sn–F---Sn bridges. This last feature is a definite indication of some covalent character, since a fully ionic structure would very probably have each F atom lying on the axis of, and equidistant from, the two Sn atoms. Moreover, a strong Sn–F vibration can be seen[14] in the infrared spectrum of $(CH_3)_3SnF$. A closely related piece of evidence is found in the properties of (neophyl)$_3$SnF (where neophyl = 2-methyl-2-phenyl-1-propyl). Unlike other triorganotin fluorides, this fluoride[15] has the remarkably low melting point of 99–100°, and is quite volatile. Apparently the considerable steric requirements of the bulky 'neophyl' groups produce a tetrahedral geometry for the fluoride, although the Mossbauer spectrum[16] suggests that there may be weak fluoride bridging. Most significantly, the fluoride is not ionic, even though this seems the most likely arrangement to reduce steric interactions. A less definitive structure is that of trimethyltin hydroxide[17] in which the $(CH_3)_3Sn$ groups are reported to be inclined at about 15° from the plane perpendicular to the chain axis. The oxygen atoms of the OH groups lie on the axis and are almost equidistant from the two tin atoms. There is, however, good infrared evidence[18] that $(CH_3)_3SnOH$ contains Sn–O–Sn bridges, and interestingly it appears[19] to have a dimeric structure, with non-planar $(CH_3)_3Sn$ groups, in benzene, chloroform or carbon tetrachloride solution. Related to this is trimethyltin cyanide, for which the crystal structure has recently been reported.[20] This contains planar $(CH_3)_3Sn$ groups with the CN groups lying along the Sn–Sn axis. While this is markedly different from the tetrahedral molecular geometry of $(CH_3)_3GeCN$, the conclusion that it is essentially an ionic structure, $(CH_3)_3Sn^+CN^-$, is based on a number of debatable assumptions. There is thus convincing structural evidence that the triorganotin fluorides and trimethyltin hydroxide do not possess fully ionic structures. The only other X-ray determination of significance here is that of dimethyltin difluoride,[21] for which the structure is an infinite two-dimensional network of tin atoms and *bridging* fluorine atoms with the methyl groups above and below the plane completing the octahedral coordination of the tin.

In addition to these detailed X-ray studies, numerous investigations of organotin and organolead compounds have been made, utilizing infrared spectroscopic methods. By such means, it is possible to determine whether R_3Sn is planar or R_2Sn is linear, according to the number of Sn–C vibrations observed in the infrared spectrum. Thus for the planar R_3Sn and linear R_2Sn groups, the Sn–C symmetric stretching vibration is predicted to be infrared inactive, so that only the Sn–C asymmetric vibration should be observed. In the same manner, it is possible to obtain information about the geometry of the anion linked to R_3Sn or R_2Sn. One would certainly anticipate that compounds such as $(CH_3)_3SnClO_4$,

$(CH_3)_3SnBF_4$, and $(CH_3)_3SnSbF_6$ would be essentially aggregates of $(CH_3)_3Sn^+$ ions and the appropriate anions. The infrared data,[22, 23] however, show that, while the $(CH_3)_3Sn$ groups are planar, the symmetry of the anion has been much reduced, apparently through cation–anion interaction. Thus, in $(CH_3)_3SnClO_4$, the ClO_4 group has C_{2v} and not T_d symmetry, and the splittings of the infrared bands are of such magnitude that the structure can best be described as a linear polymer, with planar $(CH_3)_3Sn$ bridged by ClO_4 groups. Similarly, the infrared spectrum[22] of $(CH_3)_3SnSbF_6$ suggests that the SbF_6 groups have a bridging function through *cis* F atoms, between planar $(CH_3)_3Sn$ groups. In view of this evidence for *directional* cation–anion interaction in the triorganotin derivatives of even very strong acids, the behaviour of related derivatives of weaker acids is not unexpected. Thus trimethyl- and triphenyltin nitrate[22, 14] contain coordinated nitrate groups and not free NO_3^- ions, but because of the overlap of certain infrared bands, it is not possible to decide whether NO_3 is mono- or bi-dentate, or whether the triorganotin group is planar. The triorganotin carboxylates (e.g. trimethyltin acetate) have been investigated by many workers. Infrared studies[24, 25] of the solid do not distinguish unambiguously between ionic and bridged structures, but the latter seems much more likely in view of the polymeric character of the carboxylates when dissolved in non-polar solvents.

Similar studies of diorganotin derivatives have been less extensive, but the results are quite consistent. As already mentioned, dimethyltin difluoride has a bridged rather than a purely ionic structure, and attempts to prepare dimethyltin derivatives of fluoro-anions have not been successful. Thus, the reaction[26] of dimethyltin dichloride with silver hexafluorosilicate in anhydrous methanol gave dimethyltin difluoride and silicon tetrafluoride. Similar attempts[26] to obtain $(CH_3)_2Sn(BF_4)_2$ and $(CH_3)_2Sn(SbF_6)_2$ gave mixtures of dimethyltin difluoride with the appropriate fluoro-salt. That such extensive decomposition of stable fluoro-anions should occur, indicates that $(CH_3)_2Sn$ and the anion interact very powerfully. In certain other cases, such as the chromate and carbonate, the dimethyltin derivatives could be studied spectroscopically only as intimate mixtures with silver chloride. Nevertheless, the spectroscopic evidence shows that the chromate[14] is one of the very few compounds to contain a coordinated CrO_4^{2-} group, while the carbonate shows spectroscopic features completely analogous to those of transition metal complexes containing coordinated carbonato groups. This is also consistent with the properties of dimethyltin dinitrate,[27] for which both infrared and ultraviolet spectroscopic studies indicate the presence of covalently bound nitrato groups. There is thus evidence, for both the triorgano- and diorganotin derivatives, for series of compounds from carboxylates and carbonates, through nitrates, fluorides, perchlorates, to salts of fluoro acids, which never attain full ionicity, and where the gradation in bond type is apparently relatively small.

Organolead derivatives have been much less extensively studied, largely because of their lower stability in comparison with analogous tin compounds. Triorganolead carboxylates are reported to be associated in solution and they are spectroscopically similar to their tin analogues. The infrared spectra of trimethyllead chloride and dimethyllead dichloride show[28] only the Pb–C asymmetric stretching vibration and not the corresponding symmetric vibration, so that the $(CH_3)_3Pb$ group is planar and $(CH_3)_2Pb$ is linear, both presumably with bridging chlorine atoms. Other derivatives, such as the perchlorate, nitrate, and salts of fluoro anions have not been described.

While free anions are apparently not present in the solid organotin and organolead compounds, the process of solution usually gives the 'free' anions, and solvated organometallic cations. There are, however, exceptions. Recently, it has been reported[29] that trimethyltin formate and acetate can be prepared in both soluble and insoluble forms, all of which are sublimable in high vacuum without interconversion. The reasonable suggestion is made that the soluble forms may have associated cyclic structures, while the insoluble forms are probably linear polymers. The distinction in these cases is based on the solubility, or lack of it, in a relatively non-polar solvent such as chloroform. In strong solvating media, however, the associated structures are generally broken down, particularly at low concentrations. Thus trimethyltin perchlorate in methanol solution shows[30] some coordination of ClO_4 to trimethyltin, which gradually disappears with increasing dilution.

Similarly, trimethyllead perchlorate forms a mono-pyridine complex, formulated as $(CH_3)_3Pb(C_5H_5N)^+ClO_4^-$; spectroscopic evidence suggests that this is solvated further to dipyridine compounds in pyridine solution, so that the trimethyllead group becomes planar.

5. Organometallic cations of Group V metals

In this section, only certain organoantimony compounds will be considered, since it is not possible for this article to be comprehensive. Triorganoantimony(V) derivatives, R_3SbX_2, may be ionic, or may have neutral five-coordinate molecular geometries. There is considerable evidence[31, 32] that both triaryl- and trialkylantimony dihalides have the latter structure, but only recently have the structures of compounds where $X = NO_3^-$, SO_4^{2-}, or CO_3^{2-}, etc., been investigated. Although one study[33] suggested that $(CH_3)_3Sb(NO_3)_2$ was ionic, other infrared investigations under rigorously anhydrous conditions, and where the possibility of exchange reactions in potassium bromide pellets was avoided,[34, 35] show that nitrato groups are bound to antimony. Similarly the bands attributable to CO_3 in trimethylantimony carbonate[34] are very similar to those of organic carbonates and different from those of metal carbonato-complexes, stressing the non-ionic nature of the compound. Moreover, pure trimethylantimony derivatives containing fluoro anions, such as BF_4^- or SiF_6^{2-}, could not be obtained,[34] the products being heavily contaminated with trimethylantimony difluoride. The over-all picture for these trialkylantimony(V) derivatives is therefore similar to that for the triorgano- and diorganotin compounds, namely that in the solid state, the available evidence always shows considerable cation–anion interaction to the extent that free organometallic cations cannot be present, and to the extent that anion decomposition occurs in some instances.

6. Nature of the cation–anion interactions

The experimental data summarized above include X-ray determinations, infrared and ultraviolet spectroscopic studies, and

Mössbauer spectral work. Among the X-ray structural studies, there is no positive piece of evidence in favour of the existence of free organometallic cations. There are, in contrast, several items which argue strongly against their existence, the most significant being the non-linear, non-symmetrical nature of the Sn–F - - - Sn bridge found in trimethyltin fluoride.

The infrared evidence establishes the geometry of the organo-metallic grouping, and additionally often shows considerable split-ting of otherwise degenerate bands associated with the anion. While these effects can legitimately be attributed to some covalent inter-action with the organometallic group, the fact that similar (but usually smaller) splittings can arise due to low crystal site sym-metry of the anion cannot be overlooked. In other words, the argument can be advanced that the relatively low symmetry and high polarizing effect of, say, the $(CH_3)_3Sn^+$ ion, in the crystalline solid produces these observed infrared effects. This argument can be rejected on the following grounds:

(1) Splittings of the ClO_4 infrared bands of trimethyltin per-chlorate have been observed in methanol solution where crystal site symmetry effects are absent.

(2) The infrared splittings observed for dimethyltin chromate and trimethylantimony chromate are of a different order of magnitude from those observed for ammonium chromate. The interactions involved in the former compounds are therefore con-siderably stronger than those associated with hydrogen bonding in ammonium chromate.

(3) While it is known[36] that crystal site symmetry can cause some distortion of stable anions such as $SiF_6{}^{2-}$, $BF_4{}^-$ or $PF_6{}^-$, complete decomposition to an organometallic fluoride (e.g. $(CH_3)_2SnF_2$) must surely be due to much more intense directional effects than could arise from crystal symmetry effects.

(4) The fact that more or less continuous series of compounds (e.g. dimethyltin carboxylates, carbonate, dinitrate, chromate, and difluoride) all show interaction between the anionic and organo-metallic groups is most logically attributed to one cause, namely the existence of a reasonably covalent bond. In such series, then, transition from covalency to full ionicity is never fully achieved. Certainly, in trimethyltin perchlorate, for example, the $Sn–CH_3$

bonds must be much *more* covalent than that between $(CH_3)_3Sn$ and ClO_4, but the latter is apparently considerably less than a full ionic bond. The only alternative is to describe the carboxylates and carbonates say, as covalent, but the perchlorates as 'ionic' with considerable polarization of ClO_4^- by $(CH_3)_3Sn^+$. But there is no obvious reason for postulating any such significant difference between the two types.

The most reasonable note on which to conclude is a semantic one, since the matter reduces to just one question, namely is there any difference between a 'partially covalent bond between A and B', and an ionic bond in which 'A exerts a considerable polarization effect on B'? Or are these just alternative descriptions of the same reality?

REFERENCES

1   W. J. Pope and S. J. Peachey, *Proc. chem. Soc.* 1900, **16**, 42.
2   W. J. Pope and S. J. Peachey, *Proc. chem. Soc.* 1900, **16**, 116.
3   C. A. Kraus and W. N. Greer, *J. Am. chem. Soc.* 1923, **45**, 2946.
4   F. Hein and H. Meininger, *Z. anorg. allg. Chem.* 1925, **145**, 95.
5   R. S. Tobias, *Organometal. Chem. Rev.* 1966, 1.
6   H. C. Clark and A. L. Pickard, *J. Organometal. Chem.* 1967, **8**, 427.
7   A. L. Pickard, University of Western Ontario, unpublished results.
8   H. M. Powell and D. M. Crowfoot, *Z. Kristallogr. Kristallgeom.* 1934, **87**, 370.
9   G. D. Shier and R. S. Drago, *J. Organometal. Chem.* 1966, **5**, 330.
10 G. B. Deacon, J. H. S. Green and R. S. Nyholm, *J. chem. Soc.* 1965, p. 3411.
11 G. B. Deacon and R. S. Nyholm, *J. chem. Soc.* 1965, p. 6107.
12 I. R. Beattie and P. A. Cocking, *J. chem. Soc.* 1965, p. 3860.
13 H. C. Clark, R. J. O'Brien and J. Trotter, *J. chem. Soc.* 1964, p. 2332.
14 H. C. Clark and R. G. Goel, *Inorg. Chem.* 1965, **4**, 1428.
15 W. T. Reichle, *Inorg. Chem.* 1966, **5**, 87.
16 R. H. Herber, H. A. Stockler and W. T. Reichle, *J. chem. Physics*, 1965, **42**, 2447.
17 N. Kasai, Y. Kiyoshi and R. Okawara, *J. Organometal. Chem.* 1965, **3**, 172.
18 H. Kriegsmann, H. Hoffmann and S. Pischtschan, *Z. anorg. allg. Chem.* 1962, **315**, 283.
19 R. Okawara and K. Yasuda, *J. Organometal. Chem.* 1964, **1**, 356.
20 E. O. Schlemper and D. Britton, *Inorg. Chem.* 1966, **5**, 507.
21 E. O. Schlemper and W. C. Hamilton, *Inorg. Chem.* 1966, **5**, 995.
22 H. C. Clark and R. J. O'Brien, *Inorg. Chem.* 1963, **2**, 740.
23 H. C. Clark and R. J. O'Brien, *Inorg. Chem.* 1963, **2**, 1020.
24 I. R. Beattie and T. Gilson, *J. chem. Soc.* 1961, p. 2585.

25 M. J. Janssen,   J. G. A. Luijten   and   G. J. M. van der Kerk,   J.
    Organometal. Chem. 1964, **1**, 286.
26 H. C. Clark and R. G. Goel, J. Organometal. Chem. 1967, **7**, 263.
27 C. C. Addison, W. B. Simpson and A. Walker, J. chem. Soc. 1964,
    p. 2340.
28 G. D. Shier and R. S. Drago, J. Organometal. Chem. 1966, **6**, 359.
29 P. B. Simons and W. A. G. Graham, private communication.
30 H. C. Clark, R. J. O'Brien and A. L. Pickard, J. Organometal. Chem.
    1965, **4**, 43.
31 A. F. Wells, Z. Kristallogr. Kristallgeom. 1938, **99**, 367.
32 T. N. Polynova and M. A. Porai-Koshits, J. struct. Chem. (U.S.S.R.,
    Engl. transl.), 1960, **1**, 146.
33 G. G. Long, G. O. Doak and L. D. Freedman, J. Am. chem. Soc.
    1964, **86**, 209.
34 H. C. Clark and R. G. Goel, Inorg. Chem. 1966, **5**, 998.
35 M. Shindo and R. Okawara, J. Organometal. Chem. 1966, **5**, 537.
36 J. E. Griffiths and D. E. Irish, Inorg. Chem. 1964, **3**, 1134.

2

UNUSUAL COORDINATION NUMBERS

A. J. DOWNS

1. Introduction

Any distinction between 'chemical' and 'crystallographic' co-ordination numbers is less than realistic. Hereafter, coordination number, at least in the crystalline phase, will be defined not simply as the 'number of nearest neighbours' but as the number of vertices of the coordination polyhedron surrounding a particular atom; this polyhedron encompasses all other atoms or groups whose centres are nearer that atom than any other having an equivalent environment. The preferred definition eliminates ambiguity leading, for example, to a coordination number of 6 rather than 4 for copper in crystalline cupric chloride; chemically 6 is generally considered the more significant number.

Werner's assumption that the environments of certain metal ions are characterized by particular coordination numbers and symmetries has ceased to be a useful generalization. The coordination number is by no means fixed and invariable for each atom or ion; thus copper(II) is found with coordination numbers of 4, 5 or 6, whilst actinide elements like uranium exhibit coordination numbers ranging from 5 to 20. Recent reviews[1–3] clearly illustrate, moreover, that stereochemistry is not exclusively determined by the coordination number. This point is exemplified in Table 1, which defines, broadly, the factual subject matter of this review; it is not proposed to give a catalogue of such information (which can be found elsewhere[1–4]) but rather to examine experimental methods of determining coordination numbers, and to assess some of the factors governing the appearance of unusual coordination numbers. The problem of stereochemistry is considered incidentally, being inseparable from any discussion of coordination numbers.

As data about the structures of inorganic systems accumulate, it is increasingly evident that the word 'unusual' in the present context is purely relative. An atom in chemical combination interacts

16

TABLE 1. *Systems with 'unusual' coordination numbers*

Coordination number	Idealized geometry	Examples	Point group for idealized ML_n complex	M–L stretching modes Raman-active	M–L stretching modes IR-active	M–L stretching modes R–IR coincidences
2	Linear	XeF_2, $Hg(CH_3)_2$, $Be[C(CH_3)_3]_2$	$D_{\infty h}$	1	1	0
	Bent	$(SiH_3)_2O$, $S(CN)_2$	C_{2v}	2	2	2
3	Planar equilateral triangle	BX_3, $(SiH_3)_3N$, HgI_3^-	D_{3h}	2	1	1
	Planar T-form	ClF_3, $(CH_3)_3Se^+$	C_{2v}	3	3	3
	Pyramid	$(CH_3)_3N$, PX_3	C_{3v}	2	2	2
5	Trigonal bipyramid	PF_5, $Fe(CO)_5$, $CuCl_5^{3-}$, $TiBr_3 \cdot 2NMe_3$, $[Co(Me_6tren)Br]^+$	D_{3h}	3	2	1
	Square pyramid	$Sb(C_6H_5)_5$, B_5H_9, $(Ph_3P)_3RuCl_2$, $Re_2Cl_8^{2-}$, IF_5	C_{4v}	4	3	3
7	Pentagonal bipyramid	IF_7, $[UO_2F_5]^{3-}$, $Re_3Cl_{12}^{3-}$, $[Fe(OH_2)EDTA]^-$	D_{5h}	3	2	0
	Capped trigonal prism	$[NbF_7]^{2-}$, monoclinic Sm_2O_3–Gd_2O_3, $[Mn(OH_2)EDTA]^{2-}$	C_{2v}	7	6	6
	Capped trigonal antiprism	$[ZrF_7]^{3-}$?	C_{3v}	5	5	5
	Tetragonal base–trigonal base	Monoclinic ZrO_2, $(Ph_4C_4)Fe(CO)_3$	C_s	7	7	7

8	Cube	CsCl structure	O_h	2	1	0
	Square antiprism	$Zr(acac)_4$, $TaF_8{}^{3-}$, UF_4, $[Zr_4(OH)_8(OH_2)_{16}]^{8+}$	D_{4d}	3	2	0
	Dodecahedron	$TiCl_4(diars)_2$, $[Mo(CN)_8]^{4-}$, $[Co(NO_3)_4]^{2-}$	D_{2d}	6	4	4
	Bicapped trigonal prism	$TbCl_3$	C_{2v}	8	7	7
	Bicapped trigonal antiprism	UO_2F_2?	D_{3d}	3	3	0
	Hexagonal bipyramid	$[UO_2(O_2)_3]^{4-}$, $[UO_2(O_2C.CH_3)_3]^-$	D_{6h}	3	3	0
9	Tricapped trigonal prism	$[ReH_9]^{2-}$, $[Nd(OH_2)_9]^{3+}$, $[Ln(OH)_9]^{6-}$	D_{3h}	5	3	2
	Monocapped square antiprism	$PbClF$, $[Mo_6Cl_8Br_6]^{2-}$	C_{4v}	7	5	5
	Monocapped square prism	$[Ta_6Cl_{12}(OH_2)_6]^{2+}$	C_{4v}	7	5	5
10	Bicapped square prism	—	D_{4h}	4	3	0
	Bicapped square antiprism	—	D_{4d}	4	3	0
12	Icosahedron	$La_2(SO_4)_3 \cdot 9H_2O$, $(Ce(NO_3)_6)^{3-}$	I_h	2	1	0
	Cuboctahedron	$BaNiO_3$, perovskite structure	O_h	3	1	0
	Truncated tetrahedron	β-UH_3	T_d	4	2	2

Abbreviations:

$Me_6tren = N(CH_2.CH_2.NMe_2)_3$;
EDTA = ethylenediaminetetraacetate ion;
acac = acetylacetonate ion;
diars = o-$C_6H_4(AsMe_2)_2$;
Ln = lanthanide;
X = halogen.

with its chemical environment, and the resulting arrangement depends upon a subtle compromise between the demands of the atom and its neighbours. For the purposes of discussion four- and sixfold coordination will be treated as the 'usual' arrangements simply because of their ubiquity in classical inorganic compounds. However, there are no overwhelming factors favouring these environments, and with increasing experience, particularly of electronically delocalized systems, even the numerical superiority of 4- and 6-coordinate structures is likely to be seriously eroded. Already examples of five- and eightfold coordination are sufficiently numerous to defy the label 'rare', and certain characteristics of such systems have been established.[1-3]

Comparatively little is yet known about the energetics of inter-conversion of systems with different coordination numbers. There is reason to believe that the energy differences between different systems of high coordination number (> 6) are relatively small,[1] whilst in the liquid phase several species presumably involving five-, seven- or eightfold coordination are kinetically labile.[1, 2, 5] However, it is by no means obvious that this represents the general state of affairs. Moreover, for the solid phase, to which the bulk of definitive evidence refers, rationalizations based exclusively on simple thermodynamic arguments may be altogether misleading. The constraints peculiar to this phase are always liable to produce arrangements owing more to kinetic than to thermodynamic factors; this is also true of the numerous polyhedral atomic clusters now known, which are rich in unusual coordination numbers.[1]

2. Experimental methods

As a guide to coordination number the molecular formula of a compound is far from sufficient; the literature abounds in systems such as SbF_5, $CsCdCl_3$ and $BaUF_6$ where the formulae give no hint of the true coordination number. Directly or indirectly it is often possible to deduce more about the coordination properties by such measurements as the molecular weight and electrical con-ductivity in solution, absorption spectra or magnetic susceptibility. For example, compounds of the type $M(CO)_3(TTAS)X_2$ ($M = Cr, Mo, W$; $TTAS = (o\text{-}Me_2As \cdot C_6H_4)_2AsMe$; $X = Br, I$)

behave[6] as $1:1$ electrolytes in nitromethane, and are therefore formulated as containing the 7-coordinate cations

$$[M(CO)_3(TTAS)X]^+.$$

Caution is essential in this sort of approach. First, a polydentate ligand may not use all its potential donor atoms; thus, complexes of the type CrX_3QP (X = halogen; $QP = (o\text{-}Ph_2P.C_6H_4)_3P$) apparently[7] contain 6- rather than 7-coordinated chromium(III) with bonding to only three of the four donor atoms of the phosphine. Secondly, although appreciable association or dissociation of a species in solution is usually recognizable, the effect of solvation is more insidious. In favourable circumstances spectroscopic and other methods may give some hint of the effective environment of an atom in solution, but definitive information is altogether lacking. The lifetime of an aquated ion in solution varies, for example, from a period comparable with that of an ion–water vibration (10^{-11} s), as for the alkali metal cations,[8] to 10^5–10^6 s for inert species[9] like $[Cr(OH_2)_6]^{3+}$. The highly labile systems correspond to little more than a 'sticky' collision between solute and solvent, so that it is virtually impossible to define a meaningful coordination number for the solute species; so-called 'hydration numbers' represent, at best, the average number of near-neighbour water molecules. There is a distinct possibility that a species like $[Mo(CN)_8]^{4-}$ which is 8-coordinate in the solid state is solvated in solution giving a 9- or 10-coordinate arrangement.[10] Systems characterized by lower coordination numbers in the solid or gas phase are always liable to augment these numbers in coordinating solvents; many examples suggest themselves ($HgCl_2$, HgI_3^-, PF_5, etc.), although clearcut information about the solution species is rarely available.

The most remarkable feature of the study of unusual coordination numbers is the incidence of so-called 'stereochemical non-rigidity',[1, 2, 5] notably for 5-, 7-, 8- and 9-coordinate arrangements. The processes critical to stereochemistry are those which permute identical nuclei. Should the time-scale of such processes be comparable with, or shorter than, that of an experimental observation then the array cannot be portrayed by the conventional point-group appropriate to a rigid system. The free rotation of CH_3 groups in

TABLE 2. *Approximate time-scales for structural techniques*
(*ref.* 5)

Technique	Approximate time-scale (s)
Electron diffraction	10^{-20}
Neutron diffraction	10^{-18}
X-ray diffraction	10^{-18}
U.v.-visible spectroscopy	$10^{-15}-10^{-14}$
IR-Raman spectroscopy	10^{-13}
Microwave spectroscopy	10^{-11}
Electron spin resonance	$10^{-4}-10^{-8}$*
Nuclear magnetic resonance	$10^{-1}-10^{-9}$*
Mössbauer resonance (iron)	10^{-7}
Experimental separation of isomers	$> 10^{2}$

* Time-scale depends markedly upon the system under examination.

molecules like $Hg(CH_3)_2$ provides a more familiar aspect of this problem: the period of rotation of the CH_3 groups is such that those experimental methods working on a time-scale longer than $\sim 10^{-13}$ s (see Table 2) identify only a dynamic model for the molecule. The accurate point-group of $Hg(CH_3)_2$ is that of neither the eclipsed (D_{3h}) nor staggered (D_{3d}) forms; instead its elements are *permutations* of the positions and spins of identical nuclei.[11] In systems like PF_5, IF_7, $[Mo(CN)_8]^{4-}$ and $[ReH_9]^{2-}$ the non-rigidity arises, probably, from low-energy deformations of the framework, which provide a facile means of interconverting different geometrical arrangements.[5, 12] So PF_5 is recognizable geometrically as a trigonal bipyramid by its vibrational spectra,[12, 13] although the [19]F n.m.r. spectrum indicates magnetic equivalence of the fluorine atoms. Ideally non-rigid structures should be referred to the over-all potential energy surface[14] encompassing all possible geometries for a given aggregate. Unfortunately no physical measurement is at present able to furnish reliable details of this sort, and most discussions lean heavily on guesswork. Because of stereochemical non-rigidity geometric and optical isomers of arrangements of unusual coordination number (5, 7, 8 and 9) appear to be very short-lived in solution.[1, 15] So far only one such optically active system has been successfully resolved,[16] namely a 5-coordinate vanadium (IV) complex (Fig. 1), and even here the true coordination number of the metal atom in solution could well be 6 rather than 5. In the solid phase particular isomers are usually stabilized by the steric requirements of either the ligands or the over-all packing of units, though isomerization is a possible cause of polymorphism; for example, $Th(acac)_4$.[10]

The lack of definitive structural results for the liquid phase and the low volatility of most compounds of the class considered here has caused a rather lop-sided dependence upon X-ray crystallographic data. For proper definition of the environment of an atom it is true that the full power of a three-dimensional X-ray analysis of a single crystal is often the only satisfactory method of attack; on the other hand, the strong interactions peculiar to the solid phase

Fig. 1. (a) R = R′ = R″ = Me; (b) R′ = R″ = Me, R = Ph.

may well stabilize a configuration which is not maintained in the liquid or gas phase. Hence solid phase data are not a reliable index to ground-state geometries and coordination numbers. This point is illustrated by the following examples: (a) indium in trimethylindium is apparently 5-coordinate in the solid yet 3-coordinate in the liquid and gas phase;[17] (b) tantalum achieves coordination numbers of 6, 7 and 8, respectively, in the solid complexes $CsTaF_6$, K_2TaF_7 and Na_3TaF_8, but apparently forms only TaF_6^- and TaF_7^{2-} in HF solutions;[18] (c) MX_5 systems (M = Nb, Ta, Mo; X = halogen) are associated (with octahedral coordination of the metal) in the solid but monomeric in the vapour.[2] X-ray powder techniques are decidedly limited, and even the familiar isomorphism test, as judged by similarity of cell edges, is no guarantee that two compounds are strictly isostructural.[19] The need for precise data is underlined both (a) by the similarity of some of the geometries appropriate to a particular coordination number,[1] and (b) by the possibility that relatively distant atoms constitute part of the environment of a given atom. Once ligands impose constraints on a system the distinction between idealized geometries is commonly insignificant; this is well illustrated by the history[20] of the complex terpyridyldichlorozinc(II), which has been ascribed the basic geometry first of a trigonal bipyramid and then of a square pyramid. The inclusion of relatively remote neighbours in the coordination domain of an atom is, however, a highly relevant feature. Thus, a suggestive detail of the crystal structures of numerous pentacoordinate transition metal complexes with square-pyramidal stereochemistry is the 'blocking' of what would be the sixth site of an octahedron by some part of one of the ligands (e.g. $(Ph_3P)_3RuCl_2$, where a hydrogen of one of the phenyl rings effectively occupies the sixth position of an octahedral array[19]). Whether the extra ligand is weakly bonded is usually a moot point. Even from X-ray diffraction studies, therefore, it is not always possible to assign unambiguous coordination numbers; the process is sometimes quite arbitrary since the interatomic distances range from those short enough to imply strong interaction to those long enough to be excluded from coordination; for example, LaF_3.[1]

The gas phase is the only state which provides a virtually unperturbed environment for a molecule; strictly, the preferred ground-state structure can be identified only in this state. Possible methods of structural analysis are (a) microwave spectroscopy, which demands a permanent dipole moment of the molecule; (b) electron diffraction, which has recently undergone a minor renaissance, though the debate about the $Fe(CO)_5$ molecule[21] illustrates that the reliability of the method is not universally accepted; (c) vibrational–rotational spectroscopy. These techniques are usually presumed to define a single ground-state geometry for a molecule. If the potential energy surface of the system possesses more than one minimum, however, with a comparatively small barrier to interconversion (possibly PF_5, $(SiH_3)_2O$, etc.), the exact interpretation of, say, electron diffraction data is less obvious. A detailed study of the effect of temperature on the electron diffraction pattern of a molecule like PF_5 or IF_7 might help to clarify this important issue. Unfortunately there are not many systems of unusual coordination number that can be examined in the gas phase for reasons either of stability or volatility.

Coordination properties in the liquid- or solution-state are the most difficult to define. Molecular weight, conductivity and vapour pressure measurements apart, we depend for structural information upon vibrational and electronic spectra; the inherently more powerful and direct technique of n.m.r. spectroscopy is restricted by the stereochemical non-rigidity already mentioned. There are two serious drawbacks to the use of vibrational and electronic spectroscopy to investigate the environment of an atom:

(1) The strict point-group of a system must allow for facile torsional or, possibly, deformation modes. Moreover, many ligands are highly unsymmetrical and the complexes formed therefrom possess little or no symmetry; in these circumstances the symmetry is commonly specified simply by the central (acceptor) atom and the donor atoms surrounding it (with neglect of all other substituents). This is justifiable only on empirical grounds and can, at best, be used only to distinguish between models of widely differing geometries. To deduce information about the symmetry of a unit by ignoring certain geometric properties requires the greatest caution.

(2) Failure to observe a spectroscopic feature is not a positive test of the selection rules since it may simply mean that the band is too weak to be observed. This is the basic weakness of the method. A typical use of infrared spectroscopy relates to the monomeric ruthenium complexes $(Ph_3M)_2Ru(CO)_3$ (M = P or As), which exhibit[22] only one infrared-active C–O stretching mode; this is consistent with, *but does not establish*, a trigonal bipyramid configuration with equatorial CO groups. On the other hand, the infrared spectrum of $HMn(CO)_5$ formerly led to the erroneous conclusion that the geometry of the $Mn(CO)_5$ unit is a trigonal bipyramid.[23] Nevertheless, the combination of infrared and Raman spectra usually provides a fair indication of molecular geometry, though even for simple molecules the strict selection rules are not always followed either because of molecular interactions (e.g. CS_2)[24] or dynamic effects (e.g. free rotation of methyl groups). The most baffling results are those for molecules like $(SiH_3)_2O$, which is

now known to have a bent Si–O–Si skeleton but which, according to the usual criteria of vibrational spectroscopy, behaves as a linear molecule;[25] it is strange that infrared absorption and Raman scattering, so totally different in mechanism, should lead quite independently to the same, apparently misleading conclusion.

3. Conditions affecting coordination numbers

Unquestionably the preferred coordination number adopted by an atom or ion, A, in conjunction with suitable ligands, L, depends principally upon (1) size and steric effects, (2) the electronic configuration, oxidation state and energetically significant orbitals on A, (3) electrostatic and crystal-field stabilization, (4) the nature of L. The importance of the different factors varies widely according to whether the interaction between A and L is predominantly electrostatic or covalent. In ionic compounds steric and electrostatic influences are dominant, whereas in covalent systems the requirements of bonding, whether localized or delocalized, and electronic interactions are uppermost. In reality some interpolation between the two extremes is usually required. Attention is now directed to the various factors.

3.1. Size and steric effects

Idealized coordination numbers can be calculated[26] by assuming close-packing of ligands on the surface of the central atom. If d is the metal-ligand distance and r the van der Waals radius of the ligand, the coordination number, CN, is given by

$$\mathrm{CN} = \frac{4\pi}{2\sqrt{3}} \left(\frac{d}{r}\right)^2 \left(\frac{1}{1 - (r^2/8d^2)}\right).$$

Values thus obtained, tabulated in Table 3, follow roughly the observed trends in maximum coordination number. The main difficulty about this approach concerns the 'sizes' of atoms, which cannot be specified unequivocally because of their marked sensitivity to changes of oxidation state or bonding.

To a first approximation, the disposition of an ionic aggregate is determined by its composition and by the well-known *radius-ratio*

TABLE 3. *Theoretical coordination numbers for close-packed ligands*

| Central atom | Ligand | | | |
	F	Cl	Br	I
C^{IV}	4·3	4·0	4·0	4·0
Si^{IV}	5·4	5·1	5·0	5·2
Ti^{IV}	8·0	5·8	5·6	5·5
Zn^{II}	6·9	6·4	5·9	5·1
Zr, Hf^{IV}	8·7	6·5	6·3	6·1
Sn^{IV}	9·0	6·5	6·2	6·1
La^{III}	12·1	9·6	9·0	8·5
Lu^{III}	10·0	8·3	7·9	7·5
Pt^{IV}	7·7	6·6	6·4	6·1
Hg^{II}	11·9	6·3	6·0	5·7
Th^{IV}	12·0	8·0	6·7	8·4

rule, which imposes a lower limit on the ratio of the radius of the central ion to that of the surrounding ions for each type of co-ordination polyhedron. In practice, small cations often retain an environment of high coordination number even when the radius ratio falls below the limit for that environment. However, minimization of electrostatic energy may ultimately favour distortion of the regular environment[27] with the formation of unusual co-ordination polyhedra (e.g. 7-coordinated zirconium in Baddeleyite, ZrO_2, or 5-coordinated vanadium in V_2O_5). Severe restrictions on the choice of environment are imposed by the indefinite spatial extension of a structure. For this reason cubic 8-fold coordination is stable for a crystal as a whole although for individual ions it is energetically inferior to a square-antiprismatic arrangement; under these conditions the stability of individual coordination polyhedra is 'sacrificed in the interest of long-range order'.

The effect of bulky ligands in forcing unusual environments upon a central atom is familiar: for example, di-t-butylberyllium, unlike dimethylberyllium, apparently involves di-coordinated beryllium at room temperature.[28] According to the evidence so far accumulated—for the solid phase at least—the geometric constraints imposed by multidentate ligands are no less stringent in determining the exact environment of the central atom. An ideal illustration is the 10-coordinate complex $La(OH_2)_4AH$ (H_4A = ethylenediaminetetraacetic acid) with the structure shown in Fig. 2;[29] here the size of the metal ion and the constraints of the polydentate ligand and of chelation are primarily responsible for the selection of coordination polyhedron. A tetradentate ligand such as $N(CH_2.CH_2.NMe_2)_3$ in conjunction with bivalent $3d$ transition metal ions[30] apparently favours configurations with the basic geometry of a trigonal bipyramid, which preserves the trigonal symmetry of the three terminal donor atoms (Fig. 3); on the other hand, the less-exacting steric requirements of the parent ligand $N(CH_2.CH_2.NH_2)_3$ place it on the borderline of compatibility with 5- or 6-fold coordination. Numerous other examples confirm that apparently trivial changes in ligand substituents can

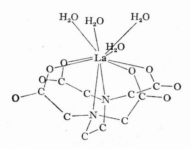

Fig. 2. Structure of $La(OH_2)_4(HEDTA)$.

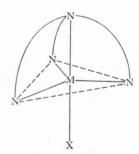

Fig. 3. M = divalent $3d$ transition metal ion; X = halogen.

have remarkable consequences on coordination behaviour. Multidentate ligands are classified[1,10] according to their rigidity, shape and 'bite' (the preferred separation of the donor atoms), and coordination geometry may well hinge on such details as the matching of the 'bite' with the edges of possible coordination polyhedra.

3.2. Nature of the central atom

It is helpful formally to treat the central atom of an aggregate as though it possessed a formal charge equal to the oxidation number. Covalent interactions transfer charge between the atom and its ligands in accordance with the Electroneutrality Principle,[31] which requires that, for stability, the charge on any atom shall not greatly exceed $\pm \frac{1}{2}e$. For a central atom in a high positive oxidation state charge-transfer may be achieved[32] either with a small number of polarizable ligands (e.g. O^{2-}) or a larger number of less polarizable ligands (e.g. F^-). Accordingly the donor character of the ligands and the acceptor character of the central atom (i.e. whether 'hard' or 'soft') are clearly significant, if not precisely defined, properties.

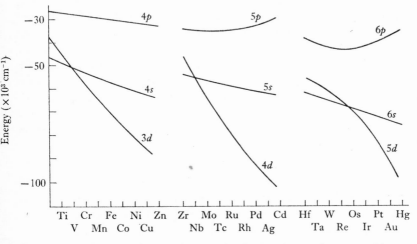

Fig. 4. Orbital ionization energies. Values for first- and second-row transition elements are from ref. 3; *approximate* values for the third-row elements have been calculated from Moore's data ('Atomic Energy Levels', National Bureau of Standards, Washington, 1958).

The acceptor ability of the central atom is a function of its oxidation number, electronic configuration and the number of potentially bonding orbitals. According to the simple valence-bond approach the coordination number of the atom is equal to the number of bonding orbitals. For transition metals in low-spin states this leads to the classical '9-orbital' rule, and the maximum coordination number is determined by the number of formally non-bonding d electrons. Both valence-bond and molecular-orbital treatments require that the bonding orbitals should have similar energies; increasing energy separation between the orbitals inevitably prejudices the attainment of the maximum coordination number. Since in a transition series the energy separation between the $(n-1)d$, ns and np orbitals tends to rise quite sharply with increasing atomic number (see Fig. 4), elements early in the series with less than four non-bonding d electrons are singularly well adapted[3] to the attainment of coordination numbers 7, 8 and 9; conversely, atoms with a full or nearly full complement of d electrons are better disposed to the attainment of relatively low co-ordination numbers (e.g. the d^{10} ions Cu^+ and Hg^{2+}). Similar arguments apply to 5-coordinate d^8 complexes like $M(triarsine)X_2$ (M = Ni, Pd, Pt; triarsine = $(Me_2As.CH_2CH_2CH_2)_2AsMe$; X = Cl, Br, I), the stability of which[33] rises as the $(n-1)d \rightarrow np$ energy separation decreases; this effect is in turn favoured by a low oxidation state and effective nuclear charge for the metal (i.e. $Ni^{II} < Co^I < Fe^0$; $Pt^{II} < Pd^{II} < Ni^{II}$). However, the plausibility of such correlations should not disguise the fact that d^0 and d^{10} metal ions of similar size (e.g. Ti^{4+}, Sn^{4+}) sometimes display comparable tendencies to adopt environments of high coordination number.

Simple valence-bond theory suggests that each orbital contributes to bonding in accordance with its radial distribution and energy. Eight hybrid orbitals can be derived, for example, from d, s and p orbitals of suitable energy; they are directed towards the corners either of a square antiprism or of a dodecahedron (corresponding to the two idealized geometries of 8-coordination), and the optimum shapes correspond remarkably closely with those observed.[3] There are, however, severe limitations to this approach. (a) Overlap calculations are highly approximate because of our limited knowledge of the real shapes of orbitals, particularly d-orbitals. (b) The relative involvement of orbitals cannot be precisely prescribed as dsp^3, d^4sp^3, etc.; in principle a complex may exist solely because of s-orbital or p-orbital or d-orbital interactions, or because of any combination of these. (c) π-Bonding is likely to be an important structural influence in some systems, e.g. transition metal oxy complexes. Participation of f orbitals deserves consideration[34] when the point-group symmetry of an aggregate possesses an irreducible representation spanned by the f-orbitals but not by s-, p- and d-orbitals; geometries which meet this requirement include the 8-coordinate cube and the 12-coordinate icosahedron. Unfortunately the complexity of solid phase interactions renders crystal structure data quite inadequate as a test for this participation; the observed crystal structures of lanthanide EDTA complexes have in fact[29] been taken to suggest that f-orbitals have no voice in determining the gross configuration.

That the number of ligands equals the number of bonding orbitals is supported by much chemical experience, but there is no causal relationship between the two. In fact, relaxation of this requirement occurs when the representations spanned by the σ-bonding orbitals cover the same irreducible representation more than once. For example, in a hypothetical L–M–L complex where the central atom and ligands each contribute one orbital, bonding may result if the molecular orbitals corresponding to the two lowest roots of the 3×3 secular determinant are occupied; one M orbital then 'does the work of two'.[34] In outline this corresponds to Rundle's 3-centre-4-electron bond.[26, 35] The same principle embodied in molecular–orbital terms extends from systems of low coordination number (e.g. I_3^-, XeF_2) to those of high coordination number (e.g. polyboranes and metal 'clusters'), and from electron-deficient aggregates at one extreme to electron-rich aggregates at the other. What is so attractive about this molecular–orbital approach is the initial neglect of energetically inferior orbitals (e.g. the d-orbitals in molecules like XeF_2 and PF_5), although the contributions of these and of π-interactions would be incorporated in a more sophisticated account.[36]

Alternatively stereochemistry can be considered to arise less from bonding influences than from electron-pair repulsions.[37] Rather surprisingly the stereochemical consequences of the molecular–orbital model, which neglects explicit electron repulsions, are usually not very different from those of the Gillespie–Nyholm model, which considers *only* such repulsions. In reality it seems more reasonable to divide systems into two categories according to whether the choice of configuration is dominated by repulsive or attractive forces,[38] which in turn are a function of the strength of bonding between a central atom and its environment. The present theoretical impasse illustrates the difficulty of applying any sort of deductive approach to the structure of inorganic compounds.

3.3. Electrostatic effects

Aggregates which are predominantly ionic can be discussed in terms of electrostatic interactions. For the stability of an ionic aggregate in which an ion M^{z+} is surrounded by n X^- ions, the

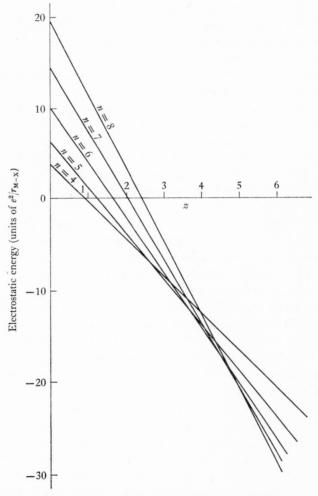

Fig. 5. Net electrostatic energies for complexes $M^{z+}(X^-)_n^{(n-z)-}$ of various co-ordination numbers. M–X distances are assumed identical and shielding by the metal ion is ignored (ref. 3).

Coulombic attraction between M^{z+} and the ligands X^- must at least balance the repulsions between the ligands. As the coordination number, n, increases so the repulsion between the ligands increases; to balance this and maintain electrostatic stability demands higher values of z, the charge on M. Figure 5 illustrates[3, 39] how the net electrostatic energy of such an aggregate varies both with n and z. According to these data high coordination numbers should be favoured by relatively high oxidation states. In practice the limit is usually set not by the energetics of the ionic system but by the increased polarizing power of the M^{z+} ion in the higher oxidation states. This leads to significant covalent contributions, which render the simple ionic model inapplicable for oxidation states outside the range $+1$ to $+4$; even within this range highly polarizable anions often lead to molecular rather than ionic systems. Calculation of electrostatic energies also reveals that the stability of a discrete unit is highly sensitive to the physical state. This is exemplified by the following enthalpies (given as ΔH in kcal/mole):[40]

$$AlF_6^{3-}(g) + F^-(g) \rightarrow AlF_7^{4-}(g) \qquad +182$$
$$AlF_6^{3-}(aq.) + F^-(aq.) \rightarrow AlF_7^{4-}(aq.) \qquad +10$$
$$AlF_6^{3-}(g) \rightarrow AlF_5^{2-}(g) + F^-(g) \qquad -59$$
$$AlF_6^{3-}(aq.) \rightarrow AlF_5^{2-}(aq.) + F^-(aq.) \qquad +39$$

If the central ion M^{z+} possesses non-bonding valence electrons these may interact strongly with the environment, producing 'crystal-field' effects. The crystal-field stabilization energies of d-block transition metal ions in fields of various geometries are depicted[3, 40] as a function of the d-electron configuration in Fig. 6. The advantages of high coordination numbers (7 or 8) for metal ions with few d electrons are self-evident. The resulting gain in stability may amount to 5–10 kcal/mole; since this is probably comparable with the net energy balance between some of the different arrangements, crystal-field stabilization energies are quite capable of exercising a structure-determining influence. For example, the stability of 5-coordinate complexes of d^8 ions is attributable[41] to the energy balance between the formation of a fifth bond and the loss of crystal-field stabilization energy accom-

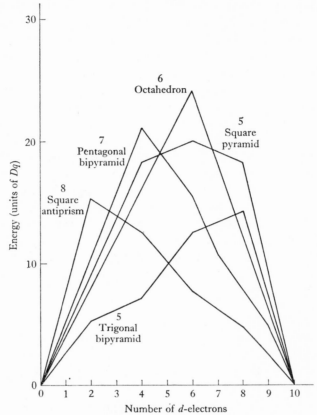

Fig. 6. Crystal-field stabilization energies for various coordination numbers. These data are based on the assumption of one-electron energy levels and low-spin states (for energy separations $> 3Dq$).

panying the change from a 4-coordinate (square-planar) to a 5-coordinate structure.

The electronic properties of certain ions are incompatible with a highly symmetrical environment; distortions result, and in extreme situations these may lead to arrangements of unusual coordination number. Two types of metal ion exemplify this feature, namely (a) those with d shells of less than spherical symmetry (Jahn–Teller distortion) and (b) those with $(n-1)d^{10}$ or $(n-1)d^{10}ns^2$ configurations possessing low-lying excited states ($(n-1)d^9ns^1$ or $(n-1)d^{10}ns^1np^1$, respectively).

In addition to the electrostatic terms so far discussed, factors normally regarded as 'secondary' may well be critical to the choice of coordination environment. For example, London dispersion attractions between ions in close proximity tend to favour high coordination numbers; these forces contribute as much as 20–30 kcal/mole to the lattice energy of a crystalline (6:6) silver halide,[42] and possibly dictate the adoption of the CsCl structure by the caesium halides (except CsF). The effect of hydrogen-bonding on the structural chemistry of the ammonium ion suggests a second factor. Likewise the multipole interactions[43] of non-spherical ions like HF_2^- and NO_3^- cannot reasonably be neglected. The solid fluoride complexes Na_3TaF_8, K_2TaF_7 and $CsTaF_6$ illustrate only too clearly[10] the nicety of choice that may exist, not only of configurational type, but even of chemical constitution.

3.4. Nature of ligand

Some properties of the *environment* of a given atom (or ion) having a major effect on the coordination number have already been indicated, namely steric factors, electronegativity and polarizability. These factors subscribe in the first place to the balance between electrostatic and covalent interactions. By causing potential bonding and non-bonding orbitals on the central atom to contract,[44] ligands of high electronegativity are capable either of promoting covalent bonding (for non-transition elements) or restricting it (for transition elements). The effects of polarizability are similarly ambivalent. Although ligands of low polarizability favour arrangements of high coordination number, high ligand polarizability does not always imply the other extreme, and 'soft' donors like As and CN^- are capable of forming aggregates of high coordination number.

Polydentate ligands generate environments only partly controlled by the central atom, and it seems that almost any geometry can be established by a suitable 'tailoring' of the ligands. If differences in direct bonding interactions are small, stabilization of a particular arrangement is determined[1, 10] by the optimum combination of (*a*) non-bonding repulsions in the primary coordination sphere, (*b*) Coulombic repulsions, and (*c*) geometrical constraints due to

E N P

the stereochemistry of the ligands. The structures[1] of such species as

$$[Fe(OH_2)EDTA]^-, \quad [M(NTA)_2]^{2-} \text{ (M = Zr}^{IV}, Hf^{IV};$$
$$NTA = \text{nitrilotriacetate ion)},$$
$$[La(OH_2)_3EDTA]^-, \quad \text{and} \quad [La(OH_2)_4(HEDTA)]$$

in the solid phase substantiate this view. Incorporation of more than one type of ligand complicates still further the selection of environment; if the energy differences between different idealized stereochemistries are small, it has been suggested[45] that the coordination number adopted is likely to be controlled as much by stoicheiometry as by more sophisticated properties.

Whereas the nature of the ligands is commonly secondary to that of the central atom in determining the preferred coordination number of the latter, there are situations where the roles are reversed. Thus, if the ligands form part of a polyhedron or cluster, the structural requirements of individual atoms are subjugated to those of the unit as a whole. Coordination numbers of

$$5 \text{ (e.g. } B_5H_9), \quad 7 \text{ (e.g. } Co_4(CO)_{10}(EtC{\equiv}CEt)),$$
$$8 \text{ (e.g. } Rh_6(CO)_{16})$$

and $\quad 9$ (e.g. $[M_6Cl_{12}(OH_2)_6]^{2+}$ $[M = Nb, Ta]$)

are encountered for individual atoms, and even atoms as small as boron are found in 7-coordinate ($B_{18}H_{22}$) and 8-coordinate ($Co_{21}V_2B_6$) environments.[1,2] Extensive multi-centre bonding is the dominating influence controlling the formation, not only of such aggregates, but also of crystalline metals, alloys and compounds such as the metal borides and silicides, which are characterized by unusually high coordination numbers. Polyhedral aggregates apart, certain ligands show a pronounced tendency to coordinate via multi-centre bonds. A simple example is the nitrate ion, which gives rise to such 8-coordinate species as $[Co(NO_3)_4]^{2-}$[,46] by acting as a bidentate ligand with a very short 'bite'. It has been proposed,[46] as a general structural principle, that a ligand which contains two chemically equivalent donor atoms much closer together than the sum of the non-bonding radii tends to interact through both atoms so that the mean positions of the pairs of atoms lie roughly at the vertices of one of the usual coordination polyhedra. For $[Co(NO_3)_4]^{2-}$ this polyhedron is a somewhat distorted tetrahedron. In the solid complex $Mg_3Ce_2(NO_3)_{12}.24H_2O$ each Ce^{IV} ion is surrounded by 6 bidentate nitrate groups such that the oxygen atoms furnish a 12-coordinate icosahedral environment for the metal ion;[47] the midpoints of the intranitrate O...O units form approximately the vertices of an octahedron. The same rationale applies to unsaturated organic groups such as C_6H_6, C_5H_5 and $CH_2{=}CH_2$, in which the donor atoms are directly bonded. The true coordination numbers of the metal atoms in the molecules $(\pi\text{-}C_5H_5)_2Fe$ and $(\pi\text{-}C_5H_5)_2WH_2$ are 10 and 12, respectively; lower values (6 and 8) suggested elsewhere, implying relatively localized metal–ligand bonding, are open to ambiguity. Clearly the occurrence of multi-centre bonding enormously enlarges the scope for 'unusual' coordination numbers, though these tend to lack chemical significance as individual atomic properties.

REFERENCES

1   E. L. Muetterties and C. M. Wright, *Q. Rev. chem. Soc.* 1967, **21**, 109.
2   E. L. Muetterties and R. A. Schunn, *Q. Rev. chem. Soc.* 1966, **20**, 245.
3   R. V. Parish, *Coordination Chem. Rev.* 1966, **1**, 439.
4   T. D. O'Brien, *The Chemistry of the Coordination Compounds* (ed. J. C. Bailar, Jr.), p. 382. New York: Reinhold, 1956.
5   E. L. Muetterties, *Inorg. Chem.* 1965, **4**, 769.
6   C. D. Cook, R. S. Nyholm and M. L. Tobe, *J. chem. Soc.* 1965, p. 4194.
7   L. M. Venanzi, *Angew. Chem., Internat. Edn*, 1964, **3**, 453.
8   O. Ya. Samoilov, *Discuss. Faraday Soc.* 1957, **24**, 141.
9   F. A. Cotton and G. Wilkinson, *Advanced Inorganic Chemistry*, 2nd ed., p. 163. London: Interscience, 1966.
10 J. L. Hoard and J. V. Silverton, *Inorg. Chem.* 1963, **2**, 235, 243. G. L. Glen, J. V. Silverton and J. L. Hoard, *ibid.* 1963, **2**, 250.
11 H. C. Longuet-Higgins, *Molec. Phys.* 1963, **6**, 445.
12 R. S. Berry, *J. chem. Phys.* 1960, **32**, 933.
13 J. E. Griffiths, R. P. Carter, Jr., and R. R. Holmes, *J. chem. Phys.* 1964, **41**, 863.
14 A. D. Liehr, *J. phys. Chem.* 1963, **67**, 471.
15 E. L. Muetterties, *Inorg. Chem.* 1967, **6**, 635.
16 K. Ramaiah, F. E. Anderson and D. F. Martin, *Inorg. Chem.* 1964, **3**, 296. K. Ramaiah and D. F. Martin, *J. inorg. nucl. Chem.* 1965, **27**, 1663.

17 E. L. Amma and R. E. Rundle, *J. Am. chem. Soc.* 1958, **80**, 4141. J. R. Hall, L. A. Woodward and E. A. V. Ebsworth, *Spectrochim. Acta*, 1964, **20**, 1249.

18 O. L. Keller, Jr., and A. Chetham-Strode, Jr., *Inorg. Chem.* 1966, **5**, 367.

19 J. A. Ibers, *A. Rev. phys. Chem.* 1965, **16**, 375.

20 Ref. no. 2, pp. 275, 298.

21 J. Donohue and A. Caron, *J. phys. Chem.* 1966, **70**, 603.

22 J. P. Collman and W. R. Roper, *J. Am. chem. Soc.* 1965, **87**, 4008.

23 S. J. La Placa, W. C. Hamilton and J. A. Ibers, *Inorg. Chem.* 1964, **3**, 1491.

24 J. C. Evans and H. J. Bernstein, *Can. J. Chem.* 1956, **34**, 1127.

25 R. C. Lord, D. W. Robinson and W. C. Schumb, *J. Am. chem. Soc.* 1956, **78**, 1327.

26 R. E. Rundle, *Surv. Progr. Chem.* 1963, **1**, 81.

27 J. D. Dunitz and L. E. Orgel, *Adv. inorg. Chem. Radiochem.* 1960, **2**, 1.

28 G. E. Coates, P. D. Roberts and A. J. Downs, *J. chem. Soc. (A)*, 1967, p. 1085.

29 M. D. Lind, B. Lee and J. L. Hoard, *J. Am. chem. Soc.* 1965, **87**, 1611, 1612.

30 M. Ciampolini and N. Nardi, *Inorg. Chem.* 1966, **5**, 41, 1150. M. Ciampolini, N. Nardi and G. P. Speroni, *Coordination Chem. Rev.* 1966, **1**, 222.

31 L. Pauling, *J. chem. Soc.* 1948, p. 1461.

32 R. S. Nyholm, *Proc. chem. Soc.* 1961, p. 273.

33 G. A. Barclay, R. S. Nyholm and R. V. Parish, *J. chem. Soc.* 1961, p. 4433.

34 S. F. A. Kettle and A. J. Smith, *J. chem. Soc. (A)*, 1967, p. 688.

35 R. E. Rundle, *J. Am. chem. Soc.* 1963, **85**, 112.

36 L. S. Bartell, *Inorg. Chem.* 1966, **5**, 1635.

37 R. J. Gillespie and R. S. Nyholm, *Q. Rev. chem. Soc.* 1957, **11**, 339. R. J. Gillespie, *J. chem. Educ.* 1963, **40**, 295; *J. chem. Soc.* 1963, pp. 4672, 4679.

38 B. F. Gray and H. O. Pritchard, *J. chem. Phys.* 1956, **25**, 779.

39 R. S. Nyholm and M. L. Tobe, *Experientia*, 1964, Suppl. 9, p. 112.

40 F. Basolo and R. G. Pearson, *Mechanisms of Inorganic Reactions*, pp. 55, 76. New York: Wiley, 1958.

41 G. Dyer and L. M. Venanzi, *J. chem. Soc.* 1965, p. 2771.

42 C. S. G. Phillips and R. J. P. Williams, *Inorganic Chemistry*, vol. 1, p. 177. Oxford, 1965.

43 T. C. Waddington, *Adv. inorg. Chem. Radiochem.* 1959, **1**, 157.

44 D. P. Craig and E. A. Magnusson, *J. chem. Soc.* 1956, p. 4895.

45 D. L. Kepert, *J. chem. Soc.* 1965, p. 4736.

46 J. G. Bergman, Jr., and F. A. Cotton, *Inorg. Chem.* 1966, **5**, 1208.

47 A. Zalkin, J. D. Forrester and D. H. Templeton, *J. chem. Phys.* 1963, **39**, 2881.

3

GALLIUM HYDRIDE AND ITS DERIVATIVES

N. N. GREENWOOD

1. Introduction

The chemistry of gallium hydride derivatives has developed rapidly since the independent discovery[1-3] of the first stable adduct $Me_3N.GaH_3$. Before that time there had been a long and unrewarding search for convenient routes to uncoordinated gallium hydride and its derivatives. Early experiments with metallic gallium and hydrogen showed little positive result[4-10] and stoicheiometric compounds were not isolated When gallium is vaporized at 2300 °C in an atmosphere of hydrogen and helium, ultraviolet absorption lines are observed at 2160–2400 Å which have been assigned[11] to the diatomic radical Ga–H. The bond length is 1·72 Å. Characteristic lines in the region 4210–4250 Å and in the visible spectrum near 5700 Å have also been assigned to Ga–H and Ga–D.[12] The most recent analysis of such spectra[13] puts the bond dissociation energy between 2·79 and 2·94 eV (about 65–67 kcal.mole⁻¹). This suggests that the bond is quite reasonably strong but that the compound GaH is unstable because of the considerable energy released on disproportionation into metallic gallium and molecular hydrogen:

$$GaH(g) \rightarrow Ga(l) + \tfrac{1}{2}H_2(g); \quad -\Delta H \simeq 57 \text{ kcal.mole}^{-1}$$

This arises from the very high heat of vaporization of gallium (71 kcal.mole⁻¹) and the high bond-dissociation energy of hydrogen (104 kcal.mole⁻¹).

Wiberg and his group were the first to report the isolation of definite hydrides of gallium. They described the preparation of a liquid thought to be digallane, Ga_2H_6, by the action of an electric discharge on a mixture of trimethylgallium and hydrogen, followed by disproportionation in the presence of triethylamine.[14-16] However, this synthesis has since been discounted,[17,18] and the properties purporting to belong to digallane, which have been repeated so often in textbooks and other secondary sources, are no longer thought to be those of the compound Ga_2H_6. Likewise, reports of a solid polymeric gallium hydride $(GaH_3)_x$, prepared by the reaction of ethereal solutions of lithium aluminium hydride or lithium gallium hydride on gallium trichloride,[19] have not been confirmed.[17]

Uncoordinated gallium hydride is now known[20] to be a viscous liquid which decomposes to gallium and hydrogen above $-15\ °C$. Its preparation and properties will be discussed in § 10 of this review. Gallium hydride forms complexes of varying stability with a wide range of ligands and these will form the subject of the next sections of this review. Typical ligands (L) are H^-, Me_3N, Me_2NH, Me_3P, Me_3As, Me_2O, Me_2S, etc. It has also been possible to synthesize stable adducts of halogeno-gallanes $L . GaH_2X$ and $L . GaHX_2$ and the properties of these compounds will be considered in § 9.

2. Tetrahydridogallates, $MGaH_4$

Lithium tetrahydridogallate, $LiGaH_4$, was first isolated by Finholt, Bond and Schlesinger[21] during the course of their classic work on lithium tetrahydridoaluminate, $LiAlH_4$. It is prepared by the reaction of a large excess of finely powdered lithium hydride with an ethereal solution of gallium chloride:

$$GaCl_3 + 4LiH \rightarrow LiGaH_4 + 3LiCl$$

The reagents are mixed at $-80\ °C$ and allowed to warm to room temperature, during which time the reaction proceeds briskly. Gallium tribromide has also been used[22, 23] and later modifications of technique[3, 24, 25] have enabled the compound to be prepared rapidly in 80–95 % yield. This is of some importance since virtually all preparative work on complexes of gallium hydride uses lithium tetrahydridogallate as an intermediate at some stage.

$LiGaH_4$ is readily soluble in diethyl ether and is frequently used in reactions without isolation. It apparently forms a complex with diethyl ether since it is difficult to remove the last traces of solvent.[26–28] Further evidence for such interaction with the solvent comes from n.m.r. studies using the ^{71}Ga nucleus.[29] Resonance was observed at 20 Mc.s^{-1} using fields of about 15·4 kG. An ethereal solution of $LiGaH_4$ showed a chemical shift of 425 ppm downfield from aqueous $GaCl_4^-$ and this was the most negative shift observed. More significantly the line was very broad ($c.$ 2000 c.s^{-1} compared with $c.$ 90 c.s^{-1} for $GaCl_4^-$). This suggests that interaction of the GaH_4^- ion with the solvent destroys the tetrahedral symmetry of the ion thereby broadening the line by quadrupole relaxation of the ^{71}Ga nucleus which has a spin $I = \frac{3}{2}$.

$LiGaH_4$ is a white solid which decomposes slowly at room temperature and more rapidly at 150 °C:

$$LiGaH_4 \rightarrow LiH + Ga + \tfrac{3}{2}H_2$$

The decomposition is accelerated autocatalytically by the presence of finely divided metallic gallium. Ethereal solutions when carefully prepared and sealed in glass ampoules are stable indefinitely (years) at 0 °C.[25] As expected, there is a steady decrease in the stability of tetrahydrido-complexes in the sequence

$$LiBH_4 > LiAlH_4 > LiGaH_4 > LiInH_4 > LiTlH_4$$

Like its aluminium analogue, $LiGaH_4$ reacts vigorously and quantitatively with water to evolve 4 moles of hydrogen. Dissolved in ether it is a strong reducing agent, though a more moderate one than either $LiBH_4$ or $LiAlH_4$.[28, 30] It evolves hydrogen with primary and secondary amines, reduces acetamide and acetonitrile to ethylamine, and aliphatic acids, aldehydes and ketones to the corresponding alcohols. However, unlike $LiAlH_4$, it does not reduce aromatic nitriles, aldehydes, ketones, or esters.[28, 31]

Knowledge of other tetrahydridogallates is fragmentary. The thallium(III) derivative[32, 33] is formed in ethereal solution at -115 °C: $$TlCl_3 + 3LiGaH_4 \rightarrow Tl(GaH_4)_3 + 3LiCl$$

It can be isolated as a white solid but decomposes above -90 °C. The silver salt was obtained[34] by reacting ethereal solutions of silver perchlorate with a slight excess of lithium tetrahydridogallate at -100 °C: $$AgClO_4 + LiGaH_4 \rightarrow AgGaH_4 + LiClO_4$$

The reaction is quantitative and the complex is said to separate as an orange-red deposit, insoluble in diethyl ether; it is unstable and decomposes rapidly in ethereal solution at -75 °C.

3. Trimethylamine adducts of GaH₃

Trimethylamine forms two gallane complexes: $Me_3N \cdot GaH_3$ and $(Me_3N)_2 \cdot GaH_3$. The $1:1$ adduct, trimethylamine-gallane, is one of the most stable complexes of gallium hydride so far prepared.[1-3] Its importance stems from its use as a synthetic intermediate and from the fact that it is one of the very few volatile compounds containing metal–hydrogen bonds which can afford spectroscopic information in the gas phase, free from the complications of solvent shifts or crystal-lattice interactions.

Trimethylamine-gallane is prepared[2,3] by the reaction of lithium tetrahydridogallate on trimethylammonium chloride in diethyl ether below -78 °C and then allowing the reaction mixture to warm to 0 °C:

$$Me_3NHCl + LiGaH_4 \rightarrow Me_3N \cdot GaH_3 + LiCl + H_2$$

It is important to avoid an excess of Me_3NHCl in this preparation (§ 9). Hydrogen is evolved quantitatively and the product can be obtained as colourless needles, m.p. $70 \cdot 5$ °C, by vacuum sublimation. The vapour pressure is approximately 2 mm at room temperature[2] and the complex is monomeric in dimethyl ether at -31 °C.[3] The complex decomposes over a period of weeks under vacuum at room temperature but is more stable under nitrogen. The trideuterio-derivative, $Me_3N \cdot GaD_3$, m.p. $68 \cdot 1$ °C, is obtained similarly from $LiGaD_4$.[4]

The molecular structure of trimethylamine-gallane has been shown to have C_{3v} symmetry by infrared[1,2] and Raman[35] spectroscopy and this has been confirmed by X-ray crystallography.[36] The lattice is rhombohedral with one molecule per unit cell; $a = 5 \cdot 91$ Å, $\alpha = 106° 25'$, d (calc) $= 1 \cdot 254$ g.cm⁻³. The space group is probably $R3m(C_{3v}^5)$. The Ga–N bond distance is $1 \cdot 97 \pm 0 \cdot 09$ Å, the N–C bond distance $1 \cdot 47 \pm 0 \cdot 06$ Å and the C–N–C angle $105 \pm 10°$. These are all consistent with earlier observations on related compounds.

The gas-phase infrared spectra of $Me_3N \cdot GaH_3$ and $Me_3N \cdot GaD_3$ are summarized in Table 1 and the corresponding Raman data on the solids are recorded in Table 2. In both Tables the numerous C–H and C–N bands due to the ligand vibrations have been omitted for clarity. It will be noticed that the Ga–H stretching modes have been separated in the Raman data, the antisym. stretch appearing as a shoulder on the intense sym. vibration; this may be the result of interactions within the crystal. The spectra are closely similar to those of the corresponding aluminium hydride complexes $Me_3N \cdot AlH_3$ and $Me_3N \cdot AlD_3$.[37]

TABLE 1. *Gas-phase infrared data for* $Me_3N \cdot GaH_3$ *and* $Me_3N \cdot GaD_3$[1,2]

$Me_3N \cdot GaH_3$ (ν cm^{-1})	$Me_3N \cdot GaD_3$ (ν cm^{-1})	ν_H/ν_D	Assignment
1853 vs	1330 vs	1·39	$\nu_1 \nu_4$ sym. and antisym. Ga–H stretch
758 vs	542 vs	1·40	ν_5 antisym. Ga–H def.
715 w	(510)	(1·40)	ν_2 sym. Ga–H def.
523 m	—	—	ν_6 N–GaH$_3$ rock
488 m ⎫	487 m ⎫		
482 m ⎬ PQR	479 m ⎬ PQR	1·00	ν_3 Ga–N stretch
475 m ⎭	473 m ⎭		

Intensity: vs = very strong, m = medium, w = weak.

TABLE 2. *Raman data for solid* $Me_3N \cdot GaH_3$ *and* $Me_3N \cdot GaD_3$[35]

$Me_3N \cdot GaH_3$	$Me_3N \cdot GaD_3$	ν_H/ν_D	Assignment
1852	1325	1·40	ν_1 sym. Ga–H stretch
1823	1356	1·35	ν_4 antisym. Ga–H stretch
730	536	1·35	ν_2 sym. Ga–H def.
521	370	1·41	Unassigned (probably ν_6)

Trimethylamine-gallane undergoes numerous ligand-displacement reactions and these are discussed in §§4, 5 and 8. Its reactions with halogenating reagents are discussed in § 9.

Trimethylamine-gallane takes up a second mole of trimethylamine at low temperatures to give the bis adduct $(Me_3N)_2 \cdot GaH_3$.[2,3] A typical tensiometric titration at -21 °C is shown in Fig. 1. Bis(trimethylamine)-gallane has a dissociation pressure of 1 mm at

$-63\,°C$ and $378\,mm$ at $0\,°C$. Detailed measurements over the temperature range $-28°$ to $0\,°C$ lead to the equation[2]

$$\log p_{mm} = 10{\cdot}867 - 2265/T$$

from which the heat of dissociation is $10{\cdot}35\,\text{kcal.mole}^{-1}$ and the entropy of dissociation is $36{\cdot}5$ eu for the reaction

$$(Me_3N)_2.GaH_3(s) \rightarrow Me_3N.GaH_3(s) + Me_3N(g)$$

Independent measurements[3] at a more restricted number of temperatures gave

$$\Delta H = 11{\cdot}5\,\text{kcal.mole}^{-1} \quad \text{and} \quad \Delta S = 41{\cdot}1\,\text{cal.deg.}^{-1}\,\text{mole}^{-1}.$$

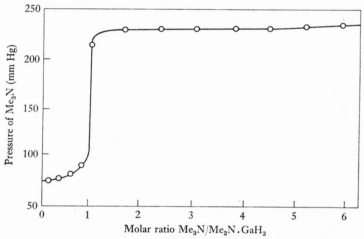

Fig. 1. Tensiometric titration of $Me_3N.GaH_3$ with Me_3N to form $(Me_3N)_2GaH_3$ at $-21\,°C$.

The infrared spectrum of the $2:1$ adduct in benzene showed seven bands in addition to those of coordinated trimethylamine.[2] It was not possible to obtain a vapour-phase spectrum because of the ready decomposition of the compound but the simplicity of the observed spectrum argues for a monomeric structure. The trigonal–bipyramidal structure in which the three hydrogen atoms are equatorial has D_{3h} symmetry; this structure, which has been established[37] for the corresponding alane $(Me_3N)_2.AlH_3$ has eight independent modes of vibration of which five are infrared active. The appearance of seven bands in the solution spectrum suggests that the forbidden modes involving the sym. Ga–H stretch and the sym. GaN_2 stretch may become allowed by interaction with the solvent. On this hypothesis and in the absence of data from the deuterium derivative, tentative assignments may be given as follows: 1850 cm^{-1} strong (antisym. Ga–H stretch); 1775 cm^{-1} medium (sym. Ga–H stretch); 780 cm^{-1} medium (GaH_3 out-of-plane deformation); 750 cm^{-1} strong (GaH_3 in-plane deformation); 527 cm^{-1} strong (N–Al–H bend); 495 cm^{-1} strong (antisym. GaN_2 stretch); 420 cm^{-1} very weak (sym. GaN_2 stretch).

4. Other amine derivatives of GaH_3

Dimethylamine quantitatively displaces trimethylamine from trimethylamine-gallane, thus showing that it is the stronger donor in this system:[38]

$$Me_3N . GaH_3(s) + Me_2NH(g) \rightarrow Me_2NH . GaH_2(l) + Me_3N(g)$$

Dimethylamine-trideutereigallane, $Me_2NH . GaD_3$, was prepared similarly. Both compounds are colourless, mobile liquids which can be purified by sublimation at reduced pressure. Dimethyl-amine-gallane was also prepared by the established reaction of lithium tetrahydridogallate on the appropriate hydrochloride in etherea. solution:[38]

$$Me_2NH_2Cl + LiGaH_4 \rightarrow Me_2NH . GaH_3 + LiCl + H_2$$

It is interesting to note that, although $LiGaH_4$ reacts smoothly with trimethylammonium chloride and dimethylammonium chloride in ether, no reaction occurs with tetramethylammonium chloride:[39]

$$Me_4NCl + LiGaH_4 \nrightarrow Me_4NGaH_4 + LiCl$$

The infrared spectrum[38] of $Me_2NH . GaH_3$ shows the characteristic intense Ga–H stretch at 1838 cm^{-1} and an even stronger Ga–H deformation at 752 cm^{-1}. The corresponding bands for $Me_2NH . GaD_3$ were at 1333 cm^{-1} and 543 cm^{-1}, corresponding to ratios ν_H/ν_D of 1·38. The N–GaH_3 rock appeared at 548 cm^{-1} and the Ga–N stretch at 527 cm^{-1}. This is somewhat higher than the value of 482 cm^{-1} found for $Me_3N . GaH_3$ (Table 1) and indicates the greater strength of the donor–acceptor bond in the secondary-amine complex.

Dimethylamine-gallane eliminates hydrogen very slowly at room temperature but, on standing under an atmosphere of dry nitrogen for three weeks, gives a quantitative yield of dimethyl-aminogallane, Me_2NGaH_2, as a white, crystalline solid:[38]

$$Me_2NH \cdot GaH_3(l) \rightarrow Me_2NGaH_2(s) + H_2(g)$$

The compound is dimeric in benzene and in this respect parallels the analogous boron compound, $(Me_2NBH_2)_2$.[40] The corresponding alane derivative is trimeric in benzene solution.[41]

The ease of intramolecular hydrogen elimination from the adducts $Me_2NH \cdot MH_3$, where M is B, Al or Ga, follows the sequence $B < Ga < Al$: $Me_2NH \cdot BH_3$ loses hydrogen only slowly below 100 °C, $Me_2NH \cdot GaH_3$ evolves hydrogen slowly at room temperature, whereas the liberation of a molar amount of hydrogen from $Me_2NH \cdot AlH_3$ is rapid at -10 °C.[42] This sequence can be explained by the relative electronegativity values of the atoms involved. Thus the hydrogen atom directly attached to nitrogen loses electron density on formation of the dative bond whereas the hydrogen atoms attached to the acceptor, M, simultaneously acquire an increased electron density. The electrical strain thereby created in the adduct molecule is relieved when elimination of one molecule of hydrogen occurs as shown below.

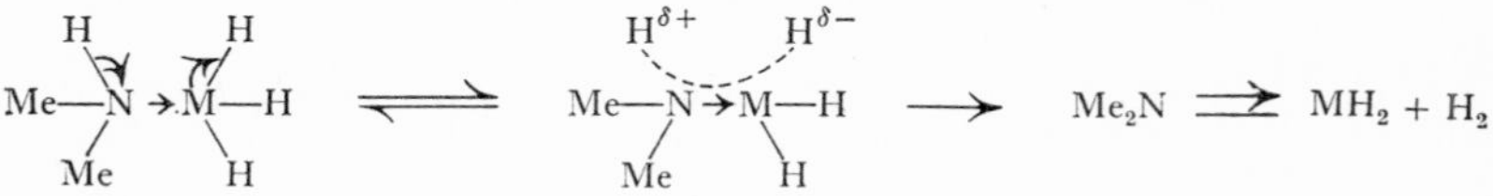

The electronegativities of the three elements, M, increase in the order $Al < Ga < B$, the values on the Allred–Rochow scale being 1·47, 1·82, 2·01 respectively, whilst that of hydrogen is 2·1. It follows that bond polarity of the type $M^{\delta+}—H^{\delta-}$, and hence hydridic character in MH_3 and ease of hydrogen elimination, should increase from B through Ga to Al as observed. Reaction may then proceed further to give cyclic dimers or trimers.

Although dimethylaminogallane is dimeric in benzene and presumably also in the solid, the gas-phase infrared spectrum at low pressures clearly shows the compound to be monomeric with C_{2v}

symmetry.[38] Gaseous Me_2NGaH_2 is thus the first stable, 3-co-ordinate derivative of gallium hydride. The data and assignments are listed in Table 3. It can be seen that the Ga–H stretches in Me_2NGaH_2 are resolved and have moved to somewhat higher wave numbers whereas the rocking mode is slightly below that in $Me_3N . GaH_3$. The most striking change, however, is the increase in the position of the Ga–N stretching vibration which moves from 527 cm^{-1} to 743 cm^{-1}; this confirms the considerable double-bond character of the $N \rightleftharpoons Ga$ bond in dimethylaminogallane.

TABLE 3. *Gas-phase infrared data for Me_2NGaH_2 and Me_2NGaD_2* [38]

Me_2NGaH_2 (ν cm^{-1})	Me_2NGaD_2 (ν cm^{-1})	ν_H/ν_D	Assignment
1912 vs	1370 vs	1·39	Sym. Ga–H stretch
1904 s	1362 vs	1·40	Antisym. Ga–H stretch
693 m	510 ms	1·36	Antisym. Ga–H def.
673 vw	.*	—	Sym. Ga–H def.
743 w	.*	—	Ga $\rightleftharpoons$ N stretch
512	.*	—	N $\rightleftharpoons$ GaH$_2$ rock

* Too weak to be detected because of the lower vapour pressure of the deuterio- complex.
  Intensity: v = very, s = strong, m = medium, w = weak.

Triethylammonium chloride reacts with an excess of lithium tetrahydridogallate to give 1 mole of hydrogen and a precipitate of lithium chloride:[39]

$$Et_3NHCl + LiGaH_4 \rightarrow Et_3N . GaH_3 + LiCl + H_2$$

Removal of the solvent from the filtrate at -20 °C gave hexagonal plate-like crystals of $Et_3N . GaH_3$, m.p. 1 °C. The complex was involatile, in contrast with $Et_3N . AlH_3$ (m.p. 18 °C)[42] and $Et_3N . BH_3$ (m.p. -3 °C)[43] both of which could be readily sublimed or distilled under reduced pressure.

The infrared spectrum of a benzene solution of $Et_3N . GaH_3$ gave the bands listed in Table 4 in addition to the bands derived from coordinated triethylamine. The results are very similar to those shown in Table 1 for the trimethylamine adduct.

TABLE 4. *Infrared data for* $Et_3N.GaH_3$ *in benzene*[39]

ν cm^{-1}	Assignment	ν cm^{-1}	Assignment
1850 vs	ν_1 sym. Ga–H stretch	1780 s	ν_4 antisym. Ga–H stretch
758 s	ν_2 sym. Ga–H def.	787 vs	ν_5 antisym. Ga–H def.
479 ms	ν_3 N–Ga stretch	553 s	ν_6 N–GaH$_3$ rock

Intensity: v = very, s = strong, m = medium.

Pyridine-gallane can be prepared as a colourless, involatile oil by a similar reaction:[39]

$$C_5H_5NHCl + LiGaH_4 \rightarrow C_5H_5N.GaH_3 + LiCl + H_2$$

Precisely 1 mole of hydrogen was evolved and the product again analysed well for all constituents. Infrared spectra were run in ether, benzene, and pyridine solutions and gave strong bands at 1830 cm^{-1} (Ga–H stretches), 760 cm^{-1} (Ga–H deformations), 545 cm^{-1} (N–GaH$_3$ rock), and 485 cm^{-1} (N–Ga stretch). It can be seen that pyridine-gallane is more closely related to the colourless involatile pyridine-borane (m.p. 11 °C) than to the highly coloured solids which result from the interaction of pyridinium chloride and lithium tetrahydridoaluminate.

Triphenylamine-gallane has been prepared in ether at -30 °C by the stoicheiometric reaction:[39]

$$Ph_3N + HCl + LiGaH_4 \rightarrow Ph_3N.GaH_3 + LiCl + H_2$$

However, at room temperature the complex in solution decomposes with further evolution of hydrogen and deposition of gallium. The dimethylaniline complex, $PhNMe_2.GaH_3$, was somewhat more stable but removal of solvent at room temperature leads to decomposition.

The results in this and the preceding section indicate that the stability of gallane adducts with nitrogen ligands decreases in the sequence

$$Me_2NH > Me_3N > C_5H_5N > Et_3N > PhNMe_2 \gg Ph_3N$$

5. Tertiary phosphine derivatives of GaH₃

One of the perennial problems in coordination chemistry is the relative donor strengths of nitrogen and phosphorus towards a given acceptor. This section discusses this problem with respect to gallium hydride and reviews the preparation, stability and properties of the GaH_3 complex with Me_3P. The reactions of GaH_3 with Et_3P and Ph_3P, as well as with Me_2PH, Et_2PH and Ph_2PH are deferred until § 6.

Trimethylphosphine-gallium hydride, m.p. 71 °C, is prepared by the reaction of lithium tetrahydridogallate on equimolar amounts of trimethyl phosphine and hydrogen chloride in diethyl ether:[44]

$$Me_3P + HCl + LiGaH_4 \rightarrow Me_3P . GaH_3 + LiCl + H_2$$

The adduct can be purified by sublimation. It decomposes slowly under vacuum, but it is quite stable under nitrogen. Trimethylphosphine-trideuteriogallane was prepared similarly from lithium gallium deuteride, trimethylphosphine, and deuterium chloride. The absence of the 2:1 adduct bis(trimethylphosphine)-gallane was shown by tensiometric titration of trimethylphosphine-gallane with trimethylphosphine at −21 °C.

In order to determine which of the two ligands, trimethylamine or trimethylphosphine, is the stronger donor towards gallium hydride, the following equilibrium was investigated by infrared spectroscopy:[44]

$$Me_3P . GaH_3(s) + Me_3N(g) \rightarrow Me_3N . GaH_3(s) + Me_3P(g)$$

The same equilibrium position was reached independently of whether trimethylamine was added to the trimethylphosphine adduct or trimethylphosphine was added to the trimethylamine adduct, and corresponded approximately to equal amounts of trimethylamine and trimethylphosphine in the gas phase. In contrast, it was found that trimethylphosphine did not displace trimethylamine from trimethylamine-alane, $Me_3N . AlH_3$; nor could the adduct trimethylphosphine-alane be prepared directly from lithium tetrahydridoaluminate, trimethylphosphine and hydrogen chloride, since this reaction resulted in reduction to aluminium metal:[44]

$$Me_3P + HCl + LiAlH_4 \rightarrow Me_3P + LiCl + Al + \tfrac{5}{2}H_2$$

These observations extend the information already available about the stability relationships among adducts of these ligands with Group III hydrides. It has been shown that trimethylphosphine is a stronger donor than trimethylamine towards borane.[45] However both trimethylamine-borane[46] and trimethylphosphine-borane[47] are thermally more stable than the corresponding alane and gallane adducts, which decompose[2] below 100 °C. These stability relationships may be summarized as follows:

$$Me_3P.BH_3 > Me_3P.GaH_3 \gg Me_3P.AlH_3$$
$$Me_3N.BH_3 < Me_3P.BH_3$$
$$Me_3N.AlH_3 \gg Me_3P.AlH_3$$
$$Me_3N.GaH_3 \sim Me_3P.GaH_3$$

Methyl proton magnetic resonance spectra in benzene have been used to determine the position of the displacement equilibrium:[44]

$$Me_3P.GaH_3 + Me_3N \rightleftharpoons Me_3N.GaH_3 + Me_3P$$

Preliminary measurements were made in order to determine the relation between the methyl proton chemical shift and composition for the two binary systems, 'gallium hydride + trimethylamine' and 'gallium hydride + trimethylphosphine'. Linear plots were obtained between chemical shift and composition and these graphs were then used as calibrations to determine the concentration of the various species in an equimolar mixture of trimethylamine and trimethylphosphine-gallane. Identical results were obtained from an equimolar mixture of trimethylphosphine and trimethylamine-gallane and lead to an equilibrium constant:

$$[Me_3N.GaH_3][Me_3P]/[Me_3P.GaH_3][Me_3N] = 2.94$$

This corresponds to a 40 % displacement of Me_3N which is therefore seen to be slightly the stronger of the two ligands.

The proton n.m.r. of the hydrogen atoms directly attached to gallium have also been observed. The τ values (in ppm with reference to Me_4Si as 10.00) are: $Me_3N.GaH_3$, 5.54; $Me_3P.GaH_3$, 5.94.[44] These can be compared with values obtained on compounds to be mentioned in other sections:[25,38] $Me_2NH.GaH_3$, 5.5; $Me_2PH.GaH_3$, 5.6; $Et_3P.GaH_3$, 6.1. All these hydridic proton resonances were broad and of low intensity and the expected quartet (^{71}Ga: I = $\frac{3}{2}$, abundance 39·8 %; ^{69}Ga: I = $\frac{3}{2}$, abundance 60·2 %) could not be resolved because of quadrupole broadening. The values occur at much lower field strengths than those found for the resonance of hydrogen bound to transition metals (15–35 ppm) but are comparable with recent values quoted for the hydride derivatives of germanium and tin.

The gas-phase infrared absorption spectrum of trimethylphosphine-gallane is weaker than that of the trimethylamine adduct (Table 1) because of the lower vapour pressure of the phosphine complex. However, strong bands were observed[44] for the Ga–H stretch at 1846 cm^{-1}, the antisym. Ga–H deformation at 750 cm^{-1} and the sym. Ga–H deformation at 704 cm^{-1}. These are very similar to the values in Table 1 and, on deuterium substitution, each band was shifted by a factor of 1·39 to 1326, 538, and 508 cm^{-1} respectively. More complete results were obtained from benzene solutions and these are listed in Table 5 together with assignments.

TABLE 5. *Infrared data for* $Me_3P.GaH_3$ *and* $Me_3P.GaD_3$
in benzene[42]

$Me_3P.GaH_3$ (ν cm^{-1})	$Me_3P.GaD_3$ (ν cm^{-1})	ν_H/ν_D	Assignment
1831 s	1314 s	1·39	ν_1 sym. Ga–H stretch
1807 s	—	—	ν_4 antisym. Ga–H stretch
846 w	745 w	—	?
752 s	540 s	1·39	ν_5 antisym. Ga–H def.
(704)	503 s	1·40	ν_2 sym. Ga–H def.
467 m	339 w	1·38	ν_6 P–GaH$_3$ rock
326 m	—	—	ν_2 P–Ga stretch

Intensity: s = strong, m = medium.

6. Other phosphine derivatives of GaH$_3$

This section considers the ligands Me$_2$PH, Et$_2$PH, Ph$_2$PH, Et$_3$P and Ph$_3$P.

Dimethylphosphine-gallane can readily be prepared in ethereal solution as follows:[38]

$$Me_2PH + HCl + LiGaH_4 \rightarrow Me_2PH.GaH_3 + LiCl + H_2$$

After filtration and removal of solvent, the product distills readily at room temperature as a colourless, mobile liquid. If left at room temperature the reaction mixture begins to evolve more gas, initially at a much slower rate than that observed for the preparative reaction but eventually at a quicker rate, indicating that a phosphino derivative is probably formed initially before decomposition sets in:

$$Me_2PH . GaH_3 \rightarrow Me_2PGaH_2 + H_2$$
$$Me_2PGaH_2 \rightarrow Me_2PH + Ga + H_2$$

This reaction scheme is consistent with results obtained when $LiGaD_4$ was substituted for $LiGaH_4$ in the preparative reaction.[38] Analysis of the evolved gases by mass spectrometry showed that a stoicheiometric amount of virtually pure HD was evolved within 1 h at room temperature as required by the equation

$$Me_2PH + HCl + LiGaD_4 \rightarrow Me_2PH . GaD_3 + LiCl + HD$$

The gas evolved in the following 2 h contained only 1 % D_2, indicating that formation of dimethylphosphinogallane preceded its decomposition to metallic gallium. This instability of dimethylphosphinogallane, Me_2PGaH_2, frustrated any attempt to isolate the compound though its existence in benzene solution was demonstrated by ¹H n.m.r. data (see later). This behaviour runs counter to the facile elimination of hydrogen from dimethylphosphine-borane, and the formation of the extremely stable, cyclic trimer, $(Me_2PBH_2)_3$. It is apparent that subsidiary bonding in a gallium-phosphorus ring must be insufficient to compensate for the reduction in σ bond strength on moving from boron to gallium. However, evidence for the existence of the parent compound, $Me_2PH . GaH_3$, is significant when set beside the difficulty experienced in preparing the corresponding alane compound $Me_2PH . AlH_3$.[48]

Since dimethylamine was found to displace trimethylamine readily from its gallane adduct (§ 4) it was of interest to determine the relative donor power of dimethylphosphine and trimethylphosphine towards gallium hydride. Gas-phase infrared spectroscopy showed that an equimolar mixture of dimethylphosphine and trimethylphosphine-gallane, $Me_3P . GaH_3$, produced no trimethylphosphine or dimethylphosphine-gallane even after 3 h at

room temperature. This, in conjunction with the equilibria discussed in §§ 4 and 5, shows that donor power towards GaH_3 is in
the order $Me_2NH > Me_3N \sim Me_3P > Me_2PH$, as might be
expected.

Gallium hydride adducts of other secondary phosphines are
even less stable than dimethylphosphine-gallane.[38] Unlike the
situation in boron chemistry, stability seems to depend markedly
on the substituent on the phosphorus atom. Thus, when diethyl-
phosphine was substituted for dimethylphosphine in the preparative reaction 1 mole of hydrogen was evolved and a colourless,
unstable oil remained after removal of solvent from the filtrate at
$-80\ °C$. This was $Et_2PH.GaH_3$.[38] Likewise with diphenylphos-
phine 1 mole of hydrogen was evolved initially at $-80\,°C$. The
product was a colourless oil, $Ph_2PH.GaH_3$, in this case stable *in
vacuo*;[38] it had a broad absorption band in the infrared at $1840\ cm^{-1}$,
a position characteristic of Ga–H stretching modes. Further elimination of hydrogen led to a complex unidentified solid. It is clear
that the P–Ga dative bond in secondary phosphine adducts with
gallium hydride is very weak and becomes weaker when higher
alkyl or aryl groups are substituted for the methyl groups in
dimethylphosphine.

TABLE 6. *Infrared data for liquid* $Me_2PH.GaH_3$[38]

ν cm^{-1}	Assignment	ν cm^{-1}	Assignment
1840 vs	ν_1 sym. Ga–H stretch	745 s	ν_2 sym. Ga–H def.
1790 vs	ν_4 antisym. Ga–H stretch	463 m	ν_6 P–GaH$_3$ rock
846 w	?	360 m	ν_3 P–Ga stretch
758 s	ν_5 antisym. Ga–H def.		

Intensity: v = very, s = strong, m = medium, w = weak.

The infrared spectrum of liquid dimethylphosphine-gallane has
been recorded[38] and bands additional to those in the parent ligand
are listed in Table 6. The strongest modes, again, are the well-
resolved Ga–H stretches and deformations. The $P.GaH_3$ portion
of the spectrum is very similar to that of $Me_3P.GaH_3$ in Table 5
even to the occurrence of a weak, unexpected band at $846\ cm^{-1}$.
The only significant difference is the value ($360\ cm^{-1}$) observed for
the P–Ga stretching mode in $Me_2PH.GaH_3$. This is somewhat

higher than the value of 326 cm^{-1} observed in the spectrum of Me$_3$P.GaH$_3$ and is nearer the value calculated from the position of the N–Ga absorption in Me$_3$N.GaH$_3$ simply on the basis of a change in mass (364 cm^{-1}) or the value of 370 cm^{-1} calculated similarly from the known position of the P–BH$_3$ stretch. Good agreement between observed and calculated wavenumbers for this P–Ga mode is also obtained for Ph$_3$P.GaH$_3$ (365 cm^{-1})[44] and Et$_3$P.GaH$_3$ (370 cm^{-1}).[25] These compounds will be considered further in the following paragraphs. The frequency of the P–H stretching vibration in free dimethylphosphine increases by 70 cm^{-1} when the ligand coordinates to GaH$_3$. This parallels an increase of 100 cm^{-1} on coordination to BH$_3$.[49]

Triethylphosphine-gallane was isolated as a colourless liquid, m.p. − 10 °C, from the following reaction which proceeded quantitatively in ethereal solutions:[25]

$$Et_3P + HCl + LiGaH_4 \rightarrow Et_3P.GaH_3 + LiCl + H_2$$

The compound is a stable, white solid at − 78 °C but decomposes above its m.p. if stored under reduced pressure. It can be manipulated at room temperature under nitrogen for several hours without decomposition provided that its isolation is completed before decomposition of the ethereal solution has begun. Once metallic gallium has started to deposit, further decomposition rapidly ensues.

The infrared spectrum of liquid Et$_3$P.GaH$_3$ resembles that of related species:[25] 1813 vs (Ga–H stretch); 750 s (antisym. Ga–H deformation); 687 vs (sym. Ga–H deformation); 474 mw (P–GaH$_3$ rock); 370 m (P–Ga stretch). The deformation modes are very close to the CH$_2$ rocking modes (765 and 748 cm^{-1}) and C–P stretching mode (690 cm^{-1}) in triethylphosphine itself, though they are of far greater intensity than these ligand bands.

Triphenylphosphine-gallane was prepared by an analogous reaction at − 5 °C.[25, 44] It is a white solid which decomposes appreciably during 3 h at room temperature. Interpretation of the infrared spectrum of a benzene solution of triphenylphosphine-gallane was complicated by the low symmetry and complex absorption spectrum of the ligand itself. However the Ga–H stretches appeared as a strong, sharp band at 1880 cm^{-1} and a broad, intense absorption at

745 cm^{-1} could be assigned to the Ga–H deformations (overlapped by the strong C–H ring deformations). The P–Ga stretching mode gave a well-defined absorption at 365 cm^{-1}.

7. Adducts of GaH$_3$ with arsines

Trimethylarsine-gallane, and triphenylarsine-gallane were too unstable to be isolated and characterized at room temperature.[25] Me$_3$As.GaH$_3$ was obtained from trimethylarsine, hydrogen chloride, and lithium tetrahydridogallate in ethereal solution at − 80 °C and could be sublimed as a white solid at low temperatures. However, it decomposed completely above 0 °C to trimethylarsine, gallium, and hydrogen. Ph$_3$As.GaH$_3$ was even less stable.

It is clear from the foregoing sections that the stabilities of Ph$_3$N.GaH$_3$, Ph$_3$P.GaH$_3$ and Ph$_3$As.GaH$_3$ are orders of magnitude less than that of Ph$_3$P.BH$_3$. Indeed phosphine-boranes are well known for their unusual stability. The sequence of donor strengths towards gallane (Ph$_3$P > Ph$_3$N > Ph$_3$As) is not unexpected. The comparability of orbital size between carbon and nitrogen leads to some delocalization of the non-bonding pair of electrons on the nitrogen thus reducing its donor strength; in fact the NC$_3$ skeleton in triphenylamine is nearly planar[50] and consequently requires more reorganization energy to create a tetrahedral environment than do its analogues. The decreased donor strength of triphenylarsine reflects trends in orbital size and in the effectiveness of d_π–d_π overlap. The low stability of Me$_3$As.GaH$_3$ again emphasizes that d_π–d_π interaction between the arsine and the metal, which is so effective in stabilizing dative bonds between arsenic and transition metals, is of little help in combating the weakness of the simple As—Ga σ bond.

8. Adducts with oxygen and sulphur donors

Gallane complexes with ligands containing oxygen or sulphur as the donor atom are less stable than those in which the donor atom is nitrogen or phosphorus. The two most studied ligands are diethyl ether[17, 22] and dimethyl sulphide.[2]

Et$_2$O.GaH$_3$ is prepared by reducing an ethereal solution of gallium trichloride with lithium tetrahydridoaluminate at 0 °C and allowing the unstable intermediate mixed hydride to decompose:[22]

$$GaCl_3 + 3LiAlH_4 \rightarrow [Ga(AlH_4)_3] + 3LiCl$$
$$x[Ga(AlH_4)_3] + xEt_2O \rightarrow xEt_2O \cdot GaH_3 + (3/x)(AlH_3)_x$$

Polymeric aluminium hydride precipitates and removal of solvent from the filtrate at 20 °C leaves lustrous crystals of diethylether-gallane, $Et_2O \cdot GaH_3$. The adduct decomposes at 35 °C:

$$Et_2O \cdot GaH_3 \rightarrow Et_2O + Ga + \tfrac{3}{2}H_2$$

Later workers[17] were unable to repeat this work and also failed to synthesize the complex by a variety of related reactions involving $LiAlH_4$, $LiGaH_4$, $GaCl_3$, $GaBr_3$, etc., in a variety of solvents. It seems probable that the conditions for the preparation are critical and that traces of mercury or gallium are able to catalyse the decomposition of $Et_2O \cdot GaH_3$. There seems little doubt, however, that the compound has been prepared on numerous occasions by the Munich group.[18]

Diethylether-gallane is reported to react with hydrazoic acid in ether at low temperatures to give gallium-triazide as colourless crystals soluble in tetrahydrofuran:[51]

$$Et_2O \cdot GaH_3 + 3HN_3 \rightarrow Ga(N_3)_3 + Et_2O + 3H_2$$

Dimethylsulphide-gallane has been synthesized from trimethyl-amine-gallane by ligand replacement:[2]

$$Me_3N \cdot GaH_3 + Me_2S + BF_3 \rightarrow Me_2S \cdot GaH_3 + Me_3N \cdot BF_3$$

The reaction is important since it effects the replacement of a strong ligand by a much weaker one; the general technique is to react the complex of the stronger ligand (Me_3N) with a strong acceptor (BF_3) using the weaker ligand (Me_2S) as solvent. The strong ligand and strong acceptor form a solid complex ($Me_3N \cdot BF_3$) which is filtered off leaving a solution of the desired complex in the weaker ligand as solvent. Removal of solvent under reduced pressure at -70 °C followed by low-temperature sublimation afforded colourless crystals of $Me_2S \cdot GaH_3$. The complex analysed correctly but decomposed completely at room temperature:[2]

$$Me_2S \cdot GaH_3 \rightarrow Me_2S + Ga + \tfrac{3}{2}H_2$$

9. Trimethylamine adducts of halogenogallanes, $GaH_{3-n}X_n$

As part of an investigation into the possibility of preparing an un-coordinated gallium hydride (see § 10) the reaction of trimethyl-amine-gallane and anhydrous hydrogen chloride was investigated.[52] Instead of displacement of trimethylamine, however, a substitution reaction occurred in which hydrogen was evolved and the new compound trimethylamine-monochlorogallane, $Me_3N.GaH_2Cl$, was isolated by vacuum sublimation. This experiment led to a systematic investigation of the preparation of halogen-substituted gallanes, and the present section discusses the various methods that have been used to synthesize these compounds, and also presents infrared data for this new group of compounds.

The trimethylamine-halogenogallanes were first prepared by treating solid trimethylamine-gallane with the stoicheiometric quantity of dry hydrogen halide gas.[52] Reaction occurred at temperatures as low as $-78\,^\circ$C to give an equivalent amount of hydrogen and the halogen derivative:

$$Me_3N.GaH_3(s) + nHX(g) \rightarrow Me_3N.GaH_{3-n}X_n(s) + nH_2(g)$$

where $n = 1$ or 2, and $X = Cl$ or Br. In experiments with an excess of hydrogen halide, 3 moles of hydrogen were evolved. A parallel series of deuterio-compounds was prepared by using trimethylamine-trideuteriogallane and deuterium chloride gas or hydrogen bromide gas in stoicheiometric quantities.[52] The action of hydrogen bromide gas on trimethylamine-trideuteriogallane liberated HD quantitatively and substituted bromine for deuterium in the adduct; no deuterium-hydrogen exchange was observed.

A second method of introducing halogen atoms into trimethyl-amine-gallane is by reaction with an excess of trimethylammonium halide in ether;[52] 2 moles of hydrogen were produced as required by the equation

$$Me_3N.GaH_3 + 2Me_3NHX \rightarrow Me_3N.GaHX_2 + 2Me_3N + 2H_2$$

for the chloro-system, but only $1\frac{1}{2}$ moles for the bromo-system. It appears that further reaction does not occur, even after prolonged periods at room temperature, and this may indicate some shielding of the third hydrogen by the two halogens already present. This

reaction explains some observations made during the preparation
of trimethylamine-gallane from lithium tetrahydridogallate (§ 3)
when it was found that the use of an excess of triethylammonium
chloride produced more hydrogen than was required by the
equation

$$Et_3NHCl + LiGaH_4 \rightarrow Et_3N.GaH_3 + LiCl + H_2$$

In a related experiment it was found that equimolar proportions
of trimethylamine-trideuteriogallane and trimethyldeuterioam-
monium chloride in ether solution gave a quantitative yield of
the monochloro-compound, $Me_3N.GaD_2Cl$ and deuterium:[52]

$$Me_3N.GaD_3 + Me_3NDCl \rightarrow Me_3N.GaD_2Cl + Me_3N + D_2$$

Analogous behaviour has not been reported in either borane or
alane chemistry.

A third method of preparation of the mixed hydride–halide
compounds involves direct reaction of trimethylamine-gallane or
-trideuteriogallane with the appropriate amount of trimethylamine-gallium trihalide in benzene.[52] The monosubstituted
derivatives can be prepared from stoicheiometric ratios according
to the equation

$$2Me_3N.GaH_3 + Me_3N.GaX_3 \rightarrow 3Me_3N.GaH_2X$$

where $X = Cl$, Br or I. For disubstitution, an excess of the trihalides appears to be necessary before all the hydride is converted
into the species $Me_3N.GaHX_2$.

The monohalogenogallane adducts were white, slightly volatile
solids which could be purified by vacuum sublimation.[52] The
dihalogenogallane adducts were involatile. $Me_3N.GaHCl_2$ melted
at 66 °C and cryoscopic measurements in benzene solution
indicated that it was slightly associated (degree of association 1·1).[52]

The vapour pressures of the substituted gallanes were too low to
permit gas-phase spectra to be recorded over the full range of
frequencies, but it was observed that when a sample of trimethylamine-monochlorogallane, $Me_3N.GaH_2Cl$, was heated in a gas
cell,[52] weak absorption bands appeared at 1900, 734 and 694 cm^{-1},
corresponding to the Ga–H stretching and deformation modes.

The infrared spectra of the trimethylamine-halogenogallanes
and trimethylamine-halogenodeuteriogallanes in benzene have

TABLE 7. *Infrared data for* $Me_3N.GaH_{3-n}X_n$ *in benzene*[52]

Me₃N.GaH₂Cl	Me₃N.GaHCl₂	Me₃N.GaH₂Br	Me₃N.GaHBr₂	Me₃N.GaH₂I	Me₃N.GaHI₂	Assignment
1905 vs 1809 sh	1968 vs 1949 vs	1907 vs	1931 s	1906 vs	1925 s	Ga–H stretch
603 m	604 vs	594 m	595 vs	583 m	585 vs	H–Ga–X def.
504 s	—	505 m	—	505 s	—	Ga–H₂ rock
491 s	520 vs	478 m	515 m	463 m	—	Ga–N stretch
345 m	370 vs 362 vs	258 m	286 vs 265 s	229 m	—	Ga–X stretch

Intensity: v = very; s = strong; m = medium; sh = shoulder.

TABLE 8. *Infrared data for* $Me_3N.GaD_{3-n}X_n$ *in benzene*[52]

Me₃N.GaD₂Cl	Me₃N.GaDCl₂	Me₃N.GaD₂Br	Me₃N.GaDBr₂	Me₃N.GaD₂I	Me₃N.GaDI₂	Assignment
1370 vs	1408 s 1393 s	1380 s	1400 s 1390 s	1373 s 1369 s	1395 sh 1380 s	Ga–D stretch
529 s 510 s	526 m 510 m	524 m 505 s	508 w	503 s	—	Ga–D def.
485 vs	484 m	479 s	481 w	476 s	—	Ga–N stretch
440 m	441 vs	434 w	440 s	439 m	436 s	D–Ga–X def.
372 w	—	340 w	—	412 w?	—	GaD₂ rock
345 m	371 vs 365 vs	253 m	282 s	227 m	—	Ga–X stretch

Intensity: v = very; s = strong; m = medium; w = weak; sh = shoulder.

been recorded and compared in detail[52] with those of the parent compounds $Me_3N.GaH_3$ and $Me_3N.GaX_3$. Those parts of the spectra which arise from the $N.GaH_{3-n}X_n$ and $N.GaD_{3-n}X_n$ groups are summarized in Tables 7 and 8. These spectra show several interesting features. The Ga–H stretching frequencies move successively to higher values on progressive substitution of halogen for hydrogen in the compounds; likewise for the corresponding Ga–D frequencies, which occur at positions moved by a factor of approximately $1/1.4$ from the normal compounds. Substitution of one halogen atom increases the Ga–H stretching frequency by about 70 cm^{-1} independent of the halogen ($X = Cl$, Br or I). Di-substitution effects a larger shift of the Ga–H stretching frequency for the dichloride (130 cm^{-1}) than for the dibromide (112 cm^{-1}) and this, in turn, is larger than for the di-iodide (106 cm^{-1}). In addition the band appears to be split in the case of the dichloride, and this may indicate partial dimerization of the compound in benzene solution. These changes in the Ga–H stretching frequency of substituted gallanes can be understood in terms of the electron-withdrawing effect of the halogen atom on the electronegativity of the gallium atom; this results in an increase in the strength of the Ga–H bond, and hence an increase in the stretching frequency. A similar shift in the Al–H stretching frequencies was noticed for the triethylamine-halogenoalane compounds,[53] $Et_3N.AlH_{3-n}X_n$ (where $X = Cl$ or Br and $n = 1$ or 2), and also for the trimethylamine-mercaptoalanes[54] $Me_3N.AlH_{3-n}(SR)_n$ (where $R = Pr^n$ or Ph and $n = 1$ or 2). Corresponding infrared data on the triethylamine-halogenoboranes has shown a similar shift to higher frequencies of the B–H stretching modes on substitution of hydrogen by halogen.[55]

The assignment of the intense bands in the range 585–605 cm^{-1} to the H–Ga–X deformation modes follows from their position and from the fact that these bands do not appear in the spectra of trimethylamine-gallane, $Me_3N.GaH_3$, or of the trimethylamine-gallium trihalides. The intensities of these bands are considerably greater in the dihalogen derivatives, but the band occurs in essentially the same position for both the mono- and di-substituted compounds. On deuteration the bands move by a factor of approximately $1/1.4$ (see Tables 7 and 8) as predicted from a mass effect. With increase in atomic number of halogen there is a cor-responding decrease in the frequency of this mode, again indicating the operation of a mass effect. A similar trend is noticed with the Ga–N stretching frequencies of the monohalogenogallanes.

The Ga–X stretching vibrations in compounds containing a single Ga–X bond have been observed for the first time in this work.[52] There is a general decrease in this stretching frequency in the compounds $Me_3N.GaH_2X$ and $Me_3N.GaD_2X$ in passing from chlorine to iodine, as might be expected from increased atomic mass of the halogen, but the position of the band remains approxi-mately the same in the normal and deuterio-derivatives of the same halogen (see Tables 7 and 8). With increase in the halogen content, however, the gallium–halogen stretching modes move to higher frequencies.

10. Uncoordinated gallium hydride and halogenogallanes

Experiments designed to liberate gallium hydride from its adducts usually lead, even at low temperatures, to immediate decomposition into gallium and hydrogen. However, gallium hydride can be pre-pared at low temperatures ($-15\,°C$) by the following displacement reaction:[20]

$$Me_3N.GaH_3(c) + BF_3(g) \rightarrow GaH_3(l) + Me_3N.BF_3(c)$$

An excess of boron trifluoride was condensed onto solid trimethyl-amine-gallane at $-196\,°C$ and allowed to warm slowly. At $-20°$ to $-15\,°C$ rapid uptake of exactly 1 mole of boron trifluoride occurred as indicated in the above equation.

Gallium hydride is a viscous liquid, m.p. $-15\,°C$, which decom-poses quantitatively at room temperature to hydrogen and gallium.[20] It is insoluble in non-coordinated solvents and hence its molecular weight could not be determined, but its viscosity and immiscibility with non-polar solvents suggests that it is to some extent polymerized.

Liquid gallium hydride shows strong, broad infrared bands at $1980\ cm^{-1}$ (Ga–H stretch) and $700\ cm^{-1}$ (Ga–H deformation). A fugitive gallium hydride species was also detected in the vapour phase by infrared spectroscopy.[20] As the reaction proceeded the sharp band of the adduct at 1853 cm^{-1} gradually disappeared and was replaced by an equally sharp band at 2000 cm^{-1}. This new band itself rapidly decreased in intensity as the gaseous gallium hydride decomposed to gallium and hydrogen.

A recent study by time-of-flight mass spectroscopy has confirmed that GaH_3 is much less stable than AlH_3.[56]

In view of the possibility that a chlorogallium hydride GaH_2Cl might be stabilized by dimerization via bridging chlorine atoms, an attempt was made to prepare this compound by an analogous reaction:[52]

$$Me_3N \cdot GaH_2Cl(c) + BF_3(g) \rightarrow GaH_2Cl(l) + Me_3N \cdot BF_3(c)$$

The reaction again resulted in the formation of an oil which decomposed rapidly under vacuum at room temperature. The infrared spectrum of the vapour showed a medium-strong band at 2000 cm^{-1}, suggesting the formation of an unstable species $(GaH_2Cl)_x$. A similar sequence occurred in benzene.

By contrast to the unstable compounds gallium hydride and monochlorogallium hydride, the compound $GaHCl_2$ can be obtained as stable, white crystals soluble in benzene or cyclohexane, in which it is dimeric.[57] The compound is prepared at $-20\ ^\circ C$ by metathesis:

$$Me_3SiH + GaCl_3 \rightarrow Me_3SiCl + HGaCl_2$$

It melts at 29 $^\circ C$ with decomposition and, when heated to 150 $^\circ C$, decomposes quantitatively into gallium dichloride and hydrogen:[57]

$$(HGaCl_2)_2 \rightarrow H_2 + Ga(GaCl_4)$$

The Ga–H stretching vibration was detected at 2018 cm^{-1}.

Dichlorogallane reacts with olefins at 0 $^\circ C$ to form alkyldichlorogallanes:[58]

$$2CH_2{=}CH_2 + (HGaCl_2)_2 \rightarrow (EtGaCl_2)_2$$

Likewise cyclohexene gave cyclohexyldichlorogallane, propylene gave n-propyldichlorogallane and n-heptene-1 gave n-heptyldichlorogallane.[58] The addition is therefore anti-Markownikov, as is hydroboration.

A similar sequence of reactions gave dibromogallanes and n-alkyldibromogallanes:[59]

$$Me_3SiH + GaBr_3 \rightarrow Me_3SiBr + HGaBr_2$$
$$2RCH{=}CH_2 + (HGaBr_2)_2 \rightarrow (RGaBr_2)_2$$

Dibromogallane is a white crystalline solid which decomposes above $-8\ ^\circ C$ with evolution of hydrogen and the alkyldibromo-gallanes are all dimers.

11. Mixed hydrides of boron and gallium

Only brief accounts of mixed hydrides have appeared in the litera-ture. Gallium borohydride, $Ga(BH_4)_3$, is much less stable than aluminium borohydride.[60, 61] When an excess of lithium tetra-hydridoborate reacts with trimethylamine-monochlorogallane in benzene or ether, an unstable oil is obtained which decomposes below room temperature:[52]

$$Me_3N . GaH_2Cl + LiBH_4 \rightarrow Me_3N . GaH_2(BH_4) + LiCl$$

The compound showed infrared absorption bands characteristic of Ga–H (1910 and $1840\ cm^{-1}$) and B–H ($2300\ cm^{-1}$) stretching frequencies. It is notable that the compound is appreciably less stable than its aluminium analogue $Me_3N . AlH_2(BH_4)$.[62]

The most stable mixed hydride of boron and gallium yet reported results from the reaction of decaborane with trimethylamine-gallane in ether at room temperature; the reaction is rapid and complete and no hydrogen is evolved:[63]

$$Me_3N . GaH_3 + B_{10}H_{14} \rightarrow [Me_3NH^+][B_{10}GaH_{16}^-]$$

The compound is obtained as a fine, white powder which analyses correctly for C, B, N and Ga. It is relatively unreactive to water and other hydroxylic solvents and can be recovered unchanged from solution in 1 M hydrochloric acid. The compound is a 1:1 electrolyte and is considered to be the first example of a 6,9-bridged derivative of the dianion $B_{10}H_{16}^{2-}$ and the first compound to contain direct boron–gallium bonds. Its formation and struc-ture can be represented in the following diagrammatic way which shows only the 5, 6, 7, 8, 9, 10 boron atoms of decaborane and the position of the B—H—B bridge bonds.[62]

$$B_{10}H_{14} \longrightarrow B_{10}H_{14}GaH_2^-$$

There is undoubtedly a rich field of mixed hydride chemistry to be developed in this area of the Periodic Table.

12. Conclusion

The large number of gallium hydride derivatives which have been prepared during the last five years has indicated a rich area of study awaiting further development. Stability sequences in various series of ligands have only been roughly established and there is considerable need for precise thermochemical and other thermodynamic data. Only then can the various factors influencing strengths of the bonds be properly assessed. Little is yet known of the complexes of bidentate and polydentate ligands.

The value of spectroscopic techniques in these studies has been demonstrated and it is clear that further development of these methods especially of nuclear magnetic resonance will yield both structural and stability data which are frequently lacking at present.

The chemical reactions of lithium tetrahydridogallate would undoubtedly repay further study and the new field of hydrogallation reactions offers a promising new route to organogallium compounds. Heterocyclic compounds of gallium have also yet to be explored and no serious kinetic studies of reactions involving gallium hydride derivatives have yet been undertaken.

REFERENCES

1   N. N. Greenwood, A. Storr and M. G. H. Wallbridge, *Proc. chem. Soc.* 1962, p. 249.
2   N. N. Greenwood, A. Storr and M. G. H. Wallbridge, *Inorg. Chem.* 1963, **2**, 1036.
3   D. F. Shriver and R. W. Parry, *Inorg. Chem.* 1963, **2**, 1039.

4 F. lecoq. de Boisbaudran, *C. r. hebd. Séanc. Acad. Sci.*, Paris, 1881, **93**, 294.
5 J. F. Corrigan, *Chem. News*, 1919, **119**, 274.
6 T. W. Richards and S. Boyer, *J. Am. chem. Soc.* 1921, **43**, 280.
7 T. W. Richards and S. Boyer, *Chem. News*, 1921, **122**, 176.
8 E. Tomkinson, *Chem. News*, 1921, **122**, 238.
9 E. Pietsch, F. Seuferling, W. Roman and H. Lehl, *Z. Elektrochem.* 1933, **39**, 577.
10 B. Siegel, *J. chem. Educ.* 1961, **38**, 496.
11 W. R. S. Garton, *Proc. phys. Soc.* 1951, A**64**, 509.
12 H. Neuhaus, *Nature, Lond.* 1957, **180**, 433; *Ark. Fys.* 1959, **14**, 551.
13 M. L. Ginter and K. K. Innes, *J. molec. Spectrosc.* 1961, **7**, 64.
14 E. Wiberg and T. Johannsen, *Naturwissenschaften*, 1941, **29**, 320.
15 E. Wiberg and T. Johannsen, *Chemie*, 1942, **55**, 88.
16 E. Wiberg, T. Johannsen and O. Stecher, *Z. anorg. Chem.* 1943, **251**, 114.
17 D. F. Shriver, R. W. Parry, N. N. Greenwood, A. Storr and M. G. H. Wallbridge, *Inorg. Chem.* 1963, **2**, 867.
18 E. Wiberg, private communication.
19 E. Wiberg and M. Schmidt, *Z. Naturf.* 1952, **7**b, 577.
20 N. N. Greenwood and M. G. H. Wallbridge, *J. chem. Soc.* 1963, p. 3912.
21 A. E. Finholt, A. C. Bond and H. I. Schlesinger, *J. Am. chem. Soc.* 1947, **69**, 1199.
22 E. Wiberg and M. Schmidt, *Z. Naturf.* 1951, **6**b, 172.
23 H. H. Hütte, British Patent 707,851 (1954); German Patent 937,823 (1956).
24 I. A. Sheka, I. S. Chaus and T. T. Mityureva, *The Chemistry of Gallium*. Amsterdam: Elsevier, 1966.
25 N. N. Greenwood and E. J. F. Ross, unpublished results, 1965.
26 H. J. Hrotstowski and M. Tanenbaum, *Physica*, 1954, **20**, 1065.
27 E. Wiberg, O. Dittmann and M. Schmidt, *Z. Naturf.* 1957, **12**b, 60.
28 E. Wiberg and H. Schmidt, *Z. Naturf.* 1951, **6**b, 171.
29 J. W. Akitt, N. N. Greenwood and A. Storr, *J. chem. Soc.* 1965, p. 4410.
30 M. Schmidt and A. Nordwig, *Chem. Ber.* 1958, **91**, 506.
31 N. G. Gaylord, *J. Chem. Educ.* 1957, **34**, 367.
32 E. Wiberg and M. Schmidt, *Z. Naturf.* 1951, **6**b, 335.
33 E. Wiberg and H. Nöth, *Z. Naturf.* 1957, **12**b, 63.
34 E. Wiberg and W. Henle, *Z. Naturf.* 1952, **7**b, 576.
35 D. F. Shriver, R. L. Amster and R. C. Taylor, *J. Am. chem. Soc.* 1962, **84**, 1321.
36 D. F. Shriver and C. E. Nordman, *Inorg. Chem.* 1963, **2**, 1298.
37 G. W. Fraser, N. N. Greenwood and B. P. Straughan, *J. chem. Soc.* 1963, p. 3742.
38 N. N. Greenwood, E. J. F. Ross and A. Storr, *J. chem. Soc.* (*A*), 1966, p. 706.
39 N. N. Greenwood and A. Storr, unpublished results, 1962.
40 A. B. Burg and C. L. Randolph, *J. Am. chem. Soc.* 1951, **73**, 953.

41 J. K. Ruff and M. F. Hawthorne, *J. Am. chem. Soc.* 1961, **83**, 535.

42 J. K. Ruff and M. F. Hawthorne, *J. Am. chem. Soc.* 1960, **82**, 2141.

43 N. N. Greenwood and J. H. Morris, *J. Chem. Soc.* 1960, p. 2922.

44 N. N. Greenwood, E. J. F. Ross and A. Storr, *J. chem. Soc.* 1965, p. 1400.

45 W. A. G. Graham and F. G. A. Stone, *J. Inorg. Nuclear Chem.* 1956, **3**, 164.

46 A. B. Burg and H. I. Schlesinger, *J. Am. chem. Soc.* 1937, **59**, 780.

47 A. B. Burg and R. I. Wagner, *J. Am. chem. Soc.* 1953, **75**, 3872.

48 A. B. Burg and K. Mödritzer, *J. inorg. nucl. Chem.* 1960, **13**, 318.

49 A. B. Burg, *Inorg. Chem.* 1964, **3**, 1325.

50 Y. Sasaki, K. Kimura and M. Kubo, *J. Chem. Phys.* 1959, **31**, 477.

51 E. Wiberg and H. Michaud, *Z. Naturf.* 1954, **9***b*, 502.

52 N. N. Greenwood and A. Storr, *J. chem. Soc.* 1965, p. 3426.

53 E. G. Hoffmann and G. Schomberg, *Z. Elektrochem.* 1957, **61**, 1110.

54 W. Marconi, A. Mazzei, F. Bonati and M. de Malde, *Z. Naturf.* 1963, **18***b*, 3.

55 J. N. G. Faulks, N. N. Greenwood and J. H. Morris, *J. inorg. nucl. Chem.* 1967, **29**, 329.

56 P. Breisacher and B. Siegel, *J. Am. chem. Soc.* 1965, **87**, 4255.

57 H. Schmidbauer, W. Findeiss and E. Gast, *Angew. Chem.* 1965, **77**, 170.

58 H. Schmidbauer and H. F. Klein, *Angew. Chem.* 1966, **78**, 306.

59 H. Schmidbauer and H. F. Klein, private communication, 1966.

60 H. I. Schlesinger, H. C. Brown and G. W. Schaeffer, *J. Am. chem. Soc.* 1943, **65**, 1786.

61 E. Wiberg, *Angew. Chem.* 1953, **65**, 26.

62 J. K. Ruff, *Inorg. Chem.* 1963, **2**, 515.

63 N. N. Greenwood and J. A. McGinnety, *Chem. Comm.* 1965, p. 331.

4

PROPERTIES OF DONOR SOLVENTS AND COORDINATION CHEMISTRY IN THEIR SOLUTIONS

V. GUTMANN

1. Introduction

The choice of a 'suitable' solvent for a particular reaction rests on the experience of the individual research worker. The choice depends upon the particular complex that is being prepared, but no reliable system is available showing relations between measurable solvent properties and the characteristics of the reactions or of the products desired.

It has been long recognized that the first step in the process of dissolution is the chemical reaction between solute A and the donor solvent D (solvation):

$$A + nD \rightleftharpoons AD_n$$

The heat of this reaction has to provide sufficient energy to overcome the lattice energy of the solid to be dissolved. Most molecules or ions in solution are solvated and all the 'complexing reactions' in solution require replacements of solvent molecules by other ligands, L. Such replacement or ligand exchange reactions will depend on the coordinating properties of both the solvent molecules D and the competing ligand L towards the ion or molecule under consideration:

$$AD + L \rightleftharpoons AL + D$$

Coordination reactions, including those of solvation, can be regarded as donor–acceptor reactions, for which the donor strength of the solvent molecules will be a decisive factor.[1]

Until recently only qualitative information was available concerning the donor strength of a solvent. Lindqvist and Zackrisson[2] have arrived at a scale of relative donor strength for a number of donor molecules, including those in use as non-aqueous solvents, on the grounds of comparative calorimetric measurements. They gave the following order:[2, 3]

$$Ph_2SeO \sim Ph_3AsO > (MeO)_3PO > Me_2SO \gg Et_2S$$
$$> Me_2CO \sim MeCOOEt \sim Et_2CO \sim Et_2O \gg Me_2SO_2$$
$$> Ph_2SO_2 \sim POCl_3 \sim SeOCl_2 > SOCl_2$$

and found that the order of donor strength of a $M{=}O$ group such as $P{=}O$ is decreased by substituents at M in the following order:[2, 3]

$$RO \geqslant R > Cl$$

It is not surprising that no further progress was made, since no quantitative data were available. Thus in an authoritative paper[4] the statement is found that 'triethylphosphate is expected to be a slightly better donor than $POCl_3$, but the similarity is such that vastly different coordination behaviour would not be expected'. The authors went so far as to question the interpretation of the results in phosphorus oxychloride only because different conclusions were obtained from their experiments in the system ferric chloride–triethylphosphate.

2. The donor number

The results of quantitative calorimetric measurements on the interactions of a number of O- and N-containing solvent molecules (D) and antimony (V) chloride as reference acceptor in an inert solvent such as dichloroethane,

$$D_{(dissolved)} + SbCl_{5(dissolved)} \rightleftharpoons D.SbCl_{5(dissolved)};$$
$$-\Delta H_{D.SbCl_5}$$

were compared with the ΔG values determined for the same equilibria obtained from spectrophotometric[5] and ^{19}F-n.m.r. measurements.[6] A linear relation was found between $-\Delta H$- and $\log K_{D.SbCl_5}$-values ($K_{D.SbCl_5}$ being the formation constants of $D.SbCl_5$) showing that the entropy contributions are equal for all solvent–acceptor interactions under consideration (Table 1). It is therefore justified from a thermodynamic point of view to consider the $-\Delta H$-values as valid expressions of the degree of interaction between solute and solvent.

TABLE I. $\Delta H_{D.SbCl_5}$- and $\log K_{D.SbCl_5}$-values for several donor solvents (D)

Donor	$-\Delta H_{D.SbCl_5}$ (kcal/mole)	Log $K_{D.SbCl_5}$
Thionyl chloride	0·4	0·3
Phosphorus oxychloride	11·7	0·7
Acetonitrile	14·1	2·8
Propane-1,2-diolcarbonate	15·1	3·0
Acetone	17·0	4·9
Water	18·0	5·3
Ether	19·2	5·5
Trimethylphosphate	23·0	9·3
Tributylphosphate	23·7	10·5

TABLE 2. ΔH-values for reactions of donor solvents with different acceptor molecules in 'inert' media

Donor	Acceptor					
	$SbCl_5$	$SbCl_3$	$SbBr_3$	C_6H_5OH	I_2	$(CH_3)_3SnCl$
$POCl_3$	11·7	.	.	2·5	.	.
CH_3CN	14·1	.	.	3·2	1·9	4·8
CH_3COOCH_3	16·4	.	.	3·2	2·5	.
$(CH_3)_2CO$	17·0	.	.	3·3	2·5	5·7
$CH_3COOC_2H_5$	17·1	.	.	4·8	.	.
$C_6H_5POCl_2$	18·5	4·4	3·7	3·4	.	.
$(C_2H_5)_2O$	19·2	.	.	5·0	4·3	.
$(CH_3O)_3PO$	23·0	6·1	5·2	5·3	.	.
$HCON(CH_3)_2$	26·6	7·4	6·8	6·1	3·7	.
DMA*	27·8	.	.	6·4	4·0	7·9
$(CH_3)_2SO$	29·8	8·8	8·3	6·5	4·4	8·2
C_5H_5N	33·1	9·9	.	8·1	7·8	(6·5)

* NN-Dimethylacetamide.

In order to study the influence of the nature of the acceptor molecule the $-\Delta H_{D.SbCl_5}$-values were compared with the $-\Delta H$-values for the reactions of the solvents with antimony(III) chloride,[7] antimony(III) bromide[7] (measured calorimetrically) and with trimethyltinchloride,[8] phenol and iodine calculated from the equilibrium constants between the solvent molecules and such acceptor molecules.[9, 10] It was found that these figures are proportional to the $-\Delta H_{\mathrm{D.SbCl_5}}$-values (Table 2) indicating that specific interactions between the solvent molecules and the acceptor molecules under consideration are small enough to be neglected.

Such relations can, however, only be expected if $1:1$ reactions are considered with analogous mechanisms involved. (Different bonding character in a particular system may lead to exceptions, as is found in the systems pyridine–iodine and pyridine–trimethyltinchloride.) Furthermore the coordinate bond formed should also have analogous ionic and covalent contributions, and secondary reactions such as ionization of the complex should not occur.

The donor number 'DN$_{\mathrm{SbCl_5}}$' has been defined as the numerical quantity of the $-\Delta H_{\mathrm{D.SbCl_5}}$-values:[9–12]

$$\mathrm{DN}_{\mathrm{SbCl_5}} \equiv -\Delta H_{\mathrm{D.SbCl_5}}$$

The donor number is therefore nearly a molecular property of the solvent, which can easily be determined by experiment. It expresses the total amount of interaction with an acceptor molecule, including the contributions both by dipole–dipole or dipole–ion interactions and by the binding effect caused by the availability of the electron pair. To some extent even certain steric properties of the solvent molecules may be contained in it (Table 3).

The donor number of acetyl chloride is smaller than that of benzoylchloride apparently owing to lack of mesomerism of the $C=O$ group in the former compound. The substitution of the methyl group in nitromethane by a phenyl group gives rise to a higher donor number for nitrobenzene than for nitromethane. A decrease of the electron density in the aromatic nucleus may be due to the $-M$ effect of the NO_2 group.[13]

The substitution of CH_3 by C_6H_5 in nitriles gives a decrease in donor number for benzonitrile. Ethylene sulphite is a weaker donor than DMSO since the cyclic alkoxy group in ethylene sulphite decreases the electron density at the sulphur atom. Sulpholane (tetramethylenesulphone) has similar donor properties.[13]

Ethylene carbonate has medium donor properties, but dichloroethylene carbonate is a very weak donor. Likewise the basicity of the $P=O$ group is small in $POCl_3$, stronger in TMP, still stronger

TABLE 3. *Donor number* DN_{SbCl_5} *and dielectric constant* ϵ
of certain solvents[12, 13]

Solvent	DN_{SbCl_5}	ϵ
1,2-Dichloroethane	.	10·1
Sulphuryl chloride	0·1	10·0
Thionyl chloride	0·4	9·2
Acetyl chloride	0·7	15·8
Benzoyl chloride	2·3	23·0
Nitromethane (NM)	2·7	35·9
Nitrobenzene (NB)	4·4	34·8
Acetic anhydride	10·5	20·7
Benzonitrile	11·9	25·2
Selenium oxychloride	12·2	46·0
Acetonitrile (AN)	14·1	38·0
Sulpholane	14·8	42·0
Propane-1,2-diolcarbonate (PDC)	15·1	69·0
Benzyl cyanide	15·1	18·4
Ethylene sulphite (ES)	15·3	41·0
iso-Butyronitrile	15·4	20·4
Propionitrile	16·1	27·7
Ethylene carbonate	16·4	89·1
Phenylphosphonic difluoride	16·4	27·9
Methylacetate	16·5	6·7
n-Butyronitrile	16·6	20·3
Acetone	17·0	20·7
Ethyl acetate	17·1	6·0
Water	18·0	81·0
Phenylphosphonic dichloride	18·5	26·0
Diethylether	19·2	4·3
Tetrahydrofurane (THF)	20·0	7·6
Diphenylphosphonic chloride	22·4	.
Trimethylphosphate (TMP)	23·0	20·6
Tributylphosphate (TBP)	23·7	6·8
Dimethylformamide (DMF)	26·6	36·1
NN-Dimethylacetamide (DMA)	27·8	38·9
Dimethylsulphoxide (DMSO)	29·8	45·0
NN-Diethylformamide	30·9	.
NN-Diethylacetamide	32·2	.
Pyridine (Py)	33·1	12·3
Hexamethylphosphoramide (HMPA)	38·2	30·0

in TBP and extremely strong in HMPA. In an analogous manner acetyl chloride has a very low donor number, acetone is much stronger, and DMA has a very high donor number.

3. The donor number of the solvent and the formation of anionic complexes

It is now of immediate interest to check the relations between the donor number and certain properties. When an acceptor compound AX_n such as $SbCl_5$ is dissolved in a donor solvent D the formation of a solvate complex occurs according to [1]:

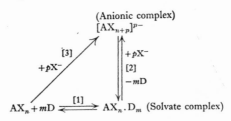

The solvate complex may then be made to react with competing ligands, such as an anion X^-, to give an anionic complex by a ligand exchange reaction according to [2]. This reaction is frequently represented according to [3] without regarding the intermediate steps [1] and [2] and considered as the formation of a complex ion, although this reaction may not take place in the absence of a suitable solvent. Thus $SbCl_5$ and KCl will not react, unless a proper solvent is used to allow the occurrence of reactions [1] and [2]. It is obvious that the formation of the anionic complex will occur only if the competing ligand has stronger donor properties than the solvent molecules.

For the assignment of a donor number to anions, such as halide or pseudohalide ions, the calorimetric measurements of the analogous reaction

$$SbCl_5 + X^- \rightleftharpoons [SbCl_5X]^-$$

cannot be used, since both halide exchange reactions may be involved and the solubilities of compounds providing X^- ions are too poor in dichloroethane.

To exclude exchange reactions vanadylacetylacetonate $VO(acac)_2$

has been used as a reference acceptor[14] which has one ligand site available for complex formation. VO(acac)$_2$ is a weaker acceptor than SbCl$_5$, so that in the place of calorimetric measurements the equilibria

$$VO(acac)_2 + X^- \rightleftharpoons [VO(acac)_2X]^-$$

were followed spectrophotometrically.

TABLE 4. *Dissociation constants of* [VO(acac)$_2$D] *and* [VO(acac)$_2$X]$^-$ *in acetonitrile and in methylene chloride at 27°*

Ligand	K_{Diss} in CH$_3$CN	K_{Diss} in CH$_2$Cl$_2$
N$_3^-$	0.8×10^{-3}	2.3×10^{-3}
NCS$^-$	1.1×10^{-2}	1.8×10^{-2}
Py	1.9×10^{-2}	2.9×10^{-2}
HMPA	1.7×10^{-2}	4.4×10^{-2}
Ph$_3$PO	2.9×10^{-2}	6.5×10^{-2}
DMSO	1.6×10^{-1}	2.2×10^{-1}
Cl$^-$	3.7×10^{-1}	4.2×10^{-1}
DMF	4.1×10^{-1}	3.9×10^{-1}
TMP	1.3	1.2
Br$^-$	$> K_{TMP}$	$> K_{TMP}$
I$^-$	$> K_{Br^-}$	$> K_{Br^-}$

Thus the following orders of increasing donor strength were obtained: in methylene chloride,

I$^-$ < Br$^-$ < TMP < DMF ~ Cl$^-$ < DMSO < Ph$_3$PO
< HMPA < Py < NCS$^-$ < N$_3^-$

and in acetonitrile,

I$^-$ < Br$^-$ < TMP < DMF $\leqslant$ Cl$^-$ < DMSO < Ph$_3$PO
< HMPA ~ Py < NCS$^-$ < N$_3^-$

These series are in agreement with the donor numbers of the solvent molecules and allow immediately qualitative predictions.

Halides, such as VOX$_2$, will easily accept further X$^-$ to give anionic complexes in solvents of low donor numbers, such as nitromethane or acetonitrile. Indeed iodo-, bromo- and chloro-complexes are easily formed in such solvents.[15] On the other hand bromo-complexes are not likely to be formed in TMP, DMF and stronger donor solvents. Chloro-complexes are accessible in TMP, but will hardly be formed in DMSO or pyridine.

A solvent of low donor number may be considered as a 'levelling' solvent, as it will allow the formation of most anionic complexes. Thus nitromethane allows the formation of anionic complexes of many transition metal ions with I^-, Br^-, Cl^-, N_3^- and NCS^- ions, in nearly stoicheiometric amounts in the last two cases.

On the other hand a solvent of high donor number may be considered a 'selective' or 'differentiating' solvent with respect to the stabilities of complex anions with different X^--ligands. Strong ligands will successfully compete with the solvent molecules for the ligand sites, but complexes with weak ligands may not be produced. Thus iodo-complexes and bromo-complexes of most class A metal ions will not be formed in DMSO.

The following displacement reaction of type [2] has been investigated for different solvents D:

$$Ph_3CCl + D.SbCl_5 \rightleftharpoons [Ph_3C][SbCl_6] + D; \quad K_{SbCl_6^-}$$

When the $K_{SbCl_6^-}$ values are compared with the respective DN_{SbCl_5} of the donor molecules D, the expected relationship

$$\log K_{SbCl_6^-} = -a.DN_{SbCl_5} + b$$

was found to exist (Table 5).

TABLE 5. *Formation constants of the chlorocomplexes from Ph_3CCl and $SbCl_5$ in several solvents at concentrations between 10^{-3} to 10^{-5} moles/l.*

Solvent	DN_{SbCl_5}	$K_{SbCl_6^-}$
$1,2\text{-}C_2H_4Cl_2$	0·1	$> 10^5$
So_2Cl_2	0·1	$> 10^5$
$SOCl_2$	0·4	$> 10^5$
C_6H_5COCl	2·3	$> 10^5$
$POCl_3$	11·7	$1·1 \times 10^2$
CH_3CN	14·1	10^5
PDC	15·1	$3·4 \times 10^2$
$C_6H_5POF_2$	16·4	$1·3 \times 10^2$
$C_6H_5POCl_2$	18·5	$3·9 \times 10^1$
$(CH_3O)_3PO$	23·0	$2·5 \times 10^0$
$HCON(CH_3)_2$	26·6	$5·0 \times 10^{-2}$
$(CH_3)_2SO$	29·8	$4·0 \times 10^{-2}$
C_5H_5N	33·1	$< 10^{-2}$

4. The donor number of the solvent and ionization of halides

If the donor properties of the solvent molecules are appreciably higher than those of the competitive ligand X, solvate bonds will be of high stabilities. Thus metal–X–bonds may be substituted by metal–solvent bonds according to reaction [4]:

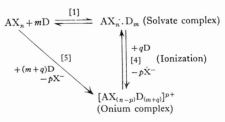

In such reactions stepwise replacement of X^- by solvent molecules may be involved. The over-all reaction is frequently represented by [5] and is termed an ionization of the compound in the solvent* with the possible formation of a completely solvated metal cation. In water this is considered as hydrolysis; for example,

$$AlCl_3 + 6H_2O \rightleftharpoons [Al(OH_2)_6]^{3+} + 3Cl^-$$

While [1]+[2] lead to the formation of a complex anion, [1]+[4] lead to ionization, which is followed by dissociation according to the dielectric constant of the solution.

Thus iodides of class A elements are ionized in TMP, DMF, DMSO and stronger donor solvents, while chlorides may be considerably ionized only in DMSO or pyridine.

5. The donor number and autocomplex-formation

When the donor properties of solvent molecules and competing ligand X^- are not vastly different, both ionization and formation of an anionic complex may occur at the same time. Reaction [4] will provide X^- units which may be consumed by other solute molecules by reaction [2]. Such a sequence of reactions is termed 'auto-complex formation'; for example,

$$2CoBr_2 + 6PDC \rightleftharpoons [Co(PDC)_6]^{2+} + [CoBr_4]^{2-}$$

* Ionization is unfortunately frequently confused with dissociation.

Such equilibria involving the formation of ions will be favoured by a high dielectric constant of the medium.

6. Other factors influencing coordination chemistry in solution

The consideration of the DN_{SbCl_5}, important as it is, can, however, never become the exclusive criterion for the choice of a solvent. Solubility properties of the solvent must also be considered as well as the dielectric constant (ϵ). A high ϵ will always be useful when the formation of a compound in solution is accompanied by the formation of ions. The contribution by ϵ to solvent properties has, however, frequently been overestimated.[12]

Other important contributions are steric considerations. If the solvent molecule is small or of a 'suitable' shape solvent co-ordination will be favoured. This is particularly true for water, which is exceeded in DN_{SbCl_5} by many other solvents, such as tri-methylphosphate, dimethylsulphoxide, dimethylformamide and pyridine. Indeed water is similar in DN_{SbCl_5} to diethylether, but the solvent properties of water are much better because of its steric properties and its high dielectric constant. Thus water ($DN_{SbCl_5} \approx 18$) would be a poor solvent for the formation of complex compounds, if it had a much higher DN_{SbCl_5}. On the other hand, certain solvents of low donor number, such as carbon tetrachloride, lack many solvent properties and can be applied only for a limited number of reactions. Acceptor solvents, such as bromine(III) fluoride[16] or arsenic(III)-chloride,[17] are excellent media for the formation of complex fluorides and chlorides respectively since there is practically no competition by reaction [1] to the coordination of the halide ions to the acceptor halides.[12]

Certain acid halides have also very low donor properties and thus will serve only as media for the formation of halide complexes. For complex chlorides liquid nitrosyl chloride is particularly useful since its donor properties at the O atoms are negligibly small. By its self-ionization the solvent nitrosyl chloride acts as a chloride ion donor[18]

$$NOCl \rightleftharpoons NO^+ + Cl^-$$

and hence provides chloride ions for the formation of chloro-complexes from many covalent chlorides; for example,

$$SbCl_5 + NOCl \rightleftharpoons [NO]^+ + [SbCl_6]^-$$
$$FeCl_3 + NOCl \rightleftharpoons [NO]^+ + [FeCl_4]^-$$

Such complexes are formed even in the absence of other chloride ion donors.

Thionyl chloride[10] and sulphuryl chloride[20] have exceedingly low donor numbers but good solvent properties for covalent chlorides. They are therefore excellent media for the formation of chloro-complexes from covalent chlorides. A stoicheiometric amount of chloride ion will usually lead to quantitative conversion of the chloride into the respective chloro-complex. On the other hand, owing to their small DN_{SbCl_5} they are poor solvents for ionic compounds and also for most transition metal halides. Alkali or alkaline earth salts of the chloro-complexes cannot be obtained in such solutions and most chloro-complexes of transition metal ions are also inaccessible. Their use is therefore restricted to the formation of chloro-complexes of representative elements with large cations.

Similar solvent properties are exhibited by acetyl or benzoyl chloride[21, 22] as solvents, but higher donor numbers are found for phosphorus oxychloride,[23] phenylphosphonic dichloride[23] and selenium oxychloride.[24] The moderately high donor number and high dielectric constant of selenium oxychloride allow both dissolution of ionic compounds, such as potassium chloride and the prompt formation of chloro-complexes.[24]

This group of solvents is very limited in their application. Other less specific solvents, such as chloroform, benzene or nitrobenzene also have low donor numbers and may therefore be useful as solvents for covalent compounds and some of their coordination reactions. Thus triphenylchloromethane has been used as a source of chloride ions to combine readily with certain metal chlorides in solutions in acetic acid,[25] nitroalkanes,[26, 27] chlorobenzenes[27] and benzene.[28] Such reactions are, however, impossible in solvents of high donor number such as dimethyl sulphoxide or tributyl phosphate.[28]

It may be concluded that solvents of low donor numbers have

severe limitations as media for complex formation due to their chemical characteristics or to restrictions in solubilities. On the other hand, solvents of high donor number will not permit certain coordination reactions because of the relative strength of the solvate bonds, thus making it difficult or impossible to replace solvent molecules in the solvate complex by other (competitive) ligands X^-.

The most useful solvents for the purpose under consideration will therefore have medium donor numbers and reasonable dielectric constants, such as acetonitrile (DN = 14, ϵ = 37), sulpholane (DN = 14·8, ϵ = 42), ethylene sulphite (DN = 15, ϵ = 41) or propane-1,2-diolcarbonate (DN = 15·1, ϵ = 69).

Many ionic compounds show reasonable solubilities and numerous bromo-, chloro- and azido-complexes have been studied in such solvents.[12, 29] It has been shown that in acetonitrile, di- and tetra-chlorides of the main-group elements are weaker chloride ion acceptors than their tri- and penta-chlorides, if a lone electron pair in the acceptor chloride is considered as a pseudo-ligand.[30] In this way $SbCl_3$ is considered a pseudo-tetrachloride and ICl_3 a pseudo-pentachloride. This is indeed the only experimental justification for counting an unshared electron pair in a compound as a ligand as has recently been proposed by Muetterties and Schunn.[31] It has further been found that the formation constant of a chloro-complex decreases with increasing donor number of the solvent, thus supporting the relationship[12]

$$\log K_{SbCl_6^-} = -a.DN_{SbCl_5} + b$$

It is now clear why only a relatively small number of iodo-complexes of class A metals have been prepared. Such complexes are hardly formed in the more common solvents, such as water, the alcohols or the ethers. By proper choice of a solvent with low donor number and still reasonable solvent properties for iodides, such as nitromethane or ethylene sulphite, such complex compounds will be accessible.

7. The coordination forms of cobalt(II) towards different X⁻ ligands in different solvents

The relationships outlined above are well demonstrated by a consideration of the complex chemistry of cobalt(II).

7.1. In nitromethane ($DN_{SbCl_5} = 2\cdot7$)

Cobalt(II) halides show moderate solubilities in nitromethane. Although the dielectric constant of the solvent is reasonably high, the solutions do not conduct an electric current. The undissociated species are nearly tetrahedral owing to the coordination of two solvent molecules: $CoX_2(NM)_2$. There are no indications of auto-complex formation. On the other hand the low donor number of the solvent molecule favours the formation of anionic complexes with all halide and pseudo-halide ions.[32-34] Such solutions are good conductors of electricity.

TABLE 6. *Coordination forms in nitromethane obtained in the presence of stoicheiometric amounts of X⁻ ligands*

X⁻	1	2	3	4	5	References
Cl⁻	−	+	+	+	−	32–34
Br⁻	−	+	+	+	−	32
I⁻	−	+	+	+	−	32, 34
N₃⁻	−	+	−	+	−	32
NCS⁻	−	+	−	+	−	32
CN⁻	−	+	−	+	+	32

The diazide and the dithiocyanate are scarcely soluble in this solvent, but they are readily dissolved on addition of cyanide and thiocyanate ions respectively to give the tetracoordinated anionic species. In the cyano-system the almost insoluble $Co(CN)_2$ is formed, and this gives a green solution on addition of two further equivalents of cyanide ions. Further addition of cyanide ions gives the pentacyano-complex ion, which is believed to contain one additional solvent molecule.[35]

Thus nitromethane is a poor ionizing solvent for dihalides and pseudohalides of cobalt(II), but it serves as an excellent medium for the formation of the respective complex anions.

7.2. In acetonitrile (DN_{SbCl_5} = 14·3)

While in the chloro- and bromo-systems tetrachlorocobaltates are readily formed in AN, the highest iodide-coordinated species in this solvent appears to be $[CoI_3(AN)]^-$. There are indications of considerable autocomplex formation of cobalt(II) iodide in acetonitrile. The decreasing stabilities of tetrahalocobaltates with increasing size of the halide ion is shown by the excess required to achieve complete formation of $[CoX_4]^{2-}$. A fourfold excess of Cl^- is necessary for $[CoCl_4]^{2-}$, a tenfold excess of Br^- for $[CoBr_4]^{2-}$, while $[CoI_4]^{2-}$ is not produced even in the presence of a 20-fold excess of iodide ions.[42] Higher stabilities are expected for thiocyanato- and azido-complexes, and indeed only slight excesses are required to obtain $[Co(N_3)_4]^{2-}$ and $[Co(NCS)_4]^{2-}$ respectively.[43]

TABLE 7. *Coordination forms in acetonitrile*

X^-	1	2	3	4	5	References
Cl^-	—	+	+	+	.	36–40
Br^-	+	+	+	+	.	38, 41
I^-	—	+	+	—	.	42
NCS^-	+	—	+	+	.	43
N_3^-	—	+	—	+	.	44

7.3. In propane-1,2-diolcarbonate (DN_{SbCl_5} = 15·1)

Although the donor numbers of AN and PDC are very similar, the differences in sterical properties of the solvent molecules seem to promote certain differences. While cobalt(II) iodide is only partly autocomplex in AN, it is completely subject to autocomplex formation in PDC; while in the former $[CoI_3]^-$ is the highest iodide coordinated species, $[CoI_4]^{2-}$ is readily formed in PDC:

$$2CoI_2 + 6PDC \underset{\text{complete}}{\rightleftharpoons} [Co(PDC)_6]^{2+} + [CoI_4]^{2-}$$

Likewise cobalt(II) bromide undergoes complete autocomplex formation in contrast to cobalt(II) chloride in this solvent. Nearly stoicheiometric amounts of bromide or chloride ions are required to achieve complete conversions into tetrabromo-cobaltate and tetra-chlorocobaltate respectively but a twofold excess of iodide ions is necessary for the $[CoI_4]^{2-}$ ion.

TABLE 8. *Coordination forms of cobalt(II) in PDC*

X^-	1	2	3	4	5	References
Cl^-	−	+	−	+	.	42
Br^-	−	−	−	+	.	41
I^-	−	−	−	+	.	42
NCS^-	+	−	+	+	.	43
N_3^-	−	+(o)	−	+	.	43
CN^-	−	+	−	−	+	43

Even the thiocyanate undergoes autocomplex formation in this solvent, and stoicheiometric amounts of thiocyanate ions are sufficient to achieve complete conversion into $[Co(NCS)_4]^{2-}$. In the azido-system an octahedral diazide $Co(N_3)_2(PDC)_4$ appears to be present in PDC.

A dicyanide and the pentacyano-complex are readily formed in PDC.

7.4. In ethylene sulphite ($DN_{SbCl_5} = 15\cdot3$)

TABLE 9. *Coordination forms of cobalt(II) in ethylene sulphite*

X^-	1	2	3	4	5	References
Cl^-	+	+	+	+	.	45
Br^-	+	+	−	+	.	45
I^-	−	−	−	+	.	45
NCS^-	+	−	+	+	.	45a
N_3^-	−	+(o)	+	+	.	45a
CN^-	−	+	−	−	+	45a

All chloride-coordinated forms possible are readily formed. In the iodo-system complete autocomplex formation of cobalt(II) iodide takes place.

A fourfold excess of Cl^--ions is necessary to produce completely $[CoCl_4]^{2-}$ in ethylene sulphite, and an 11-fold excess of bromide ions is required for the quantitative formation of $[CoBr_4]^{2-}$. The tetra-iodocomplex requires a 20-fold excess of iodide ions.[45]

The pseudohalide systems of cobalt(II) in ethylene sulphite[45a] are analogous to those in PDC. An octahedral diazide is formed and the thiocyanate is autocomplex in solution. The solubility of the dicyanide is poor, but it is readily converted into the yellow pentacyano-complex on addition of further cyanide ions. The latter shows a spectrum which is different from those in PDC or water and thus may be due to a $[Co(CN)_5D]^{3-}$ complex anion.

7.5. In acetone ($DN_{SbCl_5} = 17$)

TABLE 10. *Coordination forms of cobalt(II) in acetone*

X^-	1	2	3	4	References
Cl^-	−	+	+	+	46–48
Br^-	−	+	+	+	47, 48
I^-	−	+	+	+	47, 48
NCS^-	+	+	+	+	46, 49, 50

Acetone is similar in donor number to water and indeed reactions of cobalt(II) with halide and pseudo halide ions are easily carried out with the ultimate formation of $[CoX_4]^{2-}$.

7.6. In water ($DN_{SbCl_5} = 18$) and ethyl alcohol

TABLE 11. *Coordination forms of cobalt(II) in water*

X^-	1	2	3	4	References
Cl^-	−	+	+	+	51–54
Br^-	−	−	−	+	51, 53, 54
I^-	−	−	−	+	51, 53, 54
NCS^-	−	+ (o)	+ (o)	+	50, 55–57

The behaviour found in hydroxylic solvents is characterized by the high stabilities of the solvent–hexacoordinated metal cations. Thus

TABLE 12. *Coordination forms of cobalt(II) in ethyl alcohol*

X^-	1	2	3	4	References
Cl^-	−	+	+	+	40, 51, 58
Br^-	+	+	+	+	40, 51
I^-	−	+	+	+	40, 51
NCS^-	−	−	−	+	50

cobalt(II) bromide and cobalt(II) iodide are ionized, but addition of bromide and iodide ions respectively allows the formation of the respective tetracoordinated anions.

The dithiocyanate appears to be octahedral, as has been found in propanediolcarbonate and in ethylene sulphite.

7.7. In trimethylphosphate ($DN_{SbCl_5} = 23$)

Trimethylphosphate is a stronger donor molecule than either iodide or bromide ions. Cobalt(II) bromide and cobalt(II) iodide are extensively solvolysed in this solvent. No anionic iodo-complex is formed in TMP,[42] although a large excess of iodide ions may be present, and cobalt(II) bromide is completely solvolysed:[59]

$$CoBr_2 + 2(CH_3O)_3PO \rightarrow \begin{matrix} CH_3O & O & O & OCH_3 \\ & P & Co & P & \\ CH_3O & O & O & OCH_3 \end{matrix} + 2CH_3Br$$

Even cobalt(II) thiocyanate undergoes autocomplex formation while the diazide is stable in TMP solutions and is readily converted into tetraazidocobaltate by free azide ions. $[Co[N_3)_4]^{2-}$ is practically undissociated in TMP, while in a solvent of lower donor number, such as acetonitrile dissociation takes place.

It may be noted that $[CoCl(TMP)_5]^+$ has not been detected in TMP while both $[CoBr(TMP)_5]^+$ and $[Co(NCS)(TMP)_5]^+$ are readily formed. These species have slightly distorted octahedral structures, but all other species are tetracoordinated.

TABLE 13. *Coordination forms of cobalt(II) in TMP*

X⁻	1	2	3	4	References
Cl⁻	—	+	+	+	36, 39
Br⁻	+ (o)	—	—	—	41, 59
I⁻	—	+	—	—	42
NCS⁻	+ (o)	—	+	+	43
N₃⁻	—	+	—	+	44

7.8. In *NN*-dimethylacetamide (DMA) (DN$_{SbCl_5}$ = 27·8)

NN-Dimethylacetamide is still stronger as donor than TMP, and is approaching the donor properties of the chloride ion. Cobalt(II) chloride is at least partly ionized in this solvent, but tetrachlorocobaltate can be formed, when a fourfold excess of chloride ions is provided.[42] On the other hand no tetrabromocobaltate and no tetra-iodocobaltate have been found capable of existence in this medium. Such complexes are completely dissociated.

TABLE 14. *Coordination forms of cobalt(II) in DMA*

X⁻	1	2	3	4	References
Cl⁻	—	+	+	+	42
Br⁻	—	+	+	—	42
I⁻	—	+	—	—	42
NCS⁻	+	—	+	+	43
N₃⁻	—	+	—	+	43

A threefold excess of thiocyanate ions is sufficient to give complete conversion into tetrathiocyanatocobaltate, and with azide ions the stoicheiometric amount converts all cobalt into tetracoordinated $[Co(N_3)_4]^{2-}$, which is practically undissociated in this solvent.[43] A monothiocyanato-complex has been found to exist in DMA,

$$2Co(NCS)_2 + 6DMA \rightleftharpoons [Co(NCS)(DMA)_5]^+ \\ + [Co(NCS)_3(DMA)]^-,$$

but no analogous azide-coordinated species has been detected.

7.9. In dimethylsulphoxide (DN$_{SbCl_5}$ = 29·7)

The donor properties of DMSO are slightly higher than that of the chloride ion and thus much higher than those of bromide and iodide ions. For this reason cobalt(II) chloride undergoes auto-complex formation in DMSO, while the bromide and the iodide are completely ionized. No bromide- or iodide-coordination appears to be possible in DMSO even in the presence of a large excess of the competitive anion-ligand:

$$2CoCl_2 + 6DMSO \rightleftharpoons [Co(DMSO)_6]^{2+} + [CoCl_4]^{2-}$$
$$CoBr_2 + 6DMSO \rightleftharpoons [Co(DMSO)_6]^{2+} + 2Br^-$$
$$CoI_2 + 6DMSO \rightleftharpoons [Co(DMSO)_6]^{2+} + 2I^-$$

A 50-fold excess of chloride ion is required to convert all cobalt(II) into the tetrachlorocobaltate.

TABLE 15. *Coordination forms of cobalt(II) in DMSO*

X$^-$	1	2	3	4	References
Cl$^-$	−	+	+	+	36, 60
Br$^-$	−	−	−	−	42
I$^-$	−	−	−	−	42
NCS$^-$	+	−	−	+	43
N$_3^-$	−	+	−	+	44

Again thiocyanate gives octahedral [Co(NCS)(DMSO)$_5$]$^+$ and the dithiocyanate is subject to complete autocomplex formation in DMSO:

$$3Co(NCS)_2 + 10DMSO$$
$$\rightleftharpoons 2[Co(NCS)(DMSO)_5]^+ + [Co(NCS)_4]^{2-}$$

A large excess of thiocyanate ions is necessary to give complete conversion into [Co(NCS)$_4$]$^{2-}$, which is considerably dissociated in DMSO. Because of the stronger donor properties of the azide ion the diazide is only partly autocomplex in DMSO, and tetraazido-cobaltate is more stable than the thiocyanato-complex. A fivefold excess of azide ions is sufficient for complete formation of [Co(N$_3$)$_4$]$^{2-}$.

8. Conclusion

The relative stabilities of $[CoX_4]^{2-}$ in different solvents decrease with increasing donor number of the solvent, as can be seen from the molar ratios $X^-:Co^{2+}$ required to give nearly complete conversion into $[CoX_4]^{2-}$. While in nitromethane ($DN_{SbCl_5} = 2\cdot7$) nearly stoicheiometric quantities are sufficient, the iodo-complex requires a slight excess of iodide ions in PDC ($DN_{SbCl_5} = 15\cdot1$). In acetonitrile ($DN_{SbCl_5} = 14\cdot3$) no $[CoI_4]^{2-}$ has been found and a considerable excess of Br^- is necessary to give $[CoBr_4]^{2-}$. In TMP ($DN_{SbCl_5} = 23$) nearly stoicheiometric amounts are sufficient for the formation of the tetracoordinated thiocyanates and chlorocomplexes, but the corresponding bromide- and iodide-coordinated species cannot be formed at all in this medium. Large excesses of X^- are necessary to form $[Co(N_3)_4]^{2-}$, $[Co(NCS)_4]^{2-}$ and $[CoCl_4]^{2-}$ in DMSO ($DN_{SbCl_5} = 29\cdot7$), while the bromo-compounds and iodo-compounds are completely dissociated in this solvent.

TABLE 16. *Molar ratios required to give* $[CoX_4]^{2-}$ *in different solvents*

Solvent	$[Co(N_3)_4]^{2-}$	$[Co(NCS)_4]^{2-}$	$[CoCl_4]^{2-}$	$[CoBr_4]^{2-}$	$[CoI_4]^{2-}$
NM	4	4	4	4	5
PDC	4	4	4	4	8
ES	10	6	15	45	90
AN	8	7	16	40	∞
DMA	4	13	16	∞	∞
DMSO	20	200	200	∞	∞

REFERENCES

1 G. Briegleb, *Elektronen-Donator-Akzeptor-Komplexe*, Berlin–Göttingen–Heidelberg: Springer-Verlag, 1961.

2 I. Lindqvist and M. Zackrisson, *Acta chem. scand.* 1960, **14**, 453.

3 M. Baaz and V. Gutmann, in *Friedel-Crafts and Related Reactions* (ed. G. Olah), vol. 1, p. 367. New York and London: Interscience, 1963.

4 R. S. Drago and K. F. Purcell, *Progr. inorg. Chem.* **6**, 271 (1964), and in *Non Aqueous Solvent Systems* (ed. T. C. Waddington), p. 213. London and New York: Academic Press, 1965.

5 V. Gutmann, A. Steininger and E. Wychera, *Mh. Chem.* 1966, **97**, 460.
6 V. Gutmann, E. Wychera and F. Mairinger, *Mh. Chem.* 1966, **97**, 1265.
7 V. Gutmann, E. Wychera and S. Angelov, unpublished.
8 T. F. Bolles and R. S. Drago, *J. Am. chem. Soc.* 1966, **88**, 3921.
9 V. Gutmann and E. Wychera, *Rev. Chim. Min.* 1967, **3**, 941.
10 V. Gutmann, *Coord. Chem. Revs.* 1967, **2**, 239.
11 V. Gutmann and E. Wychera, *Inorg. Nucl. Chem. Letters*, 1966, **2**, 257.
12 V. Gutmann, *Coordination Chemistry in Non Aqueous Solutions.* Vienna, New York: Springer-Verlag, 1968.
13 V. Gutmann and A. Scherhaufer, *Mh. Chem.* 1968, **99**, 335.
14 V. Gutmann and U. Mayer, to be published in *Mh. Chem.*
15 V. Gutmann and H. Laussegger, *Mh. Chem.* 1968, **99**, 947, 963.
16 H. J. Emeléus and A. G. Sharpe, *J. Chem. Soc.* 1948, p. 2135, for reviews see: V. Gutmann, *Angew. Chem.* 1960, **62**, 312; A. G. Sharpe, *Adv. Fluorine Chem.* 1959, **1**, 29; R. D. Peacock in *Progr. in Inorg. Chem.* **2**, 192. New York: Interscience, 1960.
17 V. Gutmann, *Z. anorg. allg. Chem.* 1952, **264**, 156.
18 A. B. Burg and G. W. Campbell, *J. Am. chem. Soc.* 1948, **70**, 1964; A. B. Burg and D. E. McKenzie, *J. Am. chem. Soc.* 1952, **74**, 3143.
19 H. Spandau and E. Brunneck, *Z. anorg. Chem.* 1952, **270**, 201; 1955, **278**, 197.
20 V. Gutmann, *Mh. Chem.* 1954, **85**, 393, 404.
21 V. Gutmann and H. Tannenberger, *Mh. Chem.* 1957, **88**, 216, 292; V. Gutmann and G. Hampel, *Mh. Chem.* 1961, **92**, 1048.
22 J. Singh, R. C. Paul and S. S. Sandhu, *J. Chem. Soc.* 1959, p. 845.
23 V. Gutmann, *Öst. ChemZtg*, 1961, **62**, 326; *Halogen Chemistry* (ed. V. Gutmann), vol. II, p. 399. London and New York: Academic Press, 1967.
24 G. B. L. Smith, *Chem. Rev.* 1938, **23**, 165.
25 J. L. Cotton and A. G. Evans, *J. chem. Soc.* 1959, p. 2988.
26 J. W. Bayles, A. G. Evans and J. R. Jones, *J. chem. Soc.* 1955, p. 206.
27 J. W. Bayles, A. G. Evans and J. R. Jones, *J. chem. Soc.* 1957, p. 2020.
28 M. Baaz, V. Gutmann and J. R. Masaguer, *Mh. Chem.* 1961, **92**, 590.
29 M. Baaz, V. Gutmann and O. Kunze, *Mh. Chem.* 1962, **93**, 1142.
30 M. Baaz, V. Gutmann and O. Kunze, *Mh. Chem.* 1962, **93**, 1162.
31 E. L. Muetterties and R. A. Schunn, *Q. Rev. chem. Soc.* 1966, **20**, 245.
32 V. Gutmann and K. H. Wegleitner, *Mh. Chem.* 1968, **99**, 368.
33 S. Buffagny and T. M. Dunn, *J. chem. Soc.* 1961, p. 5105.
34 N. S. Gill and R. S. Nyholm, *J. chem. Soc.* 1959, p. 3997.
35 J. M. Pratt and P. R. Silverman, *Chem. Comm.* 1967, p. 117.
36 V. Gutmann, G. Hampel and J. R. Masaguer, *Mh. Chem.* 1963, **94**, 822.
37 F. A. Cotton, B. F. G. Johnson and R. M. Wing, *Inorg. Chem.* 1965, **4**, 502.
38 G. J. Janz, A. E. Marcinkovsky and H. V. Venkatasetty, *Electrochim. Acta*, 1963, **8**, 867.

39 M. Baaz, V. Gutmann, G. Hampel and J. R. Masaguer, *Mh. Chem.* 1962, **93**, 1416.

40 W. Libus, *Proc. VIIth Int. Conf. Coord. Chem.*, Stockholm, 1962, p. 349.

41 V. Gutmann and K. Fenkart, *Mh. Chem.* 1967, **98**, 1.

42 V. Gutmann and O. Bohunovsky, *Mh. Chem.* 1968, **99**, 740.

43 V. Gutmann and O. Bohunovsky, *Mh. Chem.* 1968, **99**, 751.

44 V. Gutmann and O. Leitmann, *Mh. Chem.* 1966, **97**, 926.

45 V. Gutmann and A. Scherhaufer, *Mh. Chem.* (in the Press).

45*a* V. Gutmann and A. Scherhaufer, *Inorg. Chim. Acta* (in the Press).

46 A. K. Babko and D. F. Drako, *Zhur. obsch. Khim.* 1949, **19**, 1809; *CA*, 1950, **44**, 1355 h.

47 M. S. Barrinok, *Zhur. obsch. Khim.* 1949, **19**, 612.

48 D. A. Fine, *J. Am. chem. Soc.* 1962, **84**, 1139.

49 A. Turco, C. Pecile and M. Nicolini, *J. chem. Soc.* 1962, p. 3008.

50 E. S. Tomula, *Z. anal. Chem.* 1931, **83**, 6.

51 M. Bobtelsky and K. Spiegler, *J. chem. Soc.* 1949, p. 143.

52 P. Job, *C. r. hebd. Séanc. Acad. Sci.*, Paris, 1937, **198**, 827.

53 W. R. Brode, *Proc. R. Soc.* A, 1928, **118**, 286.

54 W. R. Brode, *Proc. R. Soc.* A, 1928, **120**, 21.

55 M. Lehné, *Bull. Soc. chim. France*, 1951, p. 76.

56 A. de Sweemer, *Naturwet. Tijds.* 1932, **14**, 231, 237.

57 H. W. Vogel, *Ber.* 1875, **8**, 1533.

58 R. J. Macwalter and S. Barrat, *J. chem. Soc.* 1934, p. 517.

59 V. Gutmann and K. Fenkart, *Mh. Chem.* (in the Press).

60 V. Gutmann and L. Hübner, *Mh. Chem.* 1961, **92**, 1261.

5

PERFLUOROPSEUDOHALIDES AND THE CHEMISTRY OF CHLOROFLUOROMETHYLSULPHENYL COMPOUNDS

ALOIS HAAS

1. Introduction

The great interest taken in macromolecular perfluorinated hydro-carbons during the Second World War was maintained and intensified afterwards. The general aim in the earlier days was to prepare perfluorinated substances which were chemically very stable and technologically interesting, but soon scientific curiosity became the prime mover in the development of perfluoro organic chemistry. The major question then naturally arose as to whether a method of fluorination existed, which would permit the replacement of hydrogen atoms in organic molecules by fluorine while at the same time avoiding destruction of the carbon chain.

Available techniques at that time did not meet these requirements but attempts were made to find both specific and direct methods for the synthesis of organic fluorine compounds. The aim of this work was to prepare simple, reactive, fluorinated basic materials, such as CF_3I, and from them to build up more complex molecules. This led to the development of a new field of chemistry which spans classical organic and inorganic chemistry.

The expansion and exploration of this field is due mainly to the fact that perfluorinated organic groups, R_f, can be combined with a broad range of metals and non-metals such as Li, Hg, Sn, Sb, As, P, N, Se, S, etc., to give perfluoro-organometallic and -non-metallic compounds. These compounds provide a formal link between perfluoro-organic compounds and the elemental fluorides. This expansion and development is due mainly to the efforts of Professor Eméleus and his school whose work has pointed the way to future developments of perfluoro-organic chemistry of the elements.

One of the possible directions of further development is the introduction of pseudohalogens into such molecules and the study of their chemical behaviour. It is interesting to establish to what extent these reactive groups are suitable for participation in the synthesis of organic compounds. The following review outlines the present situation in perfluoropseudohalide chemistry and then deals in some detail with the chemistry of the $Cl_{3-n}F_nCS$-pseudo-halides and of the CF_3S-group.

The term 'pseudohalogen' was first used by Birkenbach and Kellermann[1] to describe a monovalent, negatively charged inorganic group, consisting of two or more electronegative elements, whose chemical and physical properties are generally similar to those of the halogens. Examples of pseudohalogens are the groups: $-CN$, $-N_3$, $-NCO$, $-SCN$, $-NCS$, $-SeCN$, $-NCSe$, $-CS_2N_3$, $-TeCN$.*

The pseudohalogen character of the perfluorinated groups CF_3- and CF_3S- was ascribed by Lagowski[2] mainly to their effective electronegativity. A more detailed study of this description by Downs[3] showed that, while some properties warranted such a classification, others did not. If the basic definition of a pseudo-halogen is extended to take into account its unsaturated character, then both CF_3- and CF_3S- are excluded. Under these conditions, both groups are treated not as pseudohalogens but rather as highly electronegative perfluorinated organic groups.

The term perfluoropseudohalide is used to describe a compound in which a pseudohalogen group is bound directly to fluorine or to a perfluorinated group. If X is a pseudohalogen then these compounds may be represented by the general formulae FX, $F_nY^mX_{m-n}$ (Y is a non-metal) and $(R_f)_nY^mX_{m-n}$ ($R_f = CF_3-$ or C_6F_5-), where $m \neq n$. In the case of four and higher valent metal-loids, divalent elements such as oxygen and sulphur may replace two fluorine atoms, so that one obtains perfluorinated pseudohalides having the general formulae

$$F_{n-2}(YZ)^{m-2}X_{m-(n+2)} \quad \text{and} \quad (R_f)_{n-2}(YZ)^{m-2}X_{m-(n+2)}.$$

* In subsequent work, Birckenbach and Huttner described further types of pseudohalogens (e.g. $-C(CN)_3$, $-N(CN)_2$, $=NCN$ and $-NCN-$) (L. Birckenbach and K. Huttner, *Ber. dtsch. chem. Ges.* 1929, **62**, 153; *Z. anorg. allg. Chem.* 1930, **190**, 1–52). These will be only briefly dealt with here as either they are included in the general formula $F_nY^mX_{m-n}$, or else no perfluorinated pseudo-halides are known to date.

FX	—CN / FCN	—N₃ / FN₃	—NCO	—SCN	—SeCN
$F_n Y^m X_{m-n}$ $Y = B$	—	—	—	—	—
C	CF_3CN	—	CF_3NCO	CF_3SCN*	CF_3SeCN*
	C_6F_5CN	$C_6F_5N_3$	—	—	—
Si	—	—	F_3SiNCO $F_2Si(NCO)_2$ $FSi(NCO)_3$	—	—
N	F_2NCN	—	—	—	—
P	F_2PCN	—	F_2PNCO $FP(NCO)_2$	F_2PNCS	—
As; O	—	—	—	—	—
S	F_5SCN	—	F_5SNCO	F_5SNCS	—
Se; Cl; Br; I	—	—	—	—	—
$(CF_3)_n Y^m X_{m-n}$ and $(C_6F_5)_n Y^m X_{m-n}$ P	$(CF_3)_2PCN$ $(C_6F_5)_2PCN$ $C_6F_5P(CN)_2$	$(CF_3)_2PN_3$	$(CF_3)_2PNCO$ $(C_6F_5)_2PNCO$ $C_6F_5P(NCO)_2$	$(CF_3)_2PNCS$ $(C_6F_5)_2PNCS$ $C_6F_5P(NCS)_2$	—
As	$(CF_3)_2AsCN$	—	$(CF_3)_2AsNCO$	$(CF_3)_2AsNCS$	—
S	CF_3SCN*	—	CF_3SNCO	CF_3SSCN	CF_3SSeCN
Se	CF_3SeCN*	—	—	—	—
$F_n(YZ)^{m-2}X_{m-(n+2)}$ $YZ = CO$ and	$FCO.CN$ $CF_3CO.CN$	$CF_3C{:}ON_3$	—	$CF_3C{:}OSCN$	—
$(CF_3)_n(YZ)^{m-2}X_{m-(n+2)}$ CS	$FC{:}S.CN$	—	—	—	—
PO	—	—	$F_2P{:}O.NCO$	—	—
SO₂	—	—	$FSO_2.NCO$	—	—

* These compounds may be considered either as cyanides of the CF₃S- and CF₃Se-groups, or else as thio- and seleno-cyanates respectively of the CF₃-group.

Compounds having these formulae have been prepared mostly where Y is carbon, phosphorus or sulphur. The perfluoropseudo-halides which have been prepared by various research groups and are scattered in the literature have been collected together in Table 1.

2. Preparation and properties of compounds having the general formulae

2.1. FX

No generally applicable methods are available for the preparation of compounds of this class, of which only FCN and FN_3 have so far been described. Fawcett and Lipscomb[4] were the first to prepare FCN in good yield by the pyrolysis of trifluorocyanuric acid at 1300° under a pressure of 50 mm. Callomon et al.[5] were able to show that the product obtained from the reaction of silver fluoride with cyanogen iodide at 220°[6] was not FCN.

Fluorine azide was first prepared by Haller[7] from the gas-phase reaction between hydrazoic acid and fluorine diluted with nitrogen. The compound may also be obtained, in a simpler and less dangerous process, by the fluorination of sodium azide.[8] The most important reaction of FN_3 is its simple thermal decomposition to give nitrogen and NF radicals which rapidly dimerize to give N_2F_2 according to

$$2FN_3 \rightarrow 2FN: + 2N_2$$
$$2FN: \rightarrow F—N{=}N—F$$

Attempts by Yakubovich and Englin[9] to prepare FNCO by fluorinating AgOCN mixed with CaF_2 were unsuccessful. At temperatures between 25° and 55 °C, COF_2 and CO_2 were obtained but no trace of any N–F containing compounds. However, the direct fluorination of potassium cyanate in fluorine diluted with nitrogen produced CF_3OF, NF_3, KF, CO_2 and COF_2. It seems likely that FNCO was formed as an intermediate in this reaction. The preparation of FNCO as the product of the fluorination of a metal cyanate (NaOCN) in fluorine diluted with helium in the presence of a polar solvent such as water is described in a patent.[10] However, no details of the chemical and physical properties of this substance were given.

As neither FSCN nor FSeCN have been described, either as monomeric or as polymeric compounds, FCN and FN_3 remain the only members of the type FX to have been prepared with certainty.

2.2. $F_n Y^m X_{m-n}$

Systematic development of the idea of 'perfluoropseudohalides' led to the compounds of the general formula $F_n Y^m X_{m-n}$ $(m \neq n)$. The members of this class are, almost exclusively, those in which Y is a non-metal. Although compounds of the first metalloid, boron, having the formulae $F_n BX_{3-n}$ $(0 < n < 3)$ are, in principle, possible, no such compound is known up to the present time. It is probable that they decompose as rapidly as the mixed halides of boron[11] by the reaction

$$BF_n X_{3-n} \rightarrow BF_3 + BX_3$$

The number of possible pseudohalides which may be derived from perfluorinated compounds of carbon is very large; however, only a few representatives have so far been synthesized. As examples of perfluoro-organopseudohalides, only the derivatives containing the CF_3- and C_6F_5-groups will be dealt with here. Other similar groups show only slight differences which are in no way fundamental. To date, only CF_3CN, CF_3NCO, CF_3SCN and CF_3SeCN among the members of the class of compounds $F_n CX_{4-n}$ $(n = 3)$ have been synthesized, and C_6F_5CN and $C_6F_5N_3$ in the class C_6F_5X.

Trifluoromethylcyanide[12] is produced when ethyl trifluoroacetate is first converted to the amide by treatment with ammonia and the elements of water are then removed with P_2O_5:

$$CF_3COOC_2H_5 \xrightarrow[-C_2H_5OH]{NH_3} CF_3CONH_2 \xrightarrow[-H_2O]{P_2O_5} CF_3CN$$

With hydrogen in the presence of a platinum catalyst, CF_3CN is reduced to 2,2,2-trifluoroethylamine. Fluorination of the compound leads, according to the reaction conditions and fluorinating agent used, to a variety of products as shown in the following reaction scheme:

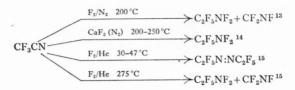

Albrecht and Husted[16] prepared the first perfluorinated organo-isocyanate in a two-step Curtius reaction

$$C_3F_7COCl \xrightarrow{NaN_3} C_3F_7CON_3 \rightarrow C_3F_7NCO + N_2$$

Barr and Haszeldine[17] prepared CF_3NCO in an analogous manner and showed that there was no recognizable difference in chemical behaviour between the CF_3- and C_3F_7-groups. Trifluoromethyl isocyanate is also prepared by the controlled hydrolysis of $CF_3N{=}CF_2$, and higher yields are obtained when $CuSO_4 . 5H_2O$ is used instead of water.[18] Both CF_3NCO and C_3F_7NCO are very reactive, thus they rapidly undergo alcohol addition to give carbamic acid esters

$$R_fNCO + C_2H_5OH \rightarrow R_fNHCO . OC_2H_5 \quad (R_f = CF_3\text{-}, C_3F_7)$$

Perfluorinated isocyanates behave differently towards water from normal isocyanates. Whereas organic isocyanates react with water to give symmetrically di-substituted ureas, C_3F_7NCO gives perfluoropropionitrile, C_2F_5CN, which, in the presence of hydrogen ions, reacts further to give $C_2F_5COONH_4$. The reaction mechanism of this chain degradation reaction is complex but it is possible to represent it in the following way:

$$C_3F_7NCO \xrightarrow{H_2O} C_3F_7\text{-}NH\text{-}COOH \xrightarrow[-CO_2]{} (C_3F_7NH_2)$$
$$\xrightarrow{H_3O^+} C_2F_5CN \xrightarrow{H_3O^+} C_2F_5CONH_2 \longrightarrow C_2F_5COONH_4$$

The compounds CF_3SCN and CF_3SeCN were prepared by Eméleus and his co-workers[19, 20] from the reaction between silver cyanide and CF_3SCl and CF_3SeCl respectively. The latter may also be prepared from CF_3SeCl and potassium cyanide.[21] Compounds of the types F_2CX_2 and FCX_3 are so far unknown. Recently the first two pseudohalide derivatives containing the perfluorophenyl group have been prepared. Perfluorobenzonitrile, C_6F_5CN, was obtained[22] from the reaction between pentafluoro-iodobenzene and CuCN in pyridine. The reaction of pentafluorophenylhydrazide with $NaNO_2$ in hydrochloric acid solution[23] yields pentafluorophenylazide, $C_6F_5N_3$, which may be separated from the pentafluoroaniline also formed by conversion of the latter to the hydrochloride. The azide, which is produced in 30% yield by this method, is a yellow oil which may be distilled under reduced pressure without decomposition. At 80 °C the azide begins to decompose, and decomposition is complete at 120 °C, 1 mole of nitrogen being given off per mole of azide. A higher yield of azide is achieved when nitrosyl chloride is used instead of $NaNO_2$ in the diazotization of the hydrazide. This reaction, which is carried out in glacial acetic acid, produces the azide in 50% yield.

The next homologue of carbon, silicon, forms all three possible isocyanates, but these are the only known examples of perfluorosilicon pseudohalides. The compounds F_3SiNCO, $F_2Si(NCO)_2$ and $FSi(NCO)_3$ are prepared by the fluorination of $Si(NCO)_4$ with antimony trifluoride;[24] they are all readily decomposed by water and methanol. Trifluorosilylisocyanate, which decomposes at 20 °C, is thermally the least stable.

Very recently a perfluorinated pseudohalide of nitrogen has been prepared by Meyers and Frank,[25] who synthesized difluorocyanamide by fluorinating a 50% aqueous solution of cyanamide, buffered with Na_2HPO_4, using elementary fluorine. The material, which is a white solid at $-196°$, melts to a clear liquid which boils at -61 °C. The best yields were obtained at a reaction temperature between 5–9 °C and at high flow rates of fluorine. F_2NCN is very stable and does not trimerize even at 120 °C. In the presence of caesium fluoride it rearranges to difluorodiazirine, thus

$$F_2NCN \xrightarrow[24\,°C]{CsF} F_2C\diagdown\diagup\begin{matrix}N\\ \|\\ N\end{matrix}$$

Anderson[26] has successfully applied the method of fluorinating non-metal pseudohalides with SbF_3 to the compounds $P(NCO)_3$ and $P(NCS)_3$. In the former case both F_2PNCO and $FP(NCO)_2$ are obtained while in the latter case only F_2PNCS was observed. All three compounds are colourless liquids having an unpleasant odour and all are remarkably sensitive to hydrolysis. F_2PNCO decomposes near its boiling point ($12\ °C$). F_2PCN was prepared[27] from the reaction

$$PF_2I + CuCN \rightarrow F_2PCN + CuI$$

The volatile, colourless compound is rather unstable and decomposes according to the scheme

$$3F_2PCN \rightarrow P(CN)_3 + 2PF_3$$

The cyanide structure appears to be confirmed particularly by the mass spectrum.

Although fluorides of sulphur having the compositions S_2F_2, SF_4 and SF_6 are known, the only pseudohalide derivatives known are those containing an SF_5-group. The first compound of this type, SF_5CN, was prepared[15] by the direct fluorination of methyl thiocyanate with fluorine diluted in nitrogen at $90\ °C$. Elegant methods for the preparation of SF_5NCO and SF_5NCS have been described by Tullock, Coffman and Muetterties.[29] SF_5NCS and SF_5NCO were prepared from the reaction of SF_5NHCF_3 with thiobenzoic and benzoic acid respectively, in the presence of sodium fluoride. Controlled hydrolysis and thiolysis of the compound SF_5NCCl_2 also afford the above-mentioned pseudohalides. The compound SF_5NCCl_2 may be prepared by the irradiation of a mixture of SF_5Cl and cyanogen chloride.

Both SF_5NCO and SF_5NCS are decomposed by bases. Benzyl alcohol reacts with the isocyanate to give the corresponding urethane:

$$F_5S.NCO + C_6H_5CH_2OH \rightarrow C_6H_5CH_2CO.ONHSF_5$$

No other perfluoro pseudohalide derivatives of these or other non-metals are known up to the present.

2.3. $F_n(YZ)^{m-2}X_{m-(n+2)}$

When two fluorine atoms on non-metals in four and higher valency states are replaced by a divalent element, for example oxygen or sulphur, then a perfluoropseudohalide of the type

$$F_n(YZ)^{m-2}X_{m-(n+2)} \quad (Z = O,\ S\ldots;\ m = 4)$$

is formed. This formulation is consistent with the relative chemical stability of the group YZ and the reactivity of these compounds being principally determined by the pseudohalogen group X. From the large number of possible members of this series, only very few examples have been prepared. Fluorocarbonylcyanide, FCOCN, and fluorothioformylcyanide, FC(:S)CN, are the only representatives of the class in which Y is carbon. When phosgene and sodium fluoride react together in liquid HCN at room temperature,[29] FCOCN is produced in 14% yield, while the main product is COF_2; HCN functions both as reagent and solvent in this reaction. If carbonyl fluoride is allowed to react with HCN in the presence of sodium fluoride (as HF trap) at 150 °C, then after 3 h a 22·5% yield of FCOCN is obtained. The corresponding thio-compound FCSCN[30] is formed by the reaction of ClFCHCN with sulphur or phosphorus pentasulphide at 350–900 °C. The excess of sulphur is trapped out and the reaction product frozen at −78 °C. At this temperature FCSCN polymerizes slowly, either with or without an initiator, to give a tough, flexible film.

Although no perfluorinated pseudohalides of pentavalent phosphorus having the general formula $F_nP^mX_{m-n}$ are known, the compound F_2PONCO has been synthesized. Antimony trifluoride will fluorinate $Cl_2P:ONCO$ in the absence of a catalyst to give $F_2P:ONCO$ in 90% yield, without any attack on the isocyanate group. A by-product of this exchange reaction is fluorochlorophosphorylisocyanate, $ClPF:ONCO$, which boils at 101–103 °C.

F_2PONCO is a colourless liquid which fumes in moist air and is considerably more stable thermally than the corresponding chlorocompound. It may be distilled at atmospheric pressure and may be stored over long periods of time without decomposition at room temperature. The isocyanate group will react with alcohols and with primary and secondary amines giving the corresponding carbamate ester or urea, thus

$$F_2PONCO + ROH \rightarrow F_2PO \cdot NHCOOR$$
$$F_2PONCO + RNH_2 \rightarrow F_2PO \cdot NHCONHR$$

As an alternative to SbF_3, silver monofluoride may also be used as fluorinating agent in the preparation of F_2PONCO.[31] The compound functions as a stabilizer for ozone.

In the case of sulphur, only FSO_2NCO is known. This compound is prepared from the corresponding chloro-compound by fluorination with sodium fluoride[32] or with anhydrous hydrogen fluoride.[34] The compound is also obtained[35] from the reaction between fluoro-sulphonic acid and sulphurylisocyanate:

$$FSO_3H + SO_2(NCO)_2 \rightarrow FSO_2NCO + HOSO_2NCO$$

Partial hydrolysis produces FSO_2NH_2 according to

$$FSO_2NCO + H_2O \rightarrow FSO_2NH_2 + CO_2$$

2.4. $(CF_3)_n(YZ)^{m-2}X_{m-(n+2)}$

When the fluorine atoms of perfluoropseudohalides of the types $F_nY^mX_{m-n}$ and $F_n(YZ)^{m-2}X_{m-(n+2)}$ are replaced by perfluorinated organic groups, compounds of the types $(R_f)_nY^mX_{m-n}$ and $(R_f)_n(YZ)^{m-2}X_{m-(n+2)}$ are obtained. We shall consider next the latter type of compound. To date, only two compounds in which Y is carbon have been prepared: they are CF_3COSCN and CF_3COCN.[35] These were prepared from the reaction between CF_3COCl and silver cyanide and silver thiocyanate respectively. Molecular weight measurements have shown that, in the gas phase, the cyanide exists as a dimer while the thiocyanate is monomeric.

2.5. $(CF_3)_nY^mX_{m-n}$

Among the perfluoropseudohalides of the type $(CF_3)_nY^mX_{m-n}$, only compounds of the fifth and sixth main group elements have been prepared. One is concerned here with pseudohalides of the bistrifluoromethyl-phosphorus $[(CF_3)_2P-]$,[36] -arsenic $[(CF_3)_2As-]$,[37] trifluoromethyl-sulphenyl $[CF_3S-]$[20] and -selenyl $[CF_3Se-]$[19] groups. These compounds may be prepared by the following general reaction:

$$(CF_3)_nY^mZ_{m-n} + (m-n)MX \rightarrow (CF_3)_nYX_{m-n} + (m-n)MZ$$
$$\text{(for } Y = P; \; Z = I, \; Y = As, \; S, \; Se; \; Z = Cl)$$

Silver pseudohalides (MX) are usually the most suitable metal salts for this reaction, and only where X is azide is LiN_3 to be preferred to the explosive silver azide. Potassium cyanide has been used with success[21] in the preparation of CF_3SeCN. In general, little is known of the chemical properties of these compounds. $(CF_3)_2PN_3$[38] is fairly stable at $0°$; however, it has been known to explode unexpectedly at $-196°$; at $20°$, it decomposes slowly and the controlled decomposition at 50–$60°$ and 37 mm gives polymeric phosphonitrilic derivatives and nitrogen thus:

$$x(CF_3)_2PN_3 \rightarrow [(CF_3)_2PN]_x + xN_2$$

The rapid development of pentafluorophenyl chemistry in recent years, largely as a result of the ready availability of reactive materials such as C_6F_5Br, has led naturally to the synthesis of compounds analogous to the CF_3-non-metal halides. With silver pseudohalides, perfluorinated compounds of the type $(C_6F_5)_n Y^m X_{m-n}$ are formed. In this way a range of compounds of the type $(C_6F_5)_n PX_{3-n}$ $(n = 1, 2$: $x = CN$, SCN and NCO) have been prepared.[39]

Extensive studies have been made of the pseudohalides of the CF_3S group, particularly of CF_3SNCO.[20] Alkaline hydrolysis of CF_3SCN first leads to a splitting of the CF_3S—C bond and the production of CF_3SH and HNCO as intermediates. Trifluoromethyl thiol is hydrolysed in the normal manner giving fluoride, sulphide and carbonate ions.[40]* The isocyanate can be identified as such:

$$CF_3SCN \xrightarrow{\text{OH}^-(\text{H}_2\text{O})} CF_3SH + NCO^-$$
$$\downarrow \text{OH}^-(\text{H}_2\text{O})$$
$$F^- + S^{-2} + CO_3^{-2}$$

The compound is stable in water at $20\ °C$, but at higher temperatures ($100\ °C$) incomplete hydrolysis occurs.

The chemistry of CF_3SSCN is determined by its thermal instability. Even at room temperature it decomposes quantitatively within a few minutes according to

$$2xCF_3SSCN \rightarrow xCF_3SSCF_3 + (SCN)_x$$

At temperatures below $-20\ °C$ the compound is stable over a period of weeks.

* The alkaline hydrolysis of CF_3S-compounds gives also polysulphides.

When CF_3SCl and silver cyanate are allowed to react in a sealed tube, in addition to CF_3SNCO, a clear, oily, less-volatile compound is formed. The latter has been characterized as the dimer of CF_3SNCO on the basis of analysis and molecular weight.[20] Hydrolysis[20] and alcoholysis[41] together with infrared and n.m.r. studies have shown that this dimer has the structure of a (bis-trifluoromethylsulphenylcarbamoyl)-isocyanate,

$$[(CF_3S)_2N.CO.NCO]$$

A cyclic dimer of CF_3SNCO is formed[42] when the pure compound is stored in a sealed tube. After a period of four to six weeks a colourless solid crystallizes out, which has the appearance of mother-of-pearl, and which can easily be recrystallized from CS_2. The compound may also be formed when CF_3SNCO is heated at 100 °C for 1 or 2 h in a sealed tube. Study of the hydrolysis, in addition to infrared and n.m.r. spectroscopy, has proved especially useful in establishing the structure of this dimer. In contrast to the linear dimer, which reacts with water to give an unsymmetrically substituted urea and CO_2, the cyclic dimer yields a symmetrically substituted urea together with CO_2. Monomeric CF_3SNCO reacts with water in a similar manner. In the presence of catalytic amounts of anhydrous sodium acetate at 100 °C in a sealed tube it gives tris-trifluoromethylsulphenyl-isocyanurate.[20, 42] The various modes of reaction of CF_3SNCO are shown in the following scheme:

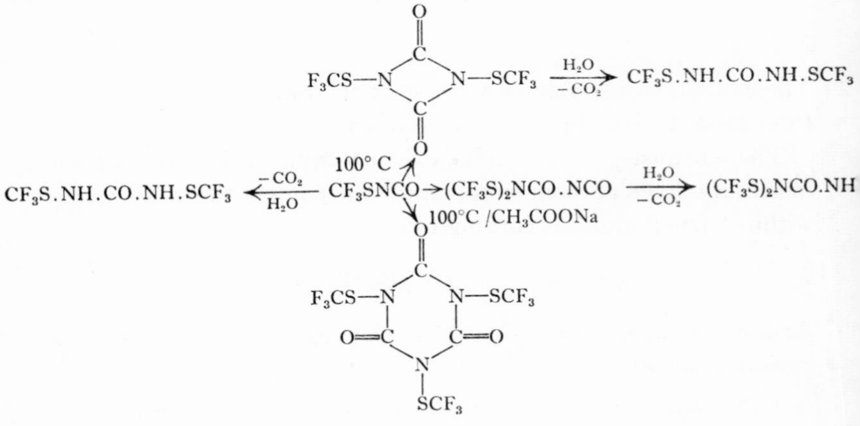

The compound $(CF_3S)_2NCO.NCO$ reacts with methanol[41] in the following manner:

$$(CF_3S)_2NCO.NCO + CH_3OH \rightarrow (CF_3S)_2N.CO.NH.COOCH_3$$

$$\downarrow H_2O$$

$$CF_3S.NH.CO.NH.COOCH_3$$

$$(I\,a)$$

The hydrolysis of CF_3SNCO, in contrast to that of CF_3NCO,[17] yields the expected disubstituted urea. This difference in behaviour may be explained by a decrease in the electron-attracting character caused by the presence of the sulphur atom, with the result that the intermediate CF_3SNH_2 is stabilized. The analogous CF_3NH_2, which must first be formed in the hydrolysis of CF_3NCO, eliminates HF immediately, giving FCN which then hydrolyses further to HF, CO_2 and NH_3. The following scheme shows this difference in behaviour more clearly

$$CF_3SNCO + H_2O \rightarrow CF_3S.NH.COOH \xrightarrow[-CO_2]{} CF_3SNH_2$$

$$\xrightarrow{CF_3SNCO} CF_3S.NH.CO.NH.SCF_3$$

$$CF_3NCO + H_2O \rightarrow CF_3NH.COOH \xrightarrow{-CO_2} CF_3NH_2$$

$$\xrightarrow{-2HF} FCN \xrightarrow{H_2O/H^+} HF + CO_2 + NH_4^+$$

The final step in the proposed reaction scheme for the hydrolysis of CF_3SNCO has been experimentally confirmed[43] by the synthesis of $CF_3S.NH.CO.NH.SCF_3$ from CF_3SNH_2 and CF_3SNCO. Thus, the hydrolysis of CF_3SNCO occurs in an analogous manner to normal isocyanates. When allowed to react with compounds containing active hydrogen atoms its behaviour is again typical of a normal isocyanate. However, exceptions do occur and phenol, thiophenol, ethyl mercaptan and triphenyl-carbinol do not add to CF_3SNCO. Certain reactions of this type carried out using CF_3SNCO are shown in the following diagram on p. 100.

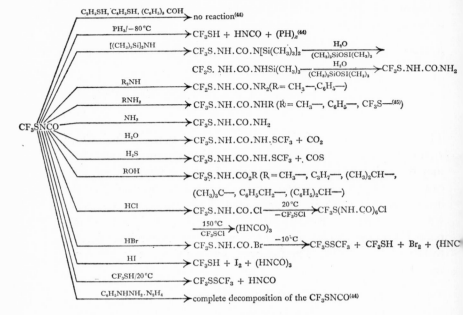

The chemical behaviour of the cyclic dimer also shows only small differences from the normal behaviour of uretidin-diones, for example

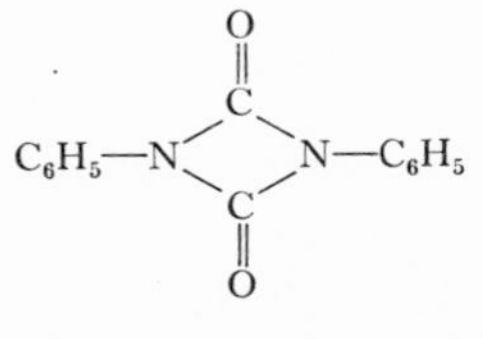

and this is shown in the diagram on p. 101.[46]

As may be seen from the diagram, the mechanism of the reaction with NH_3 and HCl is not clear. The isolated products may arise either from the intermediate uretidin-dione

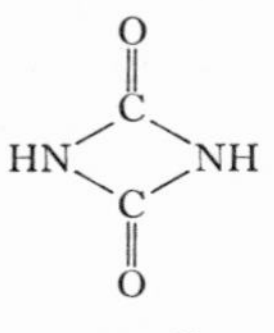

or from some linear intermediate. The primary product $CF_3S.NH.CO.N(SCF_3)^-COOCH_3$ could not be isolated from

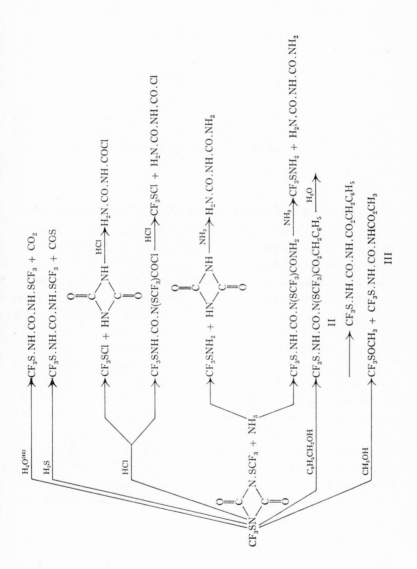

$$H_2O^{(42)} \longrightarrow CF_3S.NH.CO.NH.SCF_3 + CO_2$$

$$H_2S \longrightarrow CF_3S.NH.CO.NH.SCF_3 + COS$$

$$CF_3SCl + HN\underset{\displaystyle O=C\quad C=O}{\overbrace{\hspace{2.5em}}}NH \xrightarrow{HCl} H_2N.CO.NH.COCl$$

$$CF_3SNH.CO.N(SCF_3)COCl \xrightarrow{HCl} CF_3SCl + H_2N.CO.NH.CO.Cl$$

$$CF_3SNH_2 + HN\underset{\displaystyle O=C\quad C=O}{\overbrace{\hspace{2.5em}}}NH \xrightarrow{NH_3} H_2N.CO.NH.CO.NH_2$$

$$CF_3S.NH.CO.N(SCF_3)CONH_2 \xrightarrow{NH_3} CF_3SNH_2 + H_2N.CO.NH.CO.NH_2$$

$$CF_3S.NH.CO.N(SCF_3)CO_2CH_2C_6H_5 \xrightarrow{H_2O}$$

II

$$\longrightarrow CF_3S.NH.CO.NH.CO_2CH_2C_6H_5$$

$$CF_3SOCH_3 + CF_3S.NH.CO.NHCO_2CH_3$$

III

CF₃SN (with) C₆H₅CH₂OH, CH₃OH, HCl, $CF_3SN\underset{\displaystyle O=C\quad C=O}{\overbrace{\hspace{2.5em}}}N.SCF_3 + NH_3$

the methanolysis reaction as this material is immediately degraded into the observed products. Of the two possible formulae (a) and (b) for the final product:

$$CF_3S.NH.CO.NH.COOCH_3$$
$$(a)$$

$$H_2N.CO.N(SCF_3).COOCH_3$$
$$(b)$$

the correctness of formula (a) was established by comparison both of the infrared and n.m.r. spectra and also of the chemical and physical properties of the unambiguously synthesized compound I a on p. 99.

3. Preparation and properties of $Cl_{3-n}F_nCS$-pseudohalides

The remarkably rapid reaction of CF_3SCl with silver pseudohalides suggested the possibility of investigating the chemical behaviour of the partially fluorinated compounds of perchloromethyl mercaptan with pseudohalides. It was established that, in general, the reactivity of the compounds decreased with the decrease in the degree of fluorination in the direction $CF_3SCl \to CCl_3SCl$, which can be shown in the reaction with silver cyanate, for example. Thus, perchloromethyl mercaptan reacts with silver cyanate only in the presence of a solvent (e.g. benzene or acetonitrile) to give tris-trichloromethylsulphenyl-isocyanurate.[47] The monomeric intermediate, CCl_3SNCO, could not be isolated. Monofluorodichloromethylsulphenyl chloride reacts with AgOCN without a solvent to give FCl_2CSNCO[48] in a low yield even after a reaction time of 8–10 weeks. If, however, the reaction is carried out in a solvent, the main product is the trimer, $(FCl_2CSNCO)_3$. The monomer reacts with benzyl alcohol quantitatively to give the corresponding urethane and is hydrolysed to the disubstituted urea and carbon dioxide thus:

$$2FCl_2CSNCO + H_2O \to FCl_2CS.NH.CO.NH.SCCl_2F + CO_2$$

Difluorochloromethylsulphenyl chloride shows a very close similarity in its chemical behaviour to that of CF_3SCl. It reacts immediately with AgOCN at 20° to give $F_2ClCSNCO$, which behaves as a typical isocyanate, for example, in its reactions with water and with alcohols. When stored at room temperature over long periods (2–3 months) the isocyanate dimerses to the corresponding uretidin-dione. When, however, $F_2ClCSCl$ is allowed to react with AgOCN in benzene, the principal product is the trimer, tris-difluorochloromethylsulphenyl-isocyanurate.

The partially fluorinated sulphenyl chlorides react with selenocyanates in a similar manner to perchloromethyl mercaptan in that the product initially formed decomposes immediately into a disulphide, triselenium dicyanide and selenium dicyanide thus:

$$Cl_{3-n}F_nCSCl + KSeCN \rightarrow Cl_{3-n}F_nCSSeCN + KCl \quad (n = 0, 1, 2)$$
$$2Cl_{3-n}F_nCSSeCN \rightarrow (Cl_{3-m}F_nCS)_2 + (SeCN)_2$$
$$2(SeCN)_2 \rightarrow Se(SeCN)_2 + Se(CN)_2$$

Only from CF_3SCl is it possible to prepare a selenocyanate as a result of reaction with AgSeCN.[20] Thiocyanates are similarly unstable and, although the compounds $C_{3-n}F_nCSSCN$ can be isolated, they decompose rapidly too, according to the equation

$$2xCl_{3-n}F_nCSSCN \rightarrow x(Cl_{3-n}F_nCS)_2 + 2(SCN)_x \quad (n = 0–3)$$

The most stable derivatives are cyanides, which may be formed from the reaction of the sulphenyl chlorides with cyanides in the presence of a solvent, for example a water/ether (1:2) mixture:[49]

$$Cl_{3-n}F_nCSCl + MCN \rightarrow Cl_{3-n}F_nCSCN + MCl) \quad (n = 0, 1, 2)^{[50]}$$

Gradual changes in the chemical and physical properties with increase in the degree of fluorination are observed in the compounds formed in this reaction. Thus, the reactivity of the S—Cl bond increases with increasing fluorination and this may be explained by the high electronegativity of fluorine. When the chlorine atoms bonded to carbon in CCl_3SCl are exchanged for fluorine a shift of electrons towards the $Cl_{3-n}F_nC$-group occurs which is greater the more fluorine atoms present. This electron withdrawal causes a polarization of the S—Cl bond in the sense $S^{\delta+} \ldots Cl^{\delta-}$, thus decreasing the covalent character of this bond. Consequently the reactivity of the S—Cl bond increases with increasing degree of fluorination. Moreover, the melting point of the compounds $(Cl_{3-n}F_nCSNCO)_3$ decreases with increasing n, while the mean chemical shift $(\bar{\delta})$ increases linearly.

4. Contributions to CF₃S-chemistry

The information gained from a study of the CF_3S-pseudohalides made it worth while to synthesize other compounds containing the CF_3S-group and to study their properties. It appeared interesting to prepare compounds containing S—N bonds. The first compound of this type was prepared by Eméleus and Nabi[51] from the reaction between CF_3SCl and ammonia. Reaction between stoicheiometric proportions of the reagents at $-45\ ^\circ C$ in a sealed tube yields CF_3SNH_2. When the reaction is carried out at $20\ ^\circ C$ in the presence of an excess of CF_3SCl, $(CF_3S)_2NH$ is also produced. A better preparative method for CF_3SNH_2 consists in passing CF_3SCl through liquid ammonia at $-80\ ^\circ C$. This yields CF_3SNH_2, in addition to ammonium chloride, in large quantities ($\sim 50\%$ yield). The best route to the imine is through reaction of CF_3SNH_2 with the stoicheiometric quantity of CF_3SCl in the presence of a slight excess of trimethylamine, the latter functioning as an HCl-trap. It was also possible to prepare tris-trifluoromethylsulphenylamine for the first time by this method:[52]

$$CF_3SNH_2 + 2CF_3SCl + 2(CH_3)_3N$$
$$\rightarrow (CF_3S)_3N + 2[(CH_3)_3NH]Cl$$

$(CF_3S)_3N$ is the first perfluorinated organic compound containing an S—N bond. The compound is a colourless liquid which, in contrast to normal tertiary amines, shows no tendency to form addition compounds with BX_3 ($X = F, Cl, Br$) in the temperature range -80° to $+20\ ^\circ C$. The substance is stable to water and can be stored in the dark without decomposition. On warming and irradiation, decomposition occurs as shown by the appearance of a yellow colour. When heated under reflux, the compound is converted quantitatively to CF_3SSCF_3 and nitrogen as follows:

$$2(CF_3S)_3N \rightarrow 3CF_3SSCF_3 + N_2$$

N.m.r. spectroscopic studies[54] of samples which had been left in daylight showed the presence of another compound having a chemical shift of 50·35 ppm. At the start of irradiation the relative intensity of the signals due to CF_3SSCF_3 and to the new compound was $2:1$, so that it seems likely that the yellow compound may be $CF_3SNNSCF_3$, which arises as follows

$$2(CF_3S)_3N \to CF_3S.N{=}N.SCF_3 + CF_3SSCF_3$$

Further irradiation causes the ratio of intensities to increase in favour of the disulphide. Consequently, it has not proved possible to isolate the yellow compound.

It was expected that the reaction of CSF_2 with metal fluorides would lead to the synthesis of new compounds containing the CF_3S-group, since it was known[54] that silver and mercuric fluoride react to give $AgSCF_3$ and $Hg(SCF_3)_2$, respectively. When thiocarbonyl fluoride is allowed to react with alkali metal fluorides, either in the presence or absence of a solvent, the formation of compounds of the type $MSCF_3$ is not observed. According to the conditions polymers or dimers and trimers of CSF_2 are formed. In the absence of a solvent and using small quantities of CsF $(CSF_2:CsF = 1:60)$ at a temperature of $-78\ ^{\circ}C$, polymeric $(CSF_2)_x$ is formed. This polymer is soluble in chloroform, from which it may be re-precipitated with methanol. When the reaction is carried out at $0\ ^{\circ}C$, the trimer $(CF_3S)_2CS$ is formed very rapidly. If, however, the temperature is maintained at $-40\ ^{\circ}C$ then, in addition to the trimer, the pseudodimer of CSF_2, $CF_3S.CFS$ is also formed. The reaction occurs in a similar manner when potassium or ammonium fluoride is used; however, in these cases the reactant ratio MF/CSF_2 cannot be widely varied as in the case of CsF. The reaction with KF and with NH_4F also occurs more slowly and at higher temperatures. Chlorine adds quantitatively across the $C{=}S$ bond of both $(CF_3S)_2C{=}S$ and $CF_3S.CF{=}S$, even at $-60\ ^{\circ}C$ according to

$$CF_3S.CFS + Cl_2 \to CF_3S.CFCl.SCl$$
$$(CF_3S)_2CS + Cl_2 \to (CF_3S)_2CCl.SCl$$

5. Infrared and ^{19}F n.m.r. spectroscopic investigations of $Cl_{3-n}F_nCS$-compounds

The characterization and identification of CF_3S-derivatives is greatly facilitated by the use of infrared and n.m.r. spectroscopy. From the large number of compounds containing CF_3S-,

TABLE 2. ^{19}F-Chemical shifts and physical properties of CF_3S-compounds

Formula	Solvent (conc. %)	Chemical shift (ppm)	B.p. (m.p.) (°C)	
CF_3SCl	$FCCl_3$ (10–20)	49·8	0·7[57]	Golden yellow liquid
CF_3SCN	$FCCl_3$ (10–20)	38·8	36/(−70)	Colourless liquid
CF_3SNCO	$FCCl_3$ (10–20)	52·6	27/(−97)	Colourless liquid
CF_3SSCN	$FCCl_3$ (50–60)	45·88	(−35·5)	Colourless liquid
CF_3SSeCN	$FCCl_3$ (50–60)	42·75	119/(−67)	Yellow liquid
	$FCCl_3$ (~ 10)	58·00	—	—
CF_3SNH_2	$FCCl_3$ (~ 50)	57·67	46·5/(−89)	Colourless liquid
$(CF_3S)_2NH$	$FCCl_3$ (80–90)	55·82	73/(−47)	Colourless liquid
	$FCCl_3$ (80–90)	52·98	—	—
$(CF_3S)_3N$	$FCCl_3$ (10–20)	52·78	99·6	Colourless liquid
$(CF_3S)_2CS$	$FCCl_3$ (10–20)	42·03	110[57]	Red liquid
$CF_3SCF=S$	$FCCl_3$ (10–20)	42·97 (CF_3)	42·9/762 mm[57]	Yellow liquid
		89·55 (CF^-)		—
		18 cps (J_{F-F})	—	—
$(CF_3S)_2CCl.S.Cl$	$FCCl_3$ (~ 50)	39·11	40°/0·1 mm[55]	Yellow liquid
$CF_3S.CFCl.SCl$	$FCCl_3$ (~ 50)	38·07 (CF_3)	113[55]	Yellow liquid
		57·60 (CF)	—	—
		11 cps (J_{F-F})	—	—
CF_3SSCF_3	$FCCl_3$	48·18	34·6[57]	Colourless liquid

107

Compound	Solvent (concentration)	Shift	(value)	Appearance
CF₃S.NH.CO.NH.SCF₃	Acetone (10–20)	53·1	(182)	Colourless solid
(CF₃S)₂N.CO.NCO	CFCl₃ (10–20)	52·0	(−19)	Colourless liquid
(CF₃S)₂N.CO.NH₂	Acetone (10–20)	51·5	(43)	Colourless solid
$CF_3SN{=}C({=}O){-}C({=}O){=}NSCF_3$	Dioxan (10–20)	50·8	(79)	Colourless solid
(CF₃SNCO)₃	Acetone (10–20)	48·2 / 48·8	(111)	Colourless solid
CF₃S.NH.CO.NH.CH₃	Acetone (10–20)	53·9	(140) sublimes	Colourless solid
CF₃S.NH.CO.N(CH₃)₂	Dioxan (10–20)	53·0	(126)	Colourless solid
CF₃S.NH.CO.NHC₆H₅	Acetone (10–20)	53·3	(179)	Colourless solid
CF₃S.NH.CO.N(C₆H₅)₂	Acetone (10–20)	52·4	(157)	Colourless solid
CF₃S.NH.CO₂.CH₃	Acetone (15–20)	53·80	(40) sublimes	Colourless solid
CF₃S.NH.CO₂.C₃H₇	Acetone (15–20)	54·20	—	Colourless oil
CF₃S.NH.CO₂.CH(CH₃)₂	Acetone (15–20)	53·24	(46·5)	Colourless solid
CF₃S.NHCO₂.C(CH₃)₃	Acetone (15–20)	53·40	(38·5)	Colourless solid
CF₃S.NH.CO₂.CH₂C₆H₅	Chloroform (10–20)	55·17	(76)	Colourless solid
CF₃S.NH.CO₂.CH(C₆H₅)₂	Chloroform (15)	53·86	(88)	Colourless solid
CF₃S.(NH.CO)₂.OCH₃	Acetone (15)	52·65	(143–144)	Colourless solid
CF₃S(NH.CO)₂.OCH₂C₆H₅	Acetone (10–20)	54·20	(141)	Colourless solid
CF₃S.NH.CO.SCF₃ N.CO₂CH₂C₆H₅	External standard	51·38; 50·49	—	Colourless solid
CF₃S.NH.CO.NHSi(CH₃)₃	Chloroform (5)	54·06	(89)	Colourless solid
CF₃S.[(CH₃)₃Si]N.CO.NHSi((CH₃)₃)₃	Chloroform (5)	52·0	—	Colourless oil

CF_2ClS- and $CFCl_2S$-groups which have been synthesized, it has been possible to establish a number of group frequency assignments. This is because the sulphur atom in the bridging position serves to break down coupling between the groups bonded to it as a result of its relatively large mass. Nabi and Sheppard[56] were able to assign the vibrations of the CF_3S-group to the following frequency regions on the basis of a study of a number of trifluoromethylsulphenyl derivatives: ν_{as}(C—F) 1205–1155, ν_s(C—F) 1135–1095, $\delta_s(CF_3)$ 765–750, $\delta_{as}(CF_3)$ 510–540 and ν(C—S) 495–445 cm^{-1}. In a much more extensive study of compounds containing the CF_3S-group[41, 42] these assignments were confirmed. It has also been possible to make assignments[48] for the CF_2ClS- and $CFCl_2S$-groups. Thus, all the CF_2ClS-derivatives which have been investigated show ν_{as}(C—F) 1200–1090, ν_s(C—F) 1090–1050, ν(C—Cl) 900–850, $\delta_s(CF_2)$ 680–600, $\delta_{as}(CF_2)$ 550–500 and ν(C—S) 480–440 cm^{-1}. Similarly, for the $CFCl_2S$-group we find ν(C—F) 1055–1020, $\nu_{as}(CCl_2)$ 850–800, $\nu_s(CCl_2)$ 760–670, δ_s(C—F) 570–520 and ν(C—S) 450–430 cm^{-1}. The (C—S) stretching vibration, which normally appears at approximately 700 cm^{-1}, is considerably shifted to a lower wavelength as a result of coupling with the (C—F) deformation vibrations which belong to the same class.

As already mentioned, ^{19}F n.m.r. spectroscopy has proved of particular value in identifying $Cl_{3-n}F_nCS$-derivatives ($n = 1$–3). This is because almost all of the compounds show one single absorption line which is peculiar to, and thus characteristic of, the compound. Consideration of Tables 2 and 3 shows that all of the compounds containing the CF_3SN-group absorb in the range 51 ± 4 ppm. In a similar way, the CF_2ClSN- and $CFCl_2SN$-containing compounds show a mean chemical shift, $\bar{\delta}$, of 38 and 26 ppm, respectively; the deviation in each case being ± 3 ppm. When these values of the mean chemical shift, $\bar{\delta}$, are plotted against the degree of fluorination, n_F, a linear increase of $\bar{\delta}$ with increasing n_F is observed. The chemical shifts for the compounds CF_3S.CN, CF_3S.SCN, CF_3S.SCCN, CF_2ClS.SCN and $CFCl_2S$.CN differ so markedly from the $\bar{\delta}$ values just mentioned that one may be fairly sure that the normal form (CF_3S.CN) and *not* the iso- form (CF_3S.NCS) is present in each case. It should, however, be made

TABLE 3. ^{19}F-Chemical shifts and physical properties of $Cl_{3-n}F_nCS$-derivatives

Formula	Solvent (conc. %)	Chemical shift (ppm)	B.p. (m.p.) (°C)	
Cl_2FCSCl	$CFCl_3$ (10–20)	28·59	99[59]	Yellow liquid
Cl_2FCSCN	$CFCl_3$ (~ 50)	19·72	59/80 mm	Colourless liquid
$Cl_2FCSSCFCl_2$	$CFCl_3$ (50)	23·17	103/36 mm[60]	Colourless liquid
$(Cl_2FCSNCO)_3$	Benzene (10–20)	23·08	(160)	Colourless solid
$(Cl_2FCS.NH)_2.CO$	Acetone (10–20)	27·56	(181) decomp.	Colourless solid
$Cl_2FCS.NH.CO.O—CH_2C_6H_5$	$CFCl_3$ (10–20)	28·90	(73)	Colourless solid
ClF_2CSCl	$CFCl_3$ (10–20)	36·77	52[61]	Yellow liquid
ClF_2CSSCN	$CFCl_3$ (10–20)	31·35	Decomposes	Colourless liquid
ClF_2CSNCO	$CFCl_3$ (10–20)	38·72	—	Colourless liquid
$(ClF_2CSNCO)_2$	$CFCl_3$ (10–20)	36·86	(79)	Colourless solid
$(ClF_2CSNCO)_3$	Benzene (10–20)	35·48	(134·5)	Colourless solid
$(ClF_2CSNH)_2.CO$	Acetone (10–20)	39·87	(189)	Colourless solid
$ClF_2CS.NH.CO_2CH_3$	$CFCl_3$ (10–20)	40·28	(33·5)	Colourless solid
$ClF_2CS.NH.CO_2CH_2C_6H_5$	$CFCl_3$ (10–20)	40·47	(72)	Colourless solid
ClF_2CSSCF_2Cl	$CFCl_3$ (50)	32·21	62/95 mm[62]	Colourless liquid
ClF_2CSCN	$CFCl_3$ (50)	24·55	85	Colourless liquid

clear that other compounds, in which there is no $Cl_{3-n}F_nCSN$-group present (for example, the sulphenyl chlorides), absorb in the same region of $\overline{\delta}$.

6. Summary

The contents of Table 1 indicate the various possibilities which offer themselves to the preparative chemist in the field of perfluoropseudohalide chemistry. Only relatively few among the large number of formally possible compounds have so far been prepared. A systematic preparative investigation of the field is worth undertaking, in order to discover the relationships between structure and reactivity of these compounds and assess their practical or technological potential. A study of the chlorofluoromethylsulphenylpseudohalides and their reactions has provided a number of such relationships as the following.

(1) The reactivity of the S—Cl bond towards silver pseudohalides, of the compounds $Cl_{3-n}F_nCSCl$ ($n = 0$–3), increases with increasing 'n'.

(2) There is a linear increase in the ^{19}F n.m.r. chemical shift as the number of fluorine atoms in the group $Cl_{3-n}F_nCSN$- increases. There is also a linear decrease in melting point and boiling point of the compounds $(Cl_{3-n}F_nCSNCO)_3$ and $Cl_{3-n}F_nCSCl$, respectively, in the same direction.

(3) Assignments of characteristic absorption frequencies in the infrared for the individual groups.

(4) Preparation of fluorinated urea derivatives, carbamate esters, cyanurates, uretidin-diones, etc.

These relationships will be extended and deepened as the reactions of more, new, fluorinated organosulphenyl chlorides are studied. Such compounds are $CF_3S.CFCl.SCl$, $(CF_3S)_2CCl.SCl$, $CF_3.CFCl.SCl$ and $(CF_3)_2CCl.SCl$ which are all formally derived from $Cl_3C.SCl$ by substitution of chlorine atoms with CF_3S—, CF_3— and F—. Their reactivity should show considerable similarity to that of the $Cl_{3-n}F_nCSCl$ compounds. A further extension is possible in studying the reactions of sulphenyl chlorides such as $ClCO.SCl$, $FCO.SCl$, $ClCS.SCl$ and $FCS.SCl$, most of which have yet to be synthesized. These would yield derivatives of $Cl_{3-n}F_nCSCl$ in the broadest sense, and should permit a more profound understanding of the systematic character of this class of compound.

When the preparative work is more or less complete, kinetic studies should be carried out in order to establish quantitatively the effect of substituents in $Cl_{3-n}Y_nC.SCl$ (Y = CF_3S—, CF_3—, F—, etc.) upon the reactivity of the S—Cl bond in these compounds.

In addition to these rather theoretical considerations, one should not underestimate the practical value of these materials. It has recently been shown that, in addition to the known herbicidal activity of $CFCl_2S$-derivatives,[63] compounds containing the CF_3SN-group also show a broad spectrum of biocidal activity.[46] It is certain that this behaviour will not remain restricted to this group.

I should like to thank the Director of this Institute, Professor Dr O. Glemser, for his encouragement and Dr J. A. Connor (Technische Hochschule, Munich) for preparing the English text. My thanks are also due to the Deutsche Forschungsgemeinschaft, Fonds der Chemischen Industrie and the Farbenfabriken Bayer Leverkusen for invaluable support.

REFERENCES

1   L. Birckenbach and K. Kellermann, *Chem. Ber.* 1925, **58**, 786, 2377.
2   J. J. Lagowski, *Q. Rev. chem. Soc.* 1959, **13**, 233.
3   A. J. Downs, Ph.D. thesis, University of Cambridge, 1961.
4   F. S. Fawcett and R. D. Lipscomb, *J. Am. chem. Soc.* 1960, **86**, 1509.
5   H. J. Calloman, H. W. Thompson, F. A. Anderson and B. Bak, *J. chem. Soc.* 1953, p. 3709.
6   V. E. Coslett, *Z. anorg. allg. Chem.* 1931, **201**, 75.
7   J. F. Haller, Ph.D. thesis, Cornell University, U.S.A., 1942.
8   A. V. Pankratov, D. M. Sokolov and N. I. Savenkova, *Russ. J. inorg. Chem.* 1964, **9**, 1095. H. W. Roesky, O. Glemser and D. Bormann, *Angew. Chem.* 1964, **76**, 713; *Internat. Edit.* 1964, **3**, 701; *Chem. Ber.* 1966, **99**, 1589.
9   A. Ya. Yakubovich and M. A. Englin, *Zhur. obshchei Khim.* 1960, **30**, 2374; *Chem. Abstr.* **55**, 17336c.
10 V. Grakauskas, French Patent 1360 968 (1964); *Chem. Abstr.* **62**, 4949c.
11 F. A. Cotton and G. Wilkinson, *Advanced Inorg. Chemistry*, p. 189. Interscience Publishers, 1962.

12   F. Swarts, *Bull. Acad. roy. Belg.* 1922, **8**, 343; *Chem. Abstr.* 1923, **17**, 769.

13   R. N. Haszeldine, Research, 1951, **4**, 338.

14   G. E. Coates, J. Harris and T. Sutcliffe, *J. chem. Soc.* 1951, p. 2762.

15   J. A. Attaway, H. H. Groth and L. A. Bigelow, *J. Am. chem. Soc.* 1959, **81**, 3599.

16   A. H. Albrecht and D. R. Husted, U.S. Patent 2617817; *Chem. Abstr.* 1953, **47**, 8774.

17   D. A. Barr and R. N. Haszeldine, *J. chem. Soc.* 1956, p. 3428.

18   J. A. Young, W. S. Durrell and R. D. Dresdener, *J. Am. chem. Soc.* 1959, **81**, 1587.

19   J. W. Dale, H. J. Eméleus and R. N. Haszeldine, *J. chem. Soc.* 1958, p. 2939.

20   H. J. Eméleus and A. Haas, *J. chem. Soc.* 1963, p. 1273.

21   N. N. Yarovenko, V. N. Shemanina and G. B. Gaziera, *Zhur. obshchei Khim.* 1959, **29**, 942; *Chem. Abstr.* **54**, 2158 f.

22   W. J. Plummer and L. A. Watt, U.S. Patent 3046313 (1962), *Chem. Abstr.* **57**, 15003 d; U.S. Patent 3150163, *Chem. Abstr.* **61**, 16010 d; National Smelting Comp. Ltd. Belg. P. 630479 (1963), *Chem. Abstr.* **61**, 3033 b.

23   J. M. Birchall, R. N. Haszeldine and A. R. Parkinson, *J. chem. Soc.* 1962, p. 4966.

24   G. S. Forbes and H. H. Anderson, *J. Am. chem. Soc.* 1947, **69**, 1241.

25   M. D. Meyers and S. Frank, *Inorg. Chem.* 1966, **5**, 1455.

26   H. H. Anderson, *J. Am. chem. Soc.* 1947, **69**, 2495.

27   R. W. Rudolph, R. C. Taylor and R. W. Parry, *J. Amer. chem. Soc.* **88**, 3729.

28   C. W. Tullock, D. D. Coffmann and E. L. Muetterties, *J. Am. chem. Soc.* **86**, 357.

29   C. W. Tullock and D. D. Coffman, *J. org. Chem.* 1960, **25**, 2016.

30   St. Proskow, U.S. Patent 3026304 (1962); *Chem. Abstr.* **57**, 11032 c.

31   St. J. Kuhn and G. A. Olah, *Can. J. Chem.* 1962, **40**, 1951.

32   D. Voigt, German Patent 1043293 (1958); *Chem. Abstr.* **55**, 7298 a.

33   H. Hahn, German Patent 1083791 (1960); *Chem. Abstr.* **55**, 17509.

34   R. Appel and H. Rittersbacher, *Chem. Ber.* 1964, **97**, 849.

35   R. H. Patton and J. H. Simpson, *J. Am. chem. Soc.* 1955, **77**, 2016, 2017.

36   F. W. Bennett, H. J. Eméleus and R. N. Haszeldine, *J. chem. Soc.* 1953, p. 1565. K. Packer, Ph.D. thesis, Cambridge University, 1962.

37   H. J. Eméleus, R. N. Haszeldine and E. G. Walaschewski, *J. chem. Soc.* 1953, p. 1552.

38   G. Tesi, C. P. Haber and C. M. Douglas, *Proc. chem. Soc.* 1960, p. 219.

39   M. Fild, O. Glemser and I. Hollenberg, *Naturwissenschaften*, 1966, **53**, 130.

40   R. N. Haszeldine and J. M. Kidd, *J. chem. Soc.* 1955, p. 2901.

41   A. Haas, *Chem. Ber.* 1964, **97**, 2189.

42   A. Haas, *Chem. Ber.* 1965, **98**, 111.

43   A. Haas and M. E. Peach, unpublished results.

44 A. Haas, unpublished results.
45 A. Haas and M. E. Peach, unpublished results.
46 A. Haas and P. Schott, *Chem. Ber.* 1966, **99**, 1103.
47 A. Haas and D. Y. Oh, *Chem. Ber.* 1965, **98**, 3353.
48 A. Haas and D. Y. Oh, *Chem. Ber.* 1967, **100**, 480.
49 N. N. Yarovenko, S. P. Motornyi and L. I. Kivenskaya, *Zhur. obshchei Khim.* 1959, **29**, 3789; *Chem. Abstr.* **54**, 19479.
50 H. Brintzinger, K. Pfannstiel, H. Koddebusch and K. E. Kling, *Chem. Ber.* 1950, **83**, 87. K. A. Petrov and A. A. Neimyshewa, *Zhur. obshchei Khim.* 1959, **29**, 2165; *Chem. Abstr.* **54**, 10912; J. F. Olin, U.S. Patent 2650240; *Chem. Abstr.* **48**, 8819.
51 H. J. Emeléus and S. N. Nabi, *J. chem. Soc.* 1960, p. 1103.
52 A. Haas, M. E. Peach and P. Schott, *Angew. Chem.* 1965, **77**, 458.
53 A. Haas and P. Schott, unpublished results.
54 A. J. Downs, *J. chem. Soc.* 1962, p. 4316.
55 A. Haas and W. Klug, *Angew. Chem.* 1966, **78**, 906; *Internat. Edit.* 1966, **5**, 845.
56 S. N. Nabi and N. Sheppard, *J. chem. Soc.* 1959, p. 3439.
57 R. E. Banks and R. N. Haszeldine, *Advances in Inorganic Chemistry and Radiochemistry*, vol. III, p. 423. New York: Academic Press, 1961.
58 S. Andreades, U.S. Patent 3081350; *Chem. Abstr.* **59**, 5024.
59 E. Kober, *J. Am. chem. Soc.* 1959, **81**, 4810.
60 W. A. Sheppard and J. F. Harris, *J. Am. chem. Soc.* 1960, **82**, 5106.
61 N. N. Yarovenko, S. P. Motornyi, A. S. Vasil'eva and T. P. Gershzon, *Zhur. obshchei Khim.* 1959, **29**, 2163; *Chem. Abstr.* **54**, 9723.
62 N. N. Yarovenko, M. A. Raksha and V. N. Shemania, *Zhur. obshchei Khim.* 1960, **30**, 4069; *Chem. Abstr.* **55**, 20942.
63 E. Kühle, E. Klauke and F. Grewe, *Angew. Chem.* 1964, **76**, 807.

6

POLYFLUOROALKYL SILICON COMPOUNDS

R. N. HASZELDINE

1. Introduction

A possible route to per- and poly-fluoroalkyl derivatives of metalloids and metals became available as soon as trifluoroiodomethane and pentafluoroiodoethane were synthesized,[1]

$$CI_4 \overset{IF_5}{\rightarrow} CF_3I$$

$$C_2I_4 \overset{IF_5}{\rightarrow} CF_3.CF_2I$$

and shown to undergo carbon–iodine bond cleavage when heated or photolysed. The synthesis of the first perfluoroalkyl derivatives of a metal, $CF_3.HgI$ and $Hg(CF_3)_2$, followed soon afterwards,[2]

$$CF_3I \overset{Hg,\, u.v.}{\longrightarrow} CF_3.HgI \overset{Cd/Hg}{\longrightarrow} Hg(CF_3)_2$$

and perfluoroalkyl derivatives of phosphorus, arsenic, sulphur, etc. were obtained within a relatively short period of time.[3]

Initial attempts to extend the general procedure used for the above syntheses, i.e. photolysis of an intimate mixture of the metalloid or metal with trifluoroiodomethane, or a thermal reaction in the temperature range 180–260 °C, had failed when applied to more electropositive elements such as the alkali metals, aluminium, tin or silicon. Yet there was clearly much intriguing chemistry waiting to be investigated in per- or poly-fluoroalkyl derivatives of the elements, and encouraged by the successful results with mercury, phosphorus, sulphur, etc., a systematic and determined examination of this whole area of chemistry was begun. Now, some twenty years later, numerous per- or poly-fluoroalkyl derivatives of metalloids, non-metals, main group metals and transition metals are known, many new methods of synthesis have been developed, and many intriguing reactions have been discovered; this chapter, which does not set out to provide a comprehensive review, is devoted to selected aspects of one area of this chemistry—poly-fluoroalkyl silicon chemistry.

2. Synthesis

Four general methods have been developed for the preparation of per- or poly-fluoroalkyl silicon compounds:

(*a*) The addition of a compound containing an Si—H bond across the double bond of a per- or poly-fluoro-olefin, in the presence of ultraviolet light or a peroxide (presumed to initiate a free radical reaction), or in the presence of a transition metal catalyst (presumed to involve the intermediate formation of a compound containing a transition metal–silicon bond). These methods are by far the most useful and versatile.[4]

(*b*) The direct reaction of a per- or poly-fluoroalkyl chloride, bromide or iodide with elemental silicon in the presence of a copper catalyst.[5]

(*c*) Formation of a lithium per- or poly-fluoroalkyl, or of a per- or poly-fluoroalkyl Grignard compound, followed by its reaction with a silicon halide.[6]

(*d*) The free-radical addition of a per- or poly-fluoroalkyl bromide or iodide across the double bond of an alkenyl silicon compound.[7]

Method (*b*), the direct synthesis, has been exemplified in only sketchy fashion. The reaction of bromotrifluoromethane with a silicon–copper alloy at 400 °C is reported[5] to yield the compounds $CF_3.SiF_3$ and $CF_3.SiF_2Br$,

$$CF_3Br \xrightarrow{\text{Si/Cu, 400 °C}} CF_3.SiF_3$$

$$C_2F_5Cl \xrightarrow{\text{Si/Cu, 400–500 °C}} (C_2F_5)_2SiCl_2$$

but from what is now known (see below) of the properties of compounds which contain a $-CF_2.Si{\langle}$ group it is clear that thermal breakdown to form $F-Si{\langle}$ is very easy at temperatures below 250 °C. Low yields (5–8%) of $CF_3.SiF_3$ have been obtained by reaction of trifluoroiodomethane with silicon–copper alloy at 450–500 °C, but only when a short contact time and rapid quenching of the gas stream were employed.[8] The ready homolytic cleavage of the carbon–iodine bond in trifluoroiodomethane is doubtless an advantage here, but the major product from the direct synthesis is silicon tetrafluoride rather than $CF_3.Si$ compounds. This is not a practicable route to perfluoroalkyl silicon compounds at present, since yields are low and often zero, the reaction is difficult to control with local 'hot-spots' in the reactor, and the desired product decomposes rapidly under the conditions used. A fluidized-bed approach might be more successful. It is very doubtful that the use of compounds containing a —CF_2Cl group can lead to much more than traces of a perfluoroalkyl silicon compound since the CF_2—Cl bond cleaves homolytically only at much higher temperatures than does CF_2—I.

Method (c) is potentially of greater value, but although lithium per- or poly-fluoroalkyls and per- or poly-fluoroalkyl Grignard compounds have been shown to exist, most of them which have fluorine on carbon close to metal are at present difficult to prepare and utilize.[9, 6] Reactions such as

$$C_3F_7.MgI + Me_3SiCl \rightarrow C_3F_7.SiMe_3$$
$$n\text{-}C_3F_7Li + Et_2SiCl_2 \rightarrow n\text{-}C_3F_7.SiEt_2Cl + (n\text{-}C_3F_7)_2SiEt_2$$

can be carried out by use of a solution of the Grignard reagent prepared with difficulty from heptafluoroiodopropane and magnesium in tetrahydrofuran solution at $-40°$ to -50 °C, or by use of lithium heptafluoropropyl prepared either from the iodo-compound and lithium at -40 °C, or by the exchange reaction of heptafluoroiodopropane with lithium butyl. The alkylchlorosilane reacts almost instantly with the lithium heptafluoropropyl or heptafluoropropylmagnesium iodide, even at temperatures as low as -80 °C. Increased stabilization towards decomposition at $-20°$ to 20 °C of compounds which contain —CF_2Li or —$CF_2.MgI$ groups may well become possible by coordination and use of better aprotic solvents, but until this is demonstrated, the utility of per- and poly-fluoroalkyl lithium and Grignard compounds is low. The same comment holds for lithium and Grignard compounds that contain a CF_2 group β to the metal—for example, $CHF_2.CH_2Li$, $CHF_2.CH_2.MgI$—since unless the temperature is low they readily decompose to metal fluoride and olefin.

Once the CF_2 group is two or more carbon atoms from lithium or magnesium the organometallic reagent is much more like a conventional alkyllithium or alkylmagnesium halide, and synthesis of the corresponding polyfluoroalkyl silicon compound is reasonably straightforward; for example:

$$CF_3.CH_2.CH_2.MgCl \xrightarrow{Si(OMe)_4} CF_3.CH_2.CH_2.Si(OMe)_3$$

$$CF_3.CH_2.CH_2.MgI \xrightarrow{SiCl_4} (CF_3.CH_2.CH_2)_2SiCl_2,$$
$$(CF_3.CH_2.CH_2)_3SiCl, (CF_3.CH_2.CH_2)_4Si$$

Polyfluoroalkyl silicon compounds that contain fluorine on carbon at least two carbons removed from silicon are often best prepared by the Grignard route.

Method (a) encompasses two distinct methods for Si—H addition across the double bond of an olefin, specifically a fluoro-olefin. The first, where a noble-metal catalyst is used, for example,

$$CF_2:CFCl + SiHCl_3 \xrightarrow{Pt/C, 160\,°C} CHF_2.CFCl.SiCl_3$$

has not been studied thoroughly, and is probably capable of much extension. The direction of addition of $\rangle$Si—H to an unsymmetrical olefin may well be different from that where the reaction involves free-radical or four-centre addition (cf. the photochemical chlorotrifluoroethylene-trichlorosilane reaction below).

Hydrosilation of a fluoro-olefin presumably follows the path postulated for hydrosilation of hydrocarbon olefins,[10]

although it is often dangerous to extrapolate a reaction mechanism from hydrocarbon to fluorocarbon chemistry.

Photochemical, thermal or peroxide-initiated Si—H addition reactions have been assumed to follow a radical-chain mechanism,[4,7] for example,

$$SiHCl_3 \xrightarrow{\text{u.v., heat, or } R\cdot} SiCl_3\cdot(+H\cdot \text{ or } RH)$$

$$\dot{S}iCl_3 + C_2F_4 \rightarrow \dot{C}F_2.CF_2.SiCl_3$$

$$\xrightarrow{SiHCl_3} CHF_2.CF_2.SiCl_3 + SiCl_3\cdot$$

where R· (e.g. Ph·) is a free radical derived from a decomposing peroxide (e.g. Bz_2O_2). Such an assumption is doubtless justified, particularly when telomers of type $H.[C.C]_n.Si{<}$ are obtained as by-products; for example,

$$CF_2{:}CFCl + SiHCl_3 \xrightarrow{\text{u.v.}} CHFCl.CF_2.SiCl_3$$
$$+ H.[CFCl.CF_2]_n.SiCl_3$$

$$CF_2{:}CF_2 + MeSiHCl_2 \xrightarrow{\text{u.v.}} CHF_2.CF_2.SiMeCl_2$$
$$+ H.[CF_2.CF_2]_n.SiMeCl_2$$

$$CF_3.CH{:}CH_2 + SiH_4 \xrightarrow{\text{u.v.}} CF_3.CH_2.CH_2.SiH_3,$$
$$(CF_3.CH_2.CH_2)_2SiH_2, (CF_3.CH_2.CH_2)_3SiH$$

The yield of telomer can be minimized by use of an excess (often 5:1) of the Si—H compound to promote the chain-transfer reaction.

Kinetic evidence has established[11] that the photochemical reaction of trichlorosilane with tetrafluoroethylene is mercury-photosensitized and involves free radical intermediates as in the general mechanism given above:

$$Hg(6^1S_0) \xrightarrow{\text{u.v.}} Hg(6^3P_1)$$

$$Hg(6^3P_1) + Me_3SiH \longrightarrow Me_3Si\cdot + H\cdot + Hg(6^1S_0)$$

$$Me_3Si\cdot + C_2F_4 \longrightarrow Me_3Si.CF_2.CF_2\cdot \xrightarrow{C_2F_4} Me_3Si.(CF_2)_3.CF_2\cdot \xrightarrow{C_2F_4} \text{etc.}$$

$$\downarrow Me_3SiH \qquad\qquad \downarrow Me_3SiH \qquad \diagup Me_3SiH$$

$$Me_3Si.CF_2.CHF_2 + Me_3Si\cdot \text{ etc.} \qquad \text{Telomer}$$

Although the reaction mechanism in the above case has now been fully established, the possibility that other thermal or photochemical

Si—H addition reactions may involve a four-centre reaction in which two bonds are formed simultaneously as two are broken

must not be ignored, particularly since other four-centre reactions of Si—H compounds are well established; though at present there is no clear-cut evidence for a four-centre addition of Si—H to a fluoro-olefin, there is little evidence against it in some instances.

Most study has been devoted to polyfluoroalkyl compounds of silicon prepared by photochemical addition of Si—H across the double bond of a fluoro-olefin, and the rest of this chapter is thus restricted to polyfluoroalkyl silicon compounds prepared in this way.

3. Nucleophilic attack on silicon in polyfluoroalkylsilicon compounds

The pseudo-halogen character of a polyfluoroalkyl group, apparent in most polyfluoroalkyl derivatives of metals or metalloids, is particularly noticeable in reactions involving nucleophilic attack on silicon.

3.1. Amine complexes

The powerful inductive effect of polyfluoroalkyl, and the consequent enhanced electropositive character of silicon, enables complexes to be formed with amines. Thus 1,1,2,2-tetrafluoroethylsilane reacts with trimethylamine at room temperature to give a stable liquid $CHF_2.CF_2.SiH_3,NMe_3$, shown[12] by infrared and n.m.r. spectroscopy to be a new type of pentacoordinate silicon compound which, unlike known pentacoordinate silicon compounds, contains neither halogen nor oxygen directly attached to silicon.

3.2. Hydrolysis

Aqueous base cleaves the C—Si bond of polyfluoroalkyl silicon compounds which contain fluorine on carbon α or β to silicon; the polyfluoroalkyl group is liberated as the corresponding fluorohydrocarbon, or is converted into olefin, for example,

$$OH^- \quad Cl_3Si{-}CF_2.CHF_2 \longrightarrow \bar{C}F_2.CHF_2 \xrightarrow{H_2O} CHF_2.CHF_2$$

$$OH^- \quad Cl_3Si{-}CH_2{-}CHF{-}F \longrightarrow CH_2{:}CHF$$

Aqueous hydrolysis of a polyfluoroalkyl silicon compound that contains fluorine on α or β carbon, or base hydrolysis of a polyfluoroalkyl silicon compound that contains fluorine on carbon removed by at least two carbon atoms from silicon, does not cleave C—Si, but yields the polyfluoroalkyl-polysiloxane or -polysilsesquioxane, for example,

$$(a) \quad SiH_4 + C_2F_4 \xrightarrow{u.v.} CHF_2.CF_2.SiH_3$$

$$\xrightarrow{C_2F_4,\ u.v.} (CHF_2.CF_2)_2SiH_2$$

$$(CHF_2.CF_2)_2SiH_2 \xrightarrow[or\ COCl_2]{Cl_2\ low\ temp.} (CHF_2.CF_2)_2SiCl_2$$

$$(CHF_2.CF_2)_2SiCl_2 \xrightarrow{H_2O,\ H^+} [(CHF_2.CF_2)_2Si.O]_n$$

$$(b) \quad CHF_2.CF_2.SiH_3 \xrightarrow{COCl_2} CHF_2.CF_2.SiCl_3$$

$$\xrightarrow{H_2O,\ H^+} [CHF_2.CF_2.Si.O_{1.5}]_n$$

$$(c) \quad CF_3.CH_2.CH_2.SiCl_3 \xrightarrow{aq.\ NaOH} [CF_3.CH_2.CH_2.Si.O_{1.5}]_n$$

4. Polyfluoroalkyl silicon polymers

One of the early objectives in preparing polyfluoroalkyl silicon compounds was to use them as precursors of polyfluoroalkylpolysiloxanes and -polysilsesquioxanes. Only a few of the many polymers in which fluorine is situated α, β, or γ to silicon in

$$\overset{\gamma}{C}{-}\overset{\beta}{C}{-}\overset{\alpha}{C}{-}Si$$

will be mentioned. The hydrolytic and thermal stabilities of such polymers depend very markedly on the position of fluorine relative to silicon:[4, 13]

	Thermal stability	Hydrolytic stability
$[CHF_2.CF_2.Si.O_{1.5}]_n$	$\begin{cases} 8\% \text{ dec. at 172 °C} \\ 67\% \text{ dec. at 240 °C} \end{cases}$	$CHF_2.CHF_2$ formed
$[CHF_2.CH_2.Si.O_{1.5}]_n$	$\begin{cases} 14\% \text{ dec. at 170 °C} \\ 75\% \text{ dec. at 220 °C} \end{cases}$	$CH_2{:}CHF$ formed
$[CF_3.CH_2.CH_2.Si.O_{1.5}]_n$	$\begin{cases} 4\% \text{ dec. at 450 °C} \\ 80\% \text{ dec. at 500 °C} \end{cases}$	No reaction

Thermal stability was determined by measuring the amount of compound which could be recovered unchanged after being heated in an evacuated silica tube for 6 h. Hydrolytic stability was towards 10% aqueous sodium hydroxide at 20 °C.

Polysiloxanes and polysilsesquioxanes that contain fluorine in the α- or β-position to silicon are thus readily hydrolysed:

$$OH^- \ \backslash Si\text{—}CF_2.CHF_2 \longrightarrow \bar{C}F_2.CHF_2 \xrightarrow{\ H_2O\ } CHF_2.CHF_2$$

Even strong aqueous base fails to cleave the Si—C bond in the 3,3,3-trifluoropropyl silicon compounds but Si—O cleavage can occur in the usual way—for example the polysilsesquioxane $[CF_3.CH_2.CH_2.Si.O_{1.5}]_n$ dissolves in boiling 40% sodium hydroxide to give the sodium salt, and the polysilsesquioxane is reformed when the solution is diluted and acidified.[13]

The 3,3,3-trifluoropropyl silicone rubbers, specifically the methyl-3,3,3-trifluoropropyl polysiloxane

$$[CF_3.CH_2.CH_2.SiMe.O]_n,$$

have been developed commercially (Silastic LS-53; Dow Corning Corporation) since it shows excellent oxidation resistance and, compared with conventional dimethyl silicone rubbers, marked resistance to swelling by hydrocarbon fuels and lubricants of the type used in aircraft; its thermal stability is not improved compared with a conventional dimethyl silicone rubber. Trifluoropropyl silicone fluids are now commercially available.

The incorporation of other fluoroalkyl groups into a silicone and their influence on the properties of the polymer has not yet been studied in any detail and much investigation of industrial relevance needs to be done.

5. Thermal breakdown of polyfluoroalkyl silicon compounds

Three types of thermal decomposition have been established for these compounds.

(*a*) *Fluorine in the α-position.* Breakdown occurs by α-elimination of fluorine via internal nucleophilic attack on silicon; the carbene thus formed rearranges to olefin:

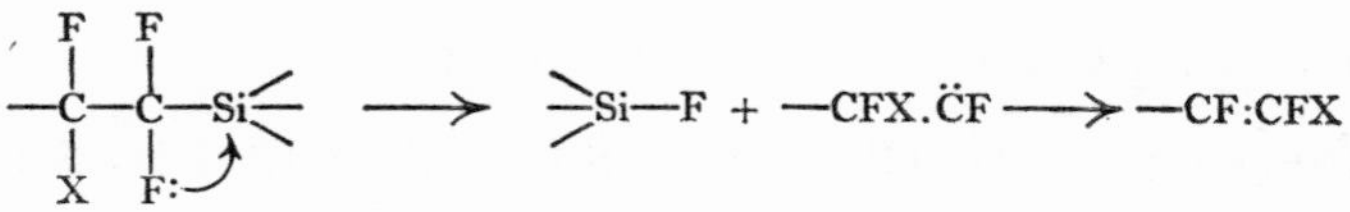

(*b*) *Fluorine in the β-position.* Olefin is formed quantitatively by β-elimination:

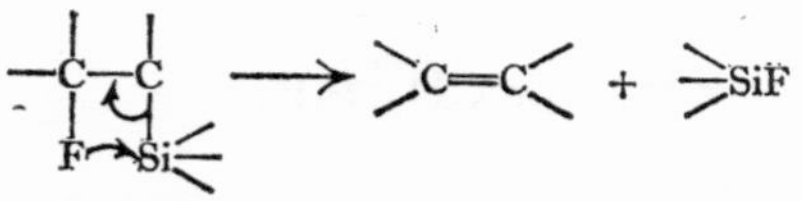

(*c*) *Fluorine in the γ-position.* The breakdown pattern is complex and involves homolytic fission of C—C and C—Si bonds:

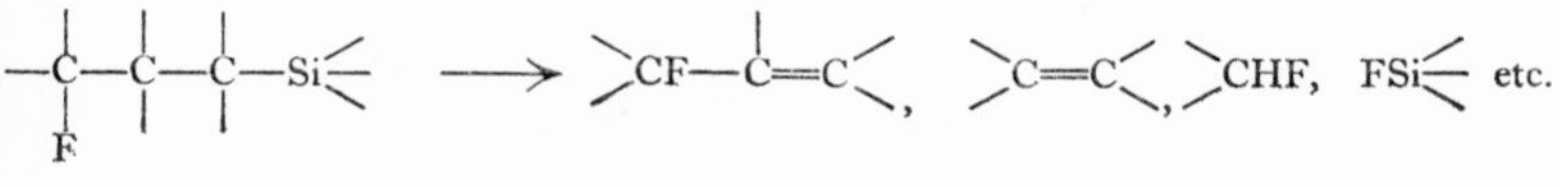

5.1. γ-Fluoro-compounds

Detailed pyrolysis studies have been carried out only on one γ-substituted polyfluoroalkyl silicon compound,

$$CF_3.CH_2.CH_2.SiF_3$$

The gas-phase thermal decomposition at 550–640 °C and initial pressures of 40–110 torr has a complex radical-chain mechanism.[14] The compounds CHF_3, $CF_3.CH:CH_2$, C_2H_4 and SiF_4 are major products, and other primary products are H_2, $CH_3.SiF_3$, $CH_2:CH.SiF_3$, $CF_2:CH_2$, $CF_3.CH_3$ and probably $CF_2:CH.CH_2.SiF_3$. The decomposition is approximately 1·5-order with respect to reactant in the initial stages, with a rate-constant given by

$$\log_{10} k_{1\cdot5}\,(\text{cm}^{\frac{3}{2}}.\text{mole}^{-\frac{1}{2}}\text{min}^{-1}) = 19\cdot6 - 74 \times 10^3/4\cdot58T$$

Reactions of the following type are thought to be involved:

$$CF_3.CH_2.CH_2.SiF_3 \overset{R\cdot}{\to} CF_3.\dot{C}H.CH_2.SiF_3$$
$$+ CF_3.CH_2.\dot{C}H.SiF_3$$

$$CF_3.\dot{C}H.CH_2.SiF_3 \to CF_3.CH:CH_2 + SiF_3\cdot$$

$$CF_3.CH_2.\dot{C}H.SiF_3 \to CH_2:CH.SiF_3 + CF_3\cdot$$

$$SiF_3\cdot + CF_3.CH_2.CH_2.SiF_3 \to SiF_4 + \dot{C}F_2.CH_2.CH_2.SiF_3$$

$$\dot{C}F_2.CH_2.CH_2.SiF_3 \left\langle \begin{array}{l} \to CF_2:CH_2 + \dot{C}H_2.SiF_3 \\ \to CF_2:CH.CH_2.SiF_3 + H\cdot \end{array} \right.$$

$R\cdot = CF_3\cdot$ or $H\cdot$ or $\dot{C}H_2.SiF_3$

The γ-fluoroalkyl silicon compounds generally are much more thermally stable than are the β- or α-fluoroalkyl compounds. Thus, the compounds $CF_3.CH_2.CH_2.SiCl_3$, $(CF_3.CH_2.CH_2)_2SiCl_2$, $(CF_3.CH_2.CH_2)_4Si$, $C_3F_7.CH_2.CH_2.SiMeCl_2$ and

$$CF_3.CHMe.CH_2.SiMeCl_2$$

and the corresponding polymers obtained from them, decompose slowly, if at all, at 300 °C, and some are stable to 400 °C.

5.2. β-Fluoro-compounds

The thermal decomposition of the polysilsesquioxane

$$[CHF_2.CH_2.Si.O_{1\cdot5}]_n$$

to give vinyl fluoride occurs at a temperature, 160–170 °C, much lower than might have been anticipated for a compound that contains a fluoroalkyl group:

$$[CHF_2.CH_2.Si.O_{1.5}]_n \rightarrow nCHF:CH_2 + \frac{n}{4}SiF_4 + \frac{3n}{4}SiO_2$$

Internal nucleophilic attack by β-fluorine on silicon was suggested to explain such ready decomposition:[13, 15]

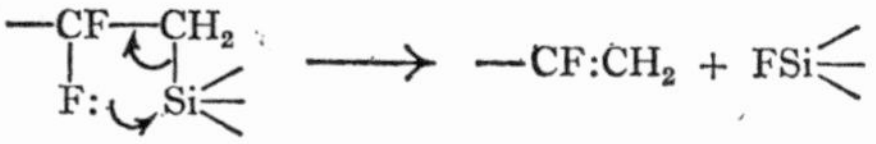

This reaction scheme has been amply substantiated by kinetic studies on related compounds.[16, 17] For example, the gas-phase thermal decomposition of 2,2-difluoroethyltrifluorosilane at 151–221 °C and pressures from 10 to 180 torr is kinetically first-order. The reaction is homogeneous and unaffected by the addition of nitric oxide or cyclohexene so that a radical-chain mechanism is not operative. The first-order rate constant

$$\log_{10}k \, (\text{s}^{-1}) = 12\cdot27 \pm 0\cdot27 - (32{,}720 \pm 530)/4\cdot576T$$

is independent of pressure. The reaction is not a radical non-chain process, since for such a process the over-all activation energy would have to equal the dissociation energy of the bond broken in the first and rate-controlling step of the reaction, and it is unlikely that the dissociation energy of the weakest bond would be much less than 70 kcal.mole^{-1} (cf. the observed activation energy of only 32·7 kcal.mole^{-1}).

The decomposition of 2,2-difluoroethyltrifluorosilane is therefore considered to be a unimolecular process involving a four-centre transition state:

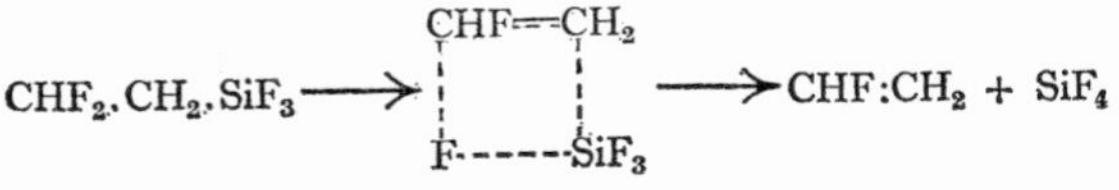

5.3. α-Fluoro-compounds

The temperature required for the onset of decomposition of the polymer $[CHF_2.CF_2.Si.O_{1.5}]_n$ is only 140°, in marked contrast to the 450° required for the polymer $[CF_3.CH_2.CH_2.Si.O_{1\cdot5}]_n$, but similar to the temperature required for the β-fluoro compound $[CHF_2.CH_2.Si.O_{1\cdot5}]_n$.[13] The earlier observations[15,18] that pyrolysis of the compounds $CHFCl.CF_2.SiCl_3$ or $CFCl_2.CF_2.SiCl_3$ at 250 °C gave mainly the olefins CHF:CFCl and CFCl:CFCl respectively were also noteworthy because of the low temperature required and the unexpected structures of the olefins formed. Simple β-elimination of fluorine would not explain the major products:

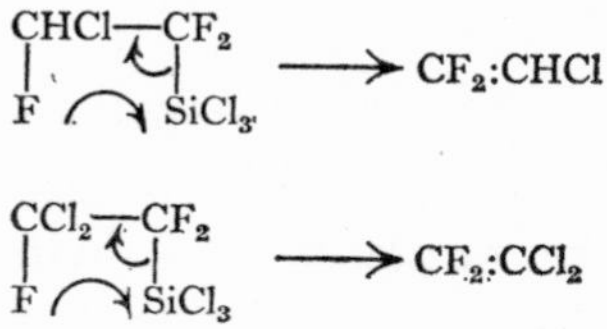

α-Elimination was therefore proposed,[15] as a route to carbenes which could rearrange to the observed olefins:

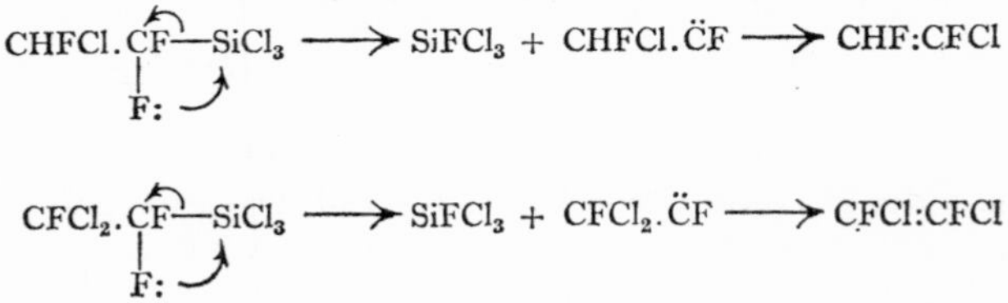

The minor (< 10%) olefinic products from these reactions, CF_2:CHCl and $CF_2:CCl_2$ respectively, may arise from fluorine shift in the intermediate carbenes:

$$CHFCl.\ddot{C}F \rightarrow CF_2:CHCl$$
$$CFCl_2.\ddot{C}F \rightarrow CF_2:CCl_2$$

or by concurrent β-elimination.

The trifluorosilyl analogue, $CFCl_2.CF_2.SiF_3$, decomposes at a lower temperature, but by a similar path:[18]

$$CFCl_2.CF_2.SiF_3 \xrightarrow{140\ ^\circ C} SiF_4$$

$$+CFCl_2.\ddot{C}F \xrightarrow{\text{Cl migration}} CFCl:CFCl \quad (91\%)$$

$$+CFCl_2.\ddot{C}F \xrightarrow[\text{F migration}]{} CF_2:CCl_2 \quad (8\%)$$

The intermediate carbene in such decomposition reactions can be trapped by addition of an excess of a hydrocarbon olefin to the reaction mixture; for example:

$$CH_2F.CF_2.SiF_3 \xrightarrow{140^\circ} CH_2F.\ddot{C}F \xrightarrow{\text{H migration}} CHF:CHF \quad (98\,\%)$$

with branch:

$$\downarrow \substack{\text{excess} \\ Me_2C:CH_2}$$

$$\begin{array}{c} Me_2C\!\!-\!\!CH_2 \\ \diagdown\!\diagup \\ CF.CH_2F \end{array} \quad (79\,\%) + CHF:CHF \quad (19\,\%)$$

Kinetic studies[17] have confirmed these ideas and established activation energies for the reactions studied. For example, the gas-phase thermal decomposition of 1,1,2,2-tetrafluoroethyltrifluoro-silane at 126–207 °C and 25–200 torr is first-order and homogeneous:

$$CHF_2.CF_2.SiF_3 \rightarrow CHF:CF_2 + SiF_4$$

The first-order rate constant is independent of pressure:

$$\log_{10} k\,(s^{-1}) = (10{\cdot}91 \pm 0{\cdot}07) - (28{,}450 \pm 140)/4{\cdot}576T$$

In unpacked vessels the intermediate carbene, $CHF_2.\ddot{C}F$, isomerizes quantitatively to trifluoroethylene; if a hydrocarbon olefin such as propene is present during reaction, the carbene adds essentially quantitatively to the double bond to give the cyclopropane:

$$CHF_2.CF_2.SiF_3 \longrightarrow CHF_2.\ddot{C}F \xrightarrow{C_3H_6} \begin{array}{c} CH_3.CH\!\!-\!\!CH_2 \\ \diagdown\!\diagup \\ CF.CHF_2 \end{array}$$

but the rate of decomposition of 1,1,2,2-tetrafluoroethyltrifluoro-silane is unchanged. This demonstrates clearly that elimination of α-fluorine and not β-fluorine is involved, and an intramolecular 3-centre transition state is therefore visualized:

$$CHF_2.CF_2.SiF_3 \longrightarrow \left[CHF_2\!\!-\!\!CF\substack{\diagup^{-F} \\ | \\ \diagdown_{SiF_3}} \right] \longrightarrow SiF_4 + CHF_2.\ddot{C}F$$

The thermal decompositions of 1,1,2,2-tetrafluoroethyltrichloro-silane and 1,1,2-trifluoroethyltrichlorosilane are particularly interesting:[18]

	First olefin	Second olefin	Third olefin
$CHF_2.CF_2.SiCl_3 \xrightarrow{225 \,°C}$	$CHF:CF_2$	$CF_2:CHCl$	$CHF:CCl_2$
	29%	21%	22%
$CH_2F.CF_2.SiCl_3 \xrightarrow{225 \,°C}$	$CHF:CHF$	$CHF:CHCl$	$CH_2:CCl_2$
	19%	7%	60%

since it will be noted that all the olefinic products contain the same number of carbon–hydrogen bonds as the original alkyl group, and that the first olefin in both cases has one less carbon–fluorine bond. The second and third olefins contain progressively one fewer carbon–fluorine bond and one more carbon–chlorine bond. The formation of trifluoroethylene and 1,2 difluoroethylene by α-elimination and rearrangement of the carbene is straightforward:

$$CHF_2.CF_2.SiCl_3 \to CHF_2.\ddot{C}F \to CHF:CF_2$$

$$CH_2F.CF_2.SiCl_3 \to CH_2F.\ddot{C}F \to CHF:CHF$$

The formation of the olefins $CF_2:CHCl$ and $CHF:CHCl$ could be accounted for by *carbene insertion* into the SiCl bond of the $SiFCl_3$ also formed, followed by a further α-elimination and carbene rearrangement:

$$CHF_2.\ddot{C}F + SiFCl_3 \to CHF_2.CFCl.SiFCl_2$$

$$CHF_2.CFCl.SiFCl_2 \xrightarrow{\alpha\text{-elimination}} SiF_2Cl_2 + CHF_2.\ddot{C}Cl$$
$$\to CF_2:CHCl$$

and $$CH_2F.\ddot{C}F + SiFCl_3 \to CH_2F.CFCl.SiFCl_2$$

$$CH_2F.CFCl.SiFCl_2 \xrightarrow{\alpha\text{-elimination}} SiF_2Cl_2 + CH_2F.\ddot{C}Cl$$
$$\to CHF:CHCl$$

Extension of this carbene-insertion postulate would lead to the formation of the third olefinic pyrolysis product by a final β- rather than α-elimination, since there would be now no α-fluorine present in the silicon compound:

$$CHF_2.\ddot{C}Cl + SiF_2Cl_2 \to CHF_2.CCl_2.SiF_2Cl$$

$$CHF_2.CCl_2.SiF_2Cl \xrightarrow{\beta\text{-elimination}} SiF_3Cl + CHF:CCl_2$$

and $\qquad$ $CH_2F . \ddot{C}Cl + SiF_2Cl_2 \rightarrow CH_2F . CCl_2 . SiF_2Cl$

$$CH_2F . CCl_2 . SiF_2Cl \xrightarrow{\beta\text{-elimination}} SiF_3Cl + CH_2 : CCl_2$$

The carbene-insertion mechanism seemed plausible at first sight, but unless the carbenes produced successively differed markedly in their reactivity and ability to insert into silicon–chlorine bonds, it would be expected that the first olefin would be produced in much higher yield than the second or third olefin—yet this was not observed. Furthermore, if the carbene-insertion mechanism was operative, pyrolysis in the presence of a large excess of silicon tetrachloride would be expected to increase the yields of the second and third olefins; in fact a control experiment gave no change in olefin ratio.

An alternative explanation was sought and found in *halogen exchange between α-carbon and the silicon*.[18] This process, which is competitive with the α-elimination process, involves *concurrent* nucleophilic attack of chlorine from the $SiCl_3$ group on the carbon with its developing electron-deficient character, and nucleophilic attack on silicon by α-fluorine, thus leading to the formation of an α-chloroalkyl silicon compound:

Carbene formation from the new chlorofluoroalkyl silicon compound gives the second olefin by α-elimination:

$$CHF_2 . CFCl . SiFCl_2 \rightarrow CHF_2 . \ddot{C}Cl \rightarrow CF_2 : CHCl$$

and the third olefin arises via concurrent exchange of a second α-fluorine on carbon for chlorine on silicon and subsequent β-elimination (see p. 130).

α-Elimination should be favoured by bulky substituents on the β-carbon which hinder the competing halogen exchange process. In accord with this the yield of the third olefin is much greater from $CH_2F . CF_2 . SiCl_3$ in which small β-substituents permit halogen exchange, than from $CHF_2 . CF_2 . SiCl_3$; models of the last compound suggest that there is appreciable steric interaction between chlorine on silicon and the β-fluorines.

9

$$CHF_2.CCl \overset{Cl}{\underset{F}{\diamond}} SiFCl \longrightarrow CHF_2.CCl_2.SiF_2Cl$$

$$CHF_2.CCl_2.SiF_2Cl \xrightarrow{\beta\text{-elimination}} CHF:CCl_2$$

Similarly, for $CH_2F.CF_2.SiCl_3$,

$$CH_2F.CF_2.SiCl_3 \begin{cases} \xrightarrow{\alpha\text{-elimination}} CH_2F.\overset{..}{C}F \longrightarrow CHF:CHF \\ \xrightarrow[\text{exchange}]{\text{halogen}} CH_2F.CFCl.SiFCl_2 \xrightarrow{\alpha\text{-elimination}} CH_2F.\overset{..}{C}Cl \\ \qquad\qquad\qquad\qquad \downarrow \text{halogen exchange} \qquad\qquad \downarrow \\ \qquad\qquad\qquad CH_2F.CCl_2.SiF_2Cl \qquad\quad CHF:CHCl \\ \qquad\qquad\qquad\qquad \downarrow \beta\text{-elimination} \\ \qquad\qquad\qquad\qquad CH_2:CCl_2 \end{cases}$$

To test the deduction that halogen exchange can occur and will be favoured by small β-substituents, the compound $CH_3.CF_2.SiCl_3$ was synthesized, after some difficulties, as follows:[19]

$$SiHCl_3 + CF_2:CCl_2 \xrightarrow{\text{u.v.}} CHCl_2.CF_2.SiCl_3$$
$$\downarrow \substack{Me_3SiH, \text{ u.v.} \\ 60\%}$$
$$CH_2Cl.CF_2.SiCl_3$$
$$\downarrow \substack{\text{n-Bu}_3SnH, \text{ u.v.} \\ 64\%}$$
$$CH_3.CF_2.SiCl_3$$

(with n-Bu_3SnH, u.v. 50% route from $CHCl_2.CF_2.SiCl_3$ to $CH_3.CF_2.SiCl_3$)

Rearrangement of 1,1-difluoroethyltrichlorosilane via halogen exchange occurred smoothly at 100°, well below the minimum temperature (225°) needed for the decomposition of a compound of type $RCF_2.SiCl_3$ via α-elimination to give a carbene:

$$CH_3.CF \overset{Cl}{\underset{F}{\diamond}} SiCl_2 \longrightarrow CH_3.CCl \overset{Cl}{\underset{F}{\diamond}} SiFCl \longrightarrow CH_3.CCl_2.SiF_2Cl$$
$$99\%$$

The yield of the rearranged product remained at $> 90\%$ for pyrolysis temperatures up to 180°, but above this temperature α-elimination began to compete with the halogen-exchange reaction.

$$CH_3.CF_2.SiCl_3 \xrightarrow{\text{α-elimination}} CH_3.\ddot{C}F \rightarrow CH_2:CHF$$

$\downarrow$ Halogen exchange

$$CH_3.CFCl.SiFCl_2 \xrightarrow{\text{α-elimination}} CH_3.\ddot{C}Cl \rightarrow CH_2:CHCl$$

$\downarrow$ Halogen exchange

$$CH_3.CCl_2.SiF_2Cl$$

At 180° the yield of vinyl fluoride was 8 % and of vinyl chloride zero. The corresponding percentage yields at 205°, 235° and 310 °C were 29, 3; 34, 53; and 31, 63 % respectively. Separate experiments with $CH_3.CCl_2.SiF_2Cl$ showed that it began to pyrolyse only at temperatures greater than 205 °C, thus providing a concurrent second route to vinyl chloride:

$$CH_3.CCl_2.SiF_2Cl \xrightarrow{\text{α-elimination}} CH_3.\ddot{C}Cl \rightarrow CH_2:CHCl$$

Pyrolysis of $CH_3.CF_2.SiCl_3$ at 400 °C showed that α-elimination to give vinyl fluoride (54 %) now predominated over exchange of one or two halogens followed by α-elimination to give vinyl chloride (35 %).

N.m.r. studies demonstrated that at 100 °C the halogen-exchange rearrangement of $CH_3.CF_2.SiCl_3$ to give $CH_3.CFCl.SiFCl_2$ was much slower than that of $CH_3.CFCl.SiFCl_2$ into

$$CH_3.CCl_2.SiF_2Cl,$$

since the concentration of $CH_3.CFCl.SiFCl_2$ present at any time was < 5 %, the limiting sensitivity of the instrument. If the second halogen exchange is easier than the first halogen exchange, the second α-elimination might be expected to be easier than the first, since both halogen exchange and α-elimination involve nucleophilic attack on silicon. If this is correct, the first halogen exchange reaction will be the rate-determining step; that the reaction may be treated to a first approximation as a first-order reaction rather than a consecutive reaction is shown by the straight-line plot from kinetic data obtained from the change in the n.m.r. peak integration curves with time. In neat solution at 100 °C the half-life of the halogen exchange reaction is approximately 21 min.[19]

Two novel features have thus emerged from these studies: the α-elimination reaction to give a carbene, and halogen exchange between carbon and silicon.

6. Carbenes from polyfluoroalkyl silicon compounds

The results described earlier establish the ease of α-elimination to give a carbene from compounds of type $RCF_2.SiX_3$, particularly when X = F. α-Chloro-compounds similarly undergo α-elimination, and a useful source of dichlorocarbene in the gas-phase is thus provided:[15, 18]

$$CH_3.SiCl_3 \xrightarrow{Cl_2, \text{ u.v.}} CCl_3.SiCl_3 \xrightarrow{SbF_3} CCl_3.SiF_3$$

$$\downarrow 250° \qquad \diagup 140\text{-}180°$$

$$\ddot{C}Cl_2$$

The halogenocarbenes, prepared by this general method,

$$\underset{\underset{Y}{|}}{\diagup}C.SiX_3 \rightarrow \diagup\ddot{C} + SiX_3Y$$

are intermediates which show a wide range of reactivity:

(a) *Reaction with carbonyl compounds.*[21] Olefin oxides, ketones or acyl halides result; for example:

$$\ddot{C}Cl_2 \begin{cases} \xrightarrow{(CF_3)_2CO} (CF_3)_2C\underset{O}{\overset{\diagdown\diagup}{-}}CCl_2 & 67\% \\[2ex] \xrightarrow{CF_3.CHO} CF_3.CH\underset{O}{\overset{\diagdown\diagup}{-}}CCl_2 & 80\% \\[2ex] \xrightarrow{(CCl_3)_2CO} CCl_3.CCl_2.CO.CCl_3 & 20\% \\[2ex] \xrightarrow{CF_3.COCl} CF_3.CO.CCl_3 & 80\% \\[2ex] \xrightarrow{CCl_3.COF} CCl_3.CFCl.COCl & 37\% \end{cases}$$

(*b*) *Reaction with nitriles or imines.*[21] Cyclic compounds or re-arrangement products have been isolated; for example:

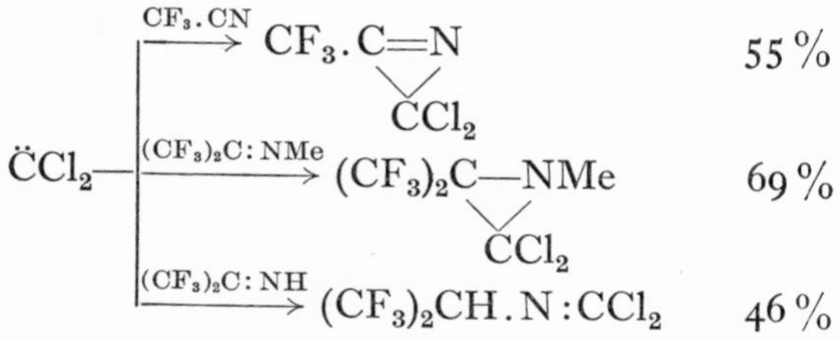

$$\ddot{C}Cl_2 - \begin{cases} \xrightarrow{CF_3.CN} CF_3.C{=}N \\ \qquad\qquad \underset{CCl_2}{\diagdown\diagup} \\ \xrightarrow{(CF_3)_2C:NMe} (CF_3)_2C{-}NMe \\ \qquad\qquad\qquad \underset{CCl_2}{\diagdown\diagup} \\ \xrightarrow{(CF_3)_2C:NH} (CF_3)_2CH.N{:}CCl_2 \end{cases}$$

55%

69%

46%

(*c*) *Carbon–hydrogen insertion reactions.*[22] 1,2,2-Trifluoroethyl-idene undergoes insertion into C–H bonds with relative rates $3° > 2° > 1°$; its rearrangement to trifluoroethylene competes with, and sometimes predominates over, C—H insertion. The yields are based on the amount of trifluoroethylidene generated during the reaction; trifluoroethylene and the cis–trans cyclo-propanes formed by reaction of trifluoroethylene with trifluoro-ethylidene are the only other organic products.

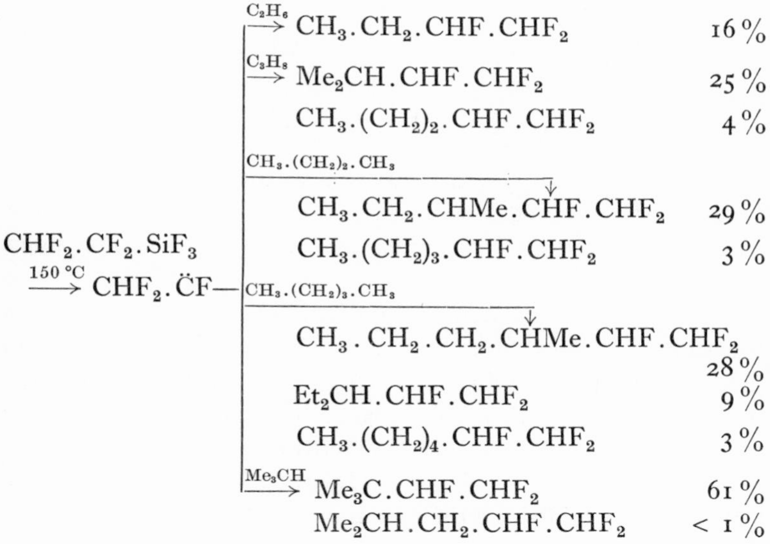

$$CHF_2.CF_2.SiF_3 \xrightarrow{150\,°C} CHF_2.\ddot{C}F - \begin{cases} \xrightarrow{C_2H_6} CH_3.CH_2.CHF.CHF_2 \\ \xrightarrow{C_3H_8} Me_2CH.CHF.CHF_2 \\ \qquad\quad CH_3.(CH_2)_2.CHF.CHF_2 \\ \overline{} \\ \qquad\quad CH_3.CH_2.CHMe.\overset{\downarrow}{C}HF.CHF_2 \\ \qquad\quad CH_3.(CH_2)_3.CHF.CHF_2 \\ \overline{} \\ \qquad\quad CH_3.CH_2.CH_2.\overset{\downarrow}{C}HMe.CHF.CHF_2 \\ \qquad\quad Et_2CH.CHF.CHF_2 \\ \qquad\quad CH_3.(CH_2)_4.CHF.CHF_2 \\ \xrightarrow{Me_3CH} Me_3C.CHF.CHF_2 \\ \qquad\quad Me_2CH.CH_2.CHF.CHF_2 \end{cases}$$

16%

25%

4%

29%

3%

28%

9%

3%

61%

$< 1\%$

(d) *Addition to olefins to give cyclopropanes*[20] (see Table).

TABLE. *Relative reactivity of* $:CCl_2$ *to olefins to give cyclopropanes*
(k_{olefin} *relative to ethylene*)

Olefin	$k_{olefin}/k_{C_2H_4}$	Olefin	$k_{olefin}/k_{C_2H_4}$
C_2H_4	1·0	$CH_2:CHCl$	10·5
$CH_2:CHF$	1·1	$CH_2:CCl_2$	45·1
$CCl_2:CCl_2$	1·2	⬡	77·3
$CH_2:CF_2$	1·8	$Me_2C:CH_2$	1,160
$CHF:CF_2$	2·7	$Me_2C:CHMe$	4,860
$CHCl:CCl_2$	4·8	$Me_2C:CMe_2$	10,210

(e) *Reactions with organometallic compounds.*[22] Carbene insertion into Si–H, and carbene capture reactions by trivalent phosphorus compounds are amongst organometallic reactions readily observed; for example,

$$Me_3SiH + CHF_2.\ddot{C}F \rightarrow Me_3Si.CHF.CHF_2 \quad 55\%$$

$$(CF_3)_3P + CHF_2.\ddot{C}F \rightarrow (CF_3)_3P:CF.CHF_2 \quad 37\%$$

7. General comment

True research meanders in a way that is impossible to predict in advance, and progress in a field can often only be ascertained properly several years after the initial discovery that opened up a new section of the subject.

The discovery of the fluorocarbon iodides and realization of their versatility in synthesis caused a tremendous surge forward in inorganic and organic fluorine chemistry. Chemists not primarily interested in the fluorine field soon realized that the new fluorine compounds becoming available had unusual properties which were relevant to their own field of study. Fluorine and fluorocarbon compounds obtained for the first time in those early days from fluorocarbon iodides and related compounds now play an integral role in research projects, academic or industrial, all over the world. The author was concerned with the original work on fluorocarbon

iodides, and with the immediate development of their organometallic chemistry; he is particularly grateful for the opportunity
at that time to work with the senior author in that Cambridge
group, and to learn and gain from his intuition, wisdom and
inspiration.

REFERENCES

1 A. A. Banks, H. J. Emeléus, R. N. Haszeldine and V. Kerrigan,
   J. chem. Soc. 1948, p. 2188.
2 H. J. Emeléus and R. N. Haszeldine, *J. chem. Soc.* 1949, pp. 2948,
   2953.
3 G. R. A. Brandt, H. J. Emeléus and R. N. Haszeldine, *J. chem. Soc.*
   1952, pp. 2198, 2549, 2552. H. J. Emeléus, R. N. Haszeldine and
   E. G. Walaschewski, *J. chem. Soc.* 1953, p. 1552. F. W. Bennett,
   H. J. Emeléus and R. N. Haszeldine, *J. chem. Soc.* 1953, p. 1565.
   H. J. Emeléus, R. N. Haszeldine and Ram Chand Paul, *J. chem.
   Soc.* 1954, p. 881, and subsequent papers.
4 A. M. Geyer and R. N. Haszeldine, *J. chem. Soc.* 1957, pp. 1038,
   3925. A. M. Geyer, R. N. Haszeldine, K. Leedham and R. J.
   Marklow, *J. chem. Soc.* 1957, p. 4472. R. N. Haszeldine and R. J.
   Marklow, *J. chem. Soc.* 1956, p. 962. R. N. Haszeldine and J. C.
   Young, *Proc. chem. Soc.* 1959, p. 394; *J. chem. Soc.* 1960, p. 4503.
   E. T. McBee, C. W. Roberts and G. W. R. Puerckhauer, *J. Am.
   chem. Soc.* 1957, **79**, 2329. A. D. Petrov, V. F. Mironov, V. A.
   Ponomarenko, S. I. Sadykh-Zade and E. A. Cherneyshev, *Izvest.
   Akad. Nauk SSSR, Otdel Khim. Nauk*, 1958, p. 954. V. A.
   Ponomarenko, B. A. Sokolov and A. D. Petrov, *Izvest. Akad.
   Nauk SSSR, Otdel. Khim. Nauk*, 1956, p. 628. P. Tarrant, G. W.
   Dyckes, R. Dunmire and G. B. Butler, *J. Am. chem. Soc.* 1957, **79**,
   6536. R. N. Haszeldine, M. J. Newlands and J. B. Plumb, *J. chem.
   Soc.* 1965, p. 2101.
5 H. J. Passino and L. C. Rubin, U.S. Patent 2,686,194 (1954). J. H.
   Simons and R. D. Dunlap, U.S. Patent 2,651,651 (1953).
6 R. N. Haszeldine, *Angew. Chem.* 1954, **66**, 693. E. T. McBee, C. W.
   Roberts, G. F. Judd and T. S. Chao, *J. Am. chem. Soc.* 1955, **77**,
   1292; Midland Silicones Ltd., British Patent 805,029 (1958). O. R.
   Pierce, E. T. McBee and R. E. Cline, *J. Am. chem. Soc.* 1953, **75**,
   5618. O. R. Pierce, R. T. McBee and G. F. Judd, *J. Am. chem.
   Soc.* 1954, **76**, 474.
7 A. M. Geyer, R. N. Haszeldine, K. Leedham and R. J. Marklow,
   J. chem. Soc. 1957, p. 4472.
8 R. N. Haszeldine, *Nature, Lond.* 1951, **168**, 1028. R. N. Haszeldine
   and K. Leedham, unpublished results.
9 R. N. Haszeldine, *J. chem. Soc.* 1952, p. 3423; *J. chem. Soc.* 1953,
   p. 1748.
10 J. W. Ryan and J. L. Speier, *J. Am. chem. Soc.* 1964, **86**, 895. A. J.
   Chalk and J. F. Harrod, *J. Am. chem. Soc.* 1965, **87**, 16. L. H.

Sommer, K. W. Michael and H. Fujimoto, *J. Am. chem. Soc.* 1967, 89, 1519.

11 R. N. Haszeldine, S. Lythgoe and P. J. Robinson, unpublished results.

12 D. I. Cook, R. Fields, R. N. Haszeldine, B. R. Iles, A. Jones and M. J. Newlands, *J. chem. Soc.* (*A*), 1966, p. 887.

13 T. N. Bell, R. N. Haszeldine, M. J. Newlands and J. B. Plumb, *J. chem. Soc.* 1965, p. 2107; *Proc. chem. Soc.* 1960, p. 147.

14 R. N. Haszeldine, P. J. Robinson and R. F. Simmons, *J. chem. Soc.* (*B*), 1967, p. 1357.

15 R. N. Haszeldine and J. C. Young, *Proc. chem. Soc.* 1959, p. 394. W. I. Bevan, R. N. Haszeldine and J. C. Young, *Chemy Ind.* 1961, p. 789.

16 R. N. Haszeldine, P. J. Robinson and R. F. Simmons, *J. chem. Soc.* 1964, p. 1890.

17 G. Fishwick, R. N. Haszeldine, C. Parkinson, P. J. Robinson and R. F. Simmons, *Chem. Comm.* 1965, p. 382. G. Fishwick, R. N. Haszeldine, P. J. Robinson and R. F. Simmons, unpublished results.

18 W. I. Bevan and R. N. Haszeldine, unpublished results.

19 W. I. Bevan, R. N. Haszeldine, J. Middleton and A. E. Tipping, unpublished results.

20 J. M. Birchall, R. A. Burton, S. G. Farrow, R. Fields, G. N. Gilmore and R. N. Haszeldine, unpublished results.

21 J. M. Birchall, R. N. Haszeldine and P. Tissington, unpublished results.

22 R. N. Haszeldine and J. G. Speight, unpublished results.

7

FLUOROALKYLMERCURIALS

J. J. LAGOWSKI

1. Introduction

The preparation of CF_3I,[1] the fluorine analogue of methyl iodide, at Cambridge in 1948 provided an entrée to the synthesis of a variety of fluorocarbon derivatives of the metals and metalloids which possess unusual chemical and physical properties. This, together with other earlier observations on the chemistry of the fluorocarbons and their derivatives, gave impetus to the exploration of a field of chemistry which has yielded results of practical as well as of theoretical interest. The first organometallic derivatives containing a fully fluorinated organic moiety, CF_3HgI and $(CF_3)_2Hg$, were prepared by Emeléus and Haszeldine.[2] Since then reports of investigations on the chemistry and properties of the fluoroalkylmercurials have appeared with increasing frequency, providing a clearer impression of the nature of these compounds. It seems particularly appropriate at this time and in this volume to indicate the progress made in the chemistry of fluoro-organo-mercurials since its origin in Professor Emeléus' laboratories nearly two decades ago.

At present, there are two interesting and important aspects of the chemistry of the fluoroalkylmercurials: (a) the use of these compounds as possible synthetic intermediates for the preparation of other fluoroalkylmetallic compounds, and (b) their coordination chemistry. Interest in the former aspects arises from the realization that the only good synthetic route to perfluoromethyl derivatives of the elements involves the reaction of CF_3I with the elements. A study of the coordinating ability of fluoroalkylmercurials is, indirectly, a study of the effect of fluoroalkyl groups on the electronic environment of the metal atom.

Although fluoroalkylmercurials are formally related to the corresponding alkylmercurials, the properties of the former are markedly different from those of the latter. Thus, a comparison of

[137]

TABLE 1. *Properties of methyl- and (trifluoromethyl)mercurials*

	M.p. (°C)	B.p. (°C)	Solubility	
			H$_2$O	Other solvents
(CH$_3$)$_2$Hg	—	92	Insol.	Miscible with most organic liquids
(CF$_3$)$_2$Hg	163	—	1·29 M	Very soluble in C$_6$H$_6$, (C$_2$H$_5$)$_2$O, or CHCl$_3$
CH$_3$HgI	152	—	1 × 10^{-3} M	Very soluble in most organic
CF$_3$HgI	112·5	—	0·155 M	solvents

the simple physical properties of (trifluoromethyl)mercurials with those of the corresponding methylmercurials (Table 1) emphasizes their differences rather than their formal relationship. In contrast with dimethylmercury, the white crystalline compound bis(trifluoromethyl)mercury is soluble in a variety of solvents including water, the latter solvent giving weakly conducting solutions. The evidence available indicates that the conductivity of bis(trifluoromethyl)mercury in aqueous solution can be attributed to the hydrolysis of an intermediate coordination compound without displacement of a trifluoromethyl group (equation 1):

$$(CF_3)_2Hg + H_2O \rightleftharpoons (CF_3)_2HgOH_2 \rightleftharpoons (CF_3)_2HgOH^- + H^+ \qquad (1)$$

The properties which reflect the electronic environment about the mercury atom are markedly affected by the successive substitution of hydrogen atoms in alkylmercurials by fluorine atoms. Thus, it would be anticipated that a perfluoroalkyl group would have a strong electron-withdrawing effect compared with the electron-releasing tendency of the corresponding alkyl groups. These suggestions are in agreement with the findings of experiments designed to estimate the relative electronegativity of perfluoroalkyl and fluoroalkyl groups (Table 2).[3] The values in Table 2 were obtained by empirically determining the functional relationship between a property in a family of compounds and the known electronegativities of the groups in these compounds (Table 3). Thus, the ionization constants of fluoro-organomercuric hydroxides and halides (equations 2 and 3) and certain infrared

$$RHgOH \rightleftharpoons RHg^+ + OH^- \qquad (2)$$

$$RHgX \rightleftharpoons RHg^+ + X^- \qquad (3)$$

TABLE 2. *Estimated electronegativities of some fluoroalkylmercurials*

Group	CF_3-	CF_3CF_2-	$(CF_3)_2CF-$	CF_3CHF-	CF_3CH_2-
Electronegativity*	3·3	3·2	3·4	2·8	2·7

* Values given on the Pauling scale in which $\chi_F = 4\cdot0$ and $\chi_{Cl} = 3\cdot0$.

TABLE 3. *Empirical relationships between the electronegativity**
and vibrational frequencies and pK

Method	Relationship
ν_{HX}†	$\chi_{R_f} = -1\cdot9 + (2\cdot20 \times 10^{-3})\nu_{HX}$
ν_{Hg-O}‡	$\chi_{R_f} = -9\cdot3 + (2\cdot59 \times 10^{-2})\nu_{Hg-O}$
pK_b	$\chi_{R_f} = -8\cdot5 + 1\cdot10\ pK_b$
pK_i	$\chi_{R_f} = 9\cdot77 + 1\cdot35\ pK_i$

* Electronegativity given on the Pauling scale.
† Mass corrected.
‡ In compounds of the type RHgOH or $(RHg)_2O$.

vibrational frequencies are relatively sensitive to the electronic nature of the alkyl groups present and can be used to estimate the relative electro negativities of these groups.

2. Coordination chemistry

The relatively high effective electronegativity of fluoroalkyl groups is consistent with the fact that fluoroalkylmercurials, in contrast to the alkylmercurials, form complex compounds. Simple sigma bonds can be formed between the ligands and mercury; if the ligand has vacant d orbitals available a $d_\pi-d_\pi$ component to the bond is possible. The latter situation occurs with the ligands in the third and higher periods of the Group V and VI elements. Equally stable

$$\begin{matrix} R \\ \diagup \\ \quad Hg \underset{\sigma}{\overset{\pi}{\rightleftharpoons}} L \quad (L = R'_3P, R'_3As, R'_2S) \\ \diagdown \\ R \end{matrix}$$

complexes of the type HgX_2L_2 (X = halide) are formed regardless of whether electron-withdrawing groups or electron-releasing groups are attached to the ligand atom L; the presence of electron-withdrawing groups on the ligands weakens the sigma component of the Hg—L bond and strengthens the pi component, while the reverse has been suggested to occur with electron-releasing groups.[4] Similar considerations should be applicable to the groups attached to the mercury atom. Electron-releasing groups, such as alkyl groups, attached to the mercury atom might be expected to weaken the sigma component of the bond whereas electron-withdrawing groups would strengthen the sigma component. It is notable that no coordination compounds of dimethylmercury have been reported, although complex compounds have been reported for alkyl-mercuric halides, some of which undergo disproportionation to give the uncomplexed bisalkylmercurial and the complexed mercuric halide.[5]

Fluoroalkylmercurials as well as fluoroalkylmercuric salts, on the other hand, form coordination compounds with halide ions and with neutral ligands. Although conductivity data[6] suggest that perfluoroalkylmercurials undergo stepwise reaction with iodide ions in aqueous solution to form $1:1$ and $1:2$ complex ions (equations 4–7),

$$(R_f)_2Hg + I^- \rightleftarrows (R_f)_2HgI^- \qquad (4)$$

$$(R_f)_2HgI^- + I^- \rightleftarrows (R_f)_2HgI_2^{2-} \qquad (5)$$

$$R_fHgI + I^- \rightleftarrows R_fHgI_2^- \qquad (6)$$

$$R_fHgI_2^- + I^- \rightleftarrows R_fHgI_3^{2-} \qquad (7)$$

where $\qquad R_f = CF_3, C_3F_7$

spectroscopic studies of these systems show only a weak interaction between mercury and halide ions.[7] The halogenoperfluoroalkyl-mercurate ions containing four-coordinate mercury were isolated as crystalline salts of ethylenediamine-transition metal complex ions (Table 4).

Coordination compounds between bis(fluoroalkyl)mercurials and a variety of neutral ligands have been detected in solution using oscillometric titration techniques[8] to follow the course of the stepwise reactions in benzene solution (equations 8 and 9). The

$$(R_f)_2Hg + L \rightleftarrows (R_f)_2HgL \qquad (8)$$

$$(R_f)_2HgL + L \rightleftarrows (R_f)_2HgL_2 \qquad (9)$$

where $\qquad R_f = CF_3, C_2F_5, (CF_3)_2CF, CF_3CFH, CF_3CH_2$

and $L = C_5H_{11}N, C_5H_5N, C_5H_5NO, (C_6H_5)_3P, (CH_3)_2{=}CO,$
$(CH_3)_2SO, C_2H_5OH, (CH_3)_2S$

TABLE 4. *Compounds containing halogenoperfluoroalkylmercurate anions*

$M[C_2H_8N_2]_x[HgR_fI_3]$			$M[C_2H_8N_2]_x[Hg(R_f)_2I_2]$		
M	x	R_f	M	x	R_f
Cu	2	CF_3	Zn	3	CF_3
Ni	3	CF_3	Ni	3	CF_3
Cd	2	CF_3			
Zn	3	C_3F_7			

inflections in the titration curves are presumably due to the formation and/or consumption of molecular species with different dipole moments, which causes a concomitant change in the dielectric constant of the solution. The infrared spectra of carbon tetrachloride solutions containing mixtures of bis(fluoroalkyl)mercurials with ethylenediamine, pyridine *N*-oxide, or tetramethylene sulphoxide indicate the formation of complexes of the type $(R_f)_2HgL$ and $(R_f)_2HgL_2$, which is in agreement with the results of oscillometric titrations.[8] Equimolar mixtures of the bismercurial and ethylenediamine show no significant change in the position of the N–H stretching mode of the amine; however, a decrease in the N–H deformation mode occurs for the same mixtures. In addition, the N–H stretching band became very sharp when the mercurials were present, in contrast to the broad bands associated with intermolecular hydrogen bonding. These observations suggest that relatively weak coordination compounds are formed between the amine and the mercurials. On the other hand, a noticeable shift in the N–O and S–O stretching frequencies was observed for mixtures containing either pyridine *N*-oxide or tetramethylene sulphoxide; the magnitude of these shifts is about half as large as those observed for the complexes formed by these ligands with transition metal ions (Table 5).

TABLE 5. *Infrared frequency shifts for fluoroalkylmercurials complexes in solution*

	ν_{NO} (cm^{-1})*	$\Delta\nu_{NO}$	ν_{SO} (cm^{-1})†	$\Delta\nu_{SO}$
$(CF_3)_2Hg$	1249	19	1019	17
$(C_2F_5)_2Hg$	1247	21	1019	17
$[(CF_3)_2CF]_2Hg$	1243	25	1016	20
$(CF_3CHF)_2Hg$	1250	18	1020	16
$(CF_3CH_2)_2Hg$	1260	8	1021	15
Ligand	1268	—	1036	—
Transition metal ions	—	39–49‡	—	30–40§

     * Pyridine *N*-oxide.   ‡ Reference no. 9.
     † Tetramethylene sulphoxide.   § Reference no. 10.

TABLE 6. *Properties of coordination compounds containing fluoroalkylmercurials*

	M.p. (°C)	Remarks
$(CF_3CF_2)_2Hg[P(C_6H_5)_3]_2$	125–30	Mol. wt. in C_6H_6 (m.p.) indicates 50% dissociation
$(CF_3CFH)_2Hg[HNC_5H_{10}]_2$	40–43	—
$(CF_3CFH)_2Hg[OS(CH_3)_2]_2$	43–44	Uncomplexed ν_{SO} observed for CCl_4 solution
$(CF_3)_2Hg[ONC_5H_5]_2$	62	Uncomplexed ν_{NO} observed in CCl_4 solution
$(CF_3CH_2)_2Hg[HNC_5H_{10}]_2$	10–15	Does not decompose on a VPC column; slowly reacts with water; b.p. 140°
$(CF_3CH_2)_2Hg(ONC_5H_5)$	—	B.p. 200°; liquid reacts rapidly with water

Isolation of the complex compounds formed between the fluoro-alkylmercurials and the neutral ligands proved difficult. In general, the compounds are very soluble in organic solvents, decompose in water, and dissociate at relatively low temperatures, suggesting that weak mercury–ligand bonds are present. These observations are in contrast with the marked stability of the complexes of bis(pentafluorophenyl)mercury.[11] However, several complex bis-(fluoroalkyl)mercurials have been isolated and characterized (Table 6).[12] Recently a potential new synthetic route for complex compounds containing fluoroalkylmercurials has been reported[13] which involves formation of coordination compounds of mercury(II) carboxylates (equation 10) followed by decarboxylation (equation 11) to

$$(R_fCO_2)_2Hg + 2L \rightleftarrows (R_fCO_2)_2HgL_2 \qquad (10)$$

$$(R_fCO_2)_2HgL_2 \rightleftarrows (R_f)_2HgL_2 + 2CO_2 \qquad (11)$$

yield the desired complex.

3. Fluoroalkylmercurials as synthetic intermediates

Although perfluoroalkylmercurials are known, there is a notable lack of information concerning the use of these compounds as intermediates in the synthesis of less readily prepared or hitherto unknown metallic derivatives; this is in contrast to the bisalkyl-mercurials, which undergo radical-transfer reactions with active metals and some metal halides to form the corresponding metallo-organic derivatives. By analogy, it might be expected that bis-(fluoroalkyl)mercurials could be used as synthetic intermediates (equations 12 and 13, where M—X represents a metal or metalloid bond

$$(R_f)_2Hg + M—X \rightarrow R_fHgX + MR_f \qquad (12)$$

$$(R_f)_2Hg + 2M' \rightarrow Hg + 2M'R_f \qquad (13)$$

and M' represents an active metal). The reactions of bis(fluoro-alkyl)mercurials with selected metals and metalloid halides have been studied with the aim of using of these mercurials as synthetic intermediates, but to date the results have been disappointing. The preparation of fluoroalkyl derivatives of aluminum,[14] zinc,[15] boron,[16] and phosphorus[17] was attempted via reactions of the types shown in equations 12 and 13 using standard synthetic techniques.

3.1. Aluminum[14]

The formation of fluoroalkylaluminum compounds could not be detected in the reactions between aluminum metal or aluminum amalgam and $(C_2F_5)_2Hg$, $(CF_3CFH)_2Hg$, or $(CF_3CH)_{22}Hg$.[14] However, in the presence of water these reactants yielded the corresponding hydrocarbon, aluminum hydroxide, mercury, and hydrogen. The first two products would be expected from the hydrolysis of a fluorocarbon derivative of aluminum, but no direct evidence for the existence of such an intermediate was observed.

Reaction readily occurred between $(CF_3CH_2)_2Hg$ and AlX_3 (X = Cl, Br) in the absence of a solvent; the fluoro-olefin $CF_2\!\!=\!\!CH_2$ was liberated in all cases together with the fluoroalkylmercuric halide or mercuric halide, depending upon the relative proportions of the reactants.

In an attempt to overcome the defluorination of a fluoroalkylaluminum compound which might be formed, the reaction of the mercurials with aluminum trichloride was conducted in a donor solvent. The solvent served the dual purpose of reducing the electron-deficient nature of aluminum, which is accentuated by the strongly electron-withdrawing nature of the fluoroalkyl group, and blocking the vacant coordination positions on the metal atom. Under these conditions, a smooth reaction occurred between $(CF_3CH_2)_2Hg$ and $AlCl_3$ in tetrahydrofuran at room temperature; after several hours colourless needles of $[Al(OC_4H_8)_6][AlCl_3F]_3$ crystallized from solution. Crystals of this complex continued to form slowly over a period of two weeks, at which time all of the aluminum in the system could be accounted for in this complex. The solution contained 2,2,2-trifluoroethylmercury(II) chloride and 1,1-difluoroethylene, the latter being extremely soluble in tetrahydrofuran.

3.2. Boron[16]

The reaction between BCl_3 and the bis(fluoroalkyl)mercurials proceeds smoothly to give quantitative yields of BF_3, together with either the fluoroalkylmercuric chloride or mercuric chloride depending upon the ratio of the reactants. The ease of reaction decreases with increasing fluorination of the alkyl group. Thus, $(CF_3CH_2)_2Hg$ reacted completely in a sealed tube while the frozen mixture warmed to room temperature, $(CF_3CHF)_2Hg$ reacted completely at 60°, and $(C_2F_5)_2Hg$ reacted only partially after being heated at 110° for several days.

3.3. Zinc[15]

Excess of zinc and $(CF_3CH_2)_2Hg$ did not react below 120 °C; however, at 135 °C 1,1-difluoroethylene was slowly liberated from this system, and the rate of liberation apparently increased with increasing temperature. No direct evidence for the formation of a CF_3CH_2—Zn bond could be obtained, but since the mercurial does not decompose alone at this temperature the results suggest that alkyl transfer does occur and that, at the temperature required for this reaction, the organozinc compound formed is thermally unstable and undergoes defluorination. An attempt was made to prepare a fluoroalkyl-metal compound by an alkyl-halogen exchange (cf. equation 12); a reaction was observed between ZnI_2 and $(CF_3CH_2)_2Hg$ at 120 °C. Under these conditions the liquid mercurial slowly disappeared accompanied by the quantitative formation of MgI_2; only 1,1-difluoroethylene, mercuric iodide, and ZnF_2 were recovered from the reaction mixture. The reaction between dichloro(1,10-phenanthroline)zinc(II) and $(CF_3CH_2)_2Hg$ also led to defluorination and liberation of olefin.

3.4. Phosphorus[16]

Attempts have been made to bring about an alkyl-halogen exchange reaction between PCl_3 and bis(fluoroalkyl)mercurials (cf. equation 12) to prepare fluoroalkyl derivatives of phosphorus; the latter have been prepared by an independent method[17] and are stable at temperatures well above 100 °C. The reactions between the bis(fluoroalkyl)mercurials and PCl_3 were conducted in sealed tubes at temperatures between 70° and 100 °C. In every case, the products consisted of PF_3 and the fluoroalkylmercuric halide, as well as the corresponding olefin. It was first thought that this reaction occurred through a complex formed between PCl_3 and the mercury atom in the bis(fluoroalkyl)mercurial, which subsequently underwent an intermolecular defluorination (see below); infrared studies in the 3.2–50μ region on these systems gave no indication of an interaction. In addition, $POCl_3$, which does not have a free electron pair on the phosphorus atom, reacted with $(CF_3CH_2)_2Hg$ to produce $CF_2{=}CH_2$, POF_3, and CF_3CH_2HgCl.

10

The results obtained thus far on the attempted preparation of fluoroalkyl derivatives of the electropositive elements indicate that the high electron-withdrawing properties of the fluoroalkyl group, coupled with the availability of an internal defluorination mechanism, yield the considerably more stable metal fluorides. The existence of such a defluorination mechanism is suggested in the results of mass spectroscopic studies on several bis(fluoroalkyl)-mercurials and fluoroalkylmercuric chlorides.[15] Parent ions were observed for monomeric species, and no evidence for association was obtained. However, very strong mass peaks representing the corresponding olefin were observed in the spectra of $(CF_3CH_2)_2Hg$, $(CF_3CFH)_2Hg$ and $(C_2F_5)_2Hg$. Excluding recombination reactions, which are unlikely to occur at such low concentrations, these results indicate an intramolecular decomposition of the mercurial and, by analogy, the other fluoroalkyl metal compounds formed by alkyl-halogen exchange reactions. The data suggest that defluorination occurs via a 4-membered ring intermediate (equation 14) in

$$(CF_3CH_2)_2Hg \rightarrow \left[\begin{array}{c} H \quad\quad H \\ F \diagdown \quad C \quad \diagup R \\ \quad C \diagdown \quad \diagup Hg \\ F \diagup \quad F \end{array} \right] \rightarrow CF_2{=}CH_2 + CF_3CH_2HgF \quad (14)$$

which a fluorine atom on the β carbon coordinates to the mercury atom. It is apparent that such decompositions would be prevented by complex formation of the metal atom with a strong ligand. Also, if this mechanism were operative the tendency towards defluorination would increase if the fluoroalkyl group were associated with an electron-deficient metal atom. It is suggested that such defluorination will always occur with electron-deficient metals capable of expanding their coordination number unless additional stability is imparted by complex formation.

It is surprising that fluoroalkylphosphorus derivatives have not been prepared by alkyl exchange with the mercurials, since these derivatives have been prepared by other methods and are known to be stable under the conditions of the experiments described here. The products of the reaction in which fluoroalkyl exchange was attempted are analogous to those obtained with the electron-deficient elements, i.e. PF_3 and the corresponding olefin, suggesting that a defluorination of the fluoroalkyl group occurs during the process of alkyl transfer. However, the data available do not permit a definitive statement of the process at this time.

We wish to acknowledge the Robert A. Welch Foundation and the National Institutes of Health for financial support of our work on the fluoroalkylmercurials.

REFERENCES

1 R. E. Banks, H. J. Emeléus, R. N. Haszeldine and V. Kerrigan, *J. chem. Soc.* 1948, p. 2188.
2 H. J. Emeléus and R. N. Haszeldine, *J. chem. Soc.* 1949, pp. 2948, 2953.
3 H. B. Powell and J. J. Lagowski, *J. chem. Soc.* 1965, p. 1392; 1962, p. 2047.
4 R. C. Cass, G. E. Coates and R. G. Hayter, *J. chem. Soc.* 1955, p. 4007.
5 G. E. Coates and A. Lauder, *J. chem. Soc.* 1965, p. 1857.
6 H. J. Emeléus and J. J. Lagowski, *Proc. chem. Soc.* 1958, p. 231; 1959, p. 1497.
7 A. J. Downs, *J. chem. Soc.* 1963, p. 5277.
8 H. B. Powell, Maung Tin Maung and J. J. Lagowski, *J. chem. Soc.* 1963, p. 4257.
9 S. Kida, J. V. Quagliano, J. A. Walmsley and S. Y. Tyree, *Spectrochim. acta*, 1963, **19**, 189.
10 D. W. Meek, D. K. Straub and R. V. Drago, *J. Am. chem. Soc.* 1960, **82**, 6013.
11 R. D. Chambers, G. E. Coates, J. G. Livingston and W. K. R. Musgrave, *J. chem. Soc.* 1962, p. 4367.
12 H. B. Powell and J. J. Lagowski, *J. chem. Soc. (A)*, 1966, p. 1282.
13 J. E. Connett and G. B. Deacon, *J. chem. Soc. (C)*, 1966, p. 1058. J. E. Connett, A. G. Davies, G. B. Deacon and J. H. S. Green, *J. chem. Soc. (C)*, 1966, p. 106.
14 A. N. Stear and J. J. Lagowski, *J. chem. Soc.* 1964, p. 5848.
15 A. Lauder, E. C. Bossert and J. J. Lagowski, Abstracts of Papers, 152nd Meeting, American Chemical Society, New York, September, 1966, O6. A. Lauder, unpublished results.
16 E. C. Bossert and J. J. Lagowski, Abstracts of Papers, 22nd Southwest Regional Meeting, American Chemical Society, Albuquerque, New Mexico, December, 1966, p. 39 A. E. C. Bossert, unpublished results.
17 F. W. Bennett, H. J. Emeléus and R. N. Haszeldine, *J. chem. Soc.* 1953, p. 1565. G. M. Bunch, H. Goldwhite and R. N. Haszeldine, *J. chem. Soc.* 1963, p. 1083.

8

CATENATION IN INORGANIC SILICON COMPOUNDS

ALAN G. MACDIARMID

1. Introduction

Since silicon lies immediately below carbon in Group IV of the Periodic Table it is particularly interesting to examine the physical, structural and chemical properties of analogous, simple carbon and silicon compounds in order to ascertain in what respects they are similar to, or different from, each other. Attempts can then be made to explain or rationalize the similarities and differences, and in this way a greater understanding of factors which cause the properties of elements and their analogous compounds to change as a group as the Periodic Table is descended, can be obtained. Where data are available, it is, of course, highly desirable to include in such a study analogous compounds of the other Group IV elements, germanium, tin and lead.

Although properties of the C—C bond have been very extensively investigated, relatively little is known about the analogous Si—Si bond; indeed there are many common misconceptions concerning the Si—Si bond. For example, it is not generally realized that silicon forms longer and more thermally stable chains of silicon atoms in certain molecular species than does carbon, if the substituent atoms are, for example, the heavier halogens or phenyl groups. It is not possible to give a comprehensive review of catenation in inorganic silicon compounds in a chapter of this size and hence only some of the more important, interesting or characteristic aspects of this subject will be described. It is also not possible to cover the more extensive topic of catenation in organosilicon compounds; however, this subject has been excellently summarized in certain reviews and books.[1-3] No attempt will be made to discuss the chemistry of the Si—X bond (X = H, halogens, C, O, N, S, etc.), which has been fully described elsewhere.[1-10]

Just as one can regard methyl and ethyl compounds as being

derived from methane and ethane respectively, by the replacement of one hydrogen atom by a given substituent, so also one can regard SiH_3-(silyl), GeH_3-(germyl), SnH_3-(stannyl), Si_2H_5-(disilanyl) and Ge_2H_5-(digermanyl) compounds as being derived from the corresponding Group IV hydride. Inorganic compounds containing the Sn_2H_5-, PbH_3- and Pb_2H_5-groups are not yet known. Disilanyl compounds may therefore be regarded as the silicon analogues of ethyl compounds, and since compounds containing the Si_2H_5-group have been more extensively studied than any other single class of inorganic polysilane, a relatively large portion of this review will be devoted to a discussion of compounds of this type. It is believed that the 'true' properties of a Si—Si bond can be obtained more readily from such species where there are a maximum number of hydrogen atoms attached to silicon, since the properties of the Si—Si bond will not then be modified by inductive or steric effects of electropositive, electronegative or bulky substituents.

2. Synthesis of the silicon–silicon bond

The methods by which a silicon–silicon bond can be synthesized may be divided into two main categories: (**2.1**) those in which there is a primary synthesis of a Si—Si bond from species containing no Si—Si bonds and (**2.2**) those in which there is a 'secondary' synthesis of a Si—Si bond from species already containing Si—Si bonds. Interconversions of polysilane derivatives in which there is no net change in the number of Si—Si bonds per molecule, between reactants and products, are described in § 3.

2.1. Primary syntheses

Many silicon compounds can be reduced by a variety of methods to give species containing Si—Si bonds. Thus, elemental silicon is produced commercially by the reduction of SiO_2 with carbon or CaC_2 in an electric furnace. Many metal silicides (which may be prepared by heating SiO_2 with excess metal, or by dissolving silicon in the molten metal) also contain Si—Si bonds.[11]

Molecular compounds such as $Si_{10}Cl_{22}$,[12] $Si_{25}Cl_{52}$[13] and

$Si_{10}Cl_{20}H_2$[14, 15] may be prepared by reducing $SiCl_4$ with hydrogen at high temperatures under carefully controlled experimental conditions. If energy is supplied in the form of an electrical discharge instead of heat then $SiCl_4$ may be converted to Si_2Cl_6 in the presence of a halogen acceptor, such as hydrogen or a metal.[16, 17] Hexaiododisilane, Si_2I_6, may be synthesized by heating SiI_4 with finely divided metallic silver at 280 °C for 6 h.[18, 19] Tribromosilane or $SiBr_4$ may even be reduced in ethereal solution by magnesium to give polymeric $(SiH)_x$[20] or $(SiBr)_x$,[21] respectively, while SiH_3I may be converted by sodium amalgam at room temperature to give 67 % yields of Si_2H_6.[22] This method has also been extended to the synthesis of $CH_3SiH_2SiH_2CH_3$ from CH_3SiH_2I.[22]

Silicon–silicon bonds may also be synthesized from monosilanes by the application of energy in the form of heat, ultraviolet light or an electrical discharge. Thus SiH_4 may be partly converted to Si_2H_6 and Si_3H_8 at 450–510 °C under appropriate experimental conditions.[23, 24] Free radical processes are probably involved. This method has been extended to the synthesis of compounds containing both Si—Si and Si—Ge bonds of general formula

$$Si_xGe_yH_{2x+2y+2}$$

by the pyrolysis of a germane with a silane.[25]

When SiH_4 containing mercury vapour (with or without added C_2H_6 or hydrogen) is irradiated with ultraviolet light from a mercury lamp under mercury resonance radiation conditions, SiH_4, Si_2H_6, Si_3H_8 and a polymeric hydride of variable composition $(SiH_{0.4-0.9})_x$ are obtained.[24, 26–28] This technique has also been extended in an interesting fashion to give SiH_3GeH_3 from a mixture of SiH_4 and GeH_4[28] and methyl-disilanes from methyl-silanes.[27]

When SiH_4 is passed through an ozonizer under appropriate conditions, it is converted to give 63 % yields of a mixture of higher silanes (66 % Si_2H_6, 23 % Si_3H_8 and 11 % higher silanes).[29, 30] If the SiH_4 is mixed with GeH_4,[31, 32] PH_3, AsH_3[33–35] or $(CH_3)_2O$,[36] then species containing both Si—Si bonds in addition to Si—Ge, Si—P, Si—As and Si—C bonds, respectively, are formed. Under the more vigorous conditions of an electric discharge, $HSiCl_3$ (or a mixture of $SiCl_4$ and hydrogen) may be converted into solid polymeric materials containing Si—Si bonds such as $(SiCl_{0.5-2.6})_x$[37] and $(SiCl_2)_x$.[38] When $HSiBr_3$ is used, Si_3Br_8 and Si_4Br_{10} are obtained.[39]

Disilane may be formed in up to 35 % yields, together with smaller quantities of Si_3H_8, by the reaction of $KSiH_3$ with SiH_3Br in 1, 2-dimethoxyethane if the Si_2H_6 is removed constantly during the reaction.[40] Eighteen per cent yields of Si_3H_8 may be obtained in an analogous reaction between $KSiH_3$ and Si_2H_5Br. If Si_2H_6 is not constantly removed from the reaction vessel, SiH_4 and $(SiH_2)_x$ are the chief reaction products, possibly due to the catalytic effect of the KBr formed during the reaction.[40] The Si_3H_8 may be produced in the former reaction by some spontaneous randomization of the SiH_3Br to SiH_2Br_2 and SiH_4 followed by reaction of the SiH_2Br_2 with $KSiH_3$.[40]

A mixture of all the known volatile silanes may be obtained by adding dilute hydrochloric acid to Mg_2Si (prepared by heating a mixture of magnesium and silicon, or more crudely by heating SiO_2 and magnesium powder in an open test-tube). Approximately 25 % of the silicon is converted to a mixture of volatile hydrides (40 % SiH_4, 30 % Si_2H_6, 15 % Si_3H_8, 10 % Si_4H_{10} and 5 % higher silanes).[41, 42] The method of preparation of the Mg_2Si greatly affects both the over-all yield of silanes and also the relative proportions of the higher and lower silanes produced,[43] and species containing Si—Si bonds may well be present in the silicide. Differing yields and relative proportions of volatile silanes are obtained when other acids, e.g. NH_4Br in liquid ammonia[44] or $N_2H_4.HCl$ in anhydrous hydrazine[45] are used. Gas chromatographic separation of the mixture of higher silanes produced by these and other methods has led to the tentative identification of many isomers of the higher silanes up to Si_8H_{18}.[30, 32, 46–51]

This type of reaction has been extended in an interesting fashion to give mixtures of compounds of general formula $Si_xGe_yH_{2x+2y+2}$ by the action of acid on Ca- or Mg—Si—Ge alloys.[49, 50] These mixtures have been resolved by vapour-phase chromatographic techniques to give species which have been tentatively identified as containing Si—Si bonds as well as Si—Ge bonds. Silicon-germanium hydrides are also formed by the action of HF on a SiO—GeO mixture.[49]

2.2. Secondary syntheses

Secondary syntheses may be described as those processes in which a reactant contains one or more Si—Si bonds and yields a product in which there are a greater or smaller number of Si—Si bonds per molecule.

Perchloropolysilanes and Si_2Br_6 may be obtained together with SiX_4 (X = Cl, Br) by passing chlorine or bromine, respectively, over heated silicon or silicon alloys. In this manner the Si—Si bond network is partly, but not completely, ruptured. Thus when a *slow* stream of chlorine is passed over a calcium silicon alloy ('calcium silicide') at 150 °C a mixture of the following composition has been obtained: 65 % $SiCl_4$, 30 % Si_2Cl_6, 4 % Si_3Cl_8, and 1 % higher perchloropolysilanes.[16] Under certain conditions, white polymeric, solid $(SiCl_2)_x$ may actually be formed.[52] Analogously, when bromine is passed over silicon or a calcium–silicon alloy, Si_2Br_6 is produced.[53, 54]

A viscous, distillable oil of empirical composition $(SiCl_{2.6})_x$[55] and also $(SiBr_2)_x$[21] may be obtained by partly rupturing the Si—Si lattice in elemental silicon when $SiCl_4$ and $SiBr_4$, respectively, are passed over heated silicon at 1200 °C. It is possible that these reactions may proceed via the formation of the short-lived free radical silene species $SiCl_2$ and $SiBr_2$, analogous to the carbenes. Indeed, some reactions of gaseous $SiCl_2$, which is produced at high temperatures, have recently been studied.[56] That silene production is indeed likely is evidenced by the fact that the extremely interesting gaseous SiF_2 species has been prepared from silicon and SiF_4 at approximately 1150 °C.[57] Many of its reactions at, or below, room temperature have been studied.[58] At low temperatures it polymerizes to translucent, rubbery $(SiF_2)_x$ and it reacts readily with HX (X = Cl, Br, I) and with BF_3 to give SiF_3SiF_2H[59] and $SiF_3(SiF_2)_xBF_2$ (x = 1, 2)[58] respectively. It adds to many unsaturated organic compounds to give a wide range of organosubstituted fluoropolysilanes. For example, with benzene, one of the products isolated is[58]

$(SiF_2)_3$

It has been suggested that one of the chemically active forms of SiF_2 at low temperatures may possibly be a diradical, $\cdot SiF_2—SiF_2 \cdot$ or $F_2Si{=}SiF_2$.[58]

Thermal decomposition of polysilanes may occur to give polysilanes of either lower or higher molecular weight. Thus Si_2H_6 is found amongst the slow, room temperature, thermal decomposition products of Si_5H_{12}.[41] Analogously, when the highly polymerized species $(SiH_2)_x$[60] is heated to 380 °C[60] or $(SiF_2)_x$ is heated to 250–350 °C[58] a mixture of volatile polysilanes or perfluoropolysilanes, respectively, is produced. Melting points and boiling points of perfluoropolysilanes up to n-Si_4F_{10} obtained by this method have been determined.[58] Also, when $(SiH)_x$ and hydrogen are exposed to ultraviolet radiation, silanes up to Si_4H_{10} are obtained.[27]

The ability of certain polysilanes to 'decompose' to give polysilanes of higher molecular weight is probably one of the most interesting chemical properties characteristic of certain compounds containing a Si—Si bond, although there is evidence that somewhat related reactions may occur with compounds such as $Cl_3Si—CCl_3$.[61] Thus, Si_2H_6 and certain substituted disilanes undergo a disproportionation type of reaction under a variety of experimental conditions in which a monosilane is eliminated and a polysilane containing more than one Si—Si bond is formed. With certain compounds a Si—Si chain can be 'grown' under relatively controllable conditions.

When Si_2H_6 in 1, 2-dimethoxyethane is treated with LiX (X = H, D, Cl, Br, I) at room temperature the Si_2H_6 is completely consumed,[40] namely

$$x Si_2H_6 \rightarrow x SiH_4 + (SiH_2)_x$$

When LiH or LiD is employed, it is believed that the Si—Si bonds in the $(SiH_2)_x$ first formed are cleaved in a reaction analogous to that between Si_2H_6 and KH (see § 5) to give species such as Si_3H_7Li, since treatment of the non-volatile reaction products with acid yields some Si_3H_8.[62, 63] Detailed studies of the reactions using various deuterated species[63] have not as yet led to an unequivocal reaction mechanism although a reactive silene, SiH_2, intermediary has been suggested.

Hexachlorodisilane, Si_2Cl_6, and certain alkylpolychlorodisilanes also undergo disproportionation reactions in the presence of amines such as $(CH_3)_3N$ or quaternary ammonium salts, such as $[(CH_3)_4N]Cl$.[61, 64–67] Thus, in the presence of catalytic amounts of $(CH_3)_3N$, Si_2Cl_6 (and also Si_3Cl_8)[65] disproportionates at temperatures above $-46\ °C$ to yield Si_6Cl_{14},[64, 65] namely

$$5Si_2Cl_6 \rightarrow Si_6Cl_{14} + 4SiCl_4$$

When Si_6Cl_{14} is treated with more $(CH_3)_3N$, Si_5Cl_{12} is obtained. One[64] of two[61] mechanisms suggested for this reaction involves initial coordination of the $(CH_3)_3N$ to a silicon atom of the Si_2Cl_6, namely

followed by elimination of $SiCl_4$. This process can, of course, continue to give the observed product after removal of the $(CH_3)_3N$ catalyst. Although very little information is available, it appears that Si_2F_6,[68] Si_2Br_6[69] and Si_2I_6[19] may undergo analogous reactions on heating at high temperatures or for prolonged periods in the absence of catalyst. In another type of disproportionation reaction it has been found that when $(CH_3)_3Si—Si(CH_3)_2CN$ is refluxed at 175 °C, in the absence of catalyst, $(CH_3)_3SiCN$ is eliminated and $(CH_3)_3Si[Si(CH_3)_2]_xCN$ $(x = 2, 3, 4, 7)$ is formed;[70] when $(CH_3O)Si(CH_3)_2—Si(CH_3)_2(OCH_3)$ is heated with $LiOCH_3$, $(CH_3)_2Si(OCH_3)_2$ is eliminated with the formation of

$$CH_3O[Si(CH_3)_2]_xOCH_3$$

$(x = 3, 4, 5)$.[71] There is strong evidence that the latter reaction proceeds via a $(CH_3)_2Si$-free radical species.[71]

3. Parent polysilanes and their derivatives

The parent polysilanes and their derivatives may exist as molecular compounds of general formula Si_nX_{2n+2} such as Si_3H_8, Si_5Cl_{12}, etc., or as highly polymerized species such as $(SiH_2)_x$, $(SiF_2)_x$,

$(SiCl)_x$, etc. Extensive cross-linking by Si—Si bonds will, of course, be present in many of the highly polymerized polysilanes. This section will be devoted to the synthesis of polysilanes and their derivatives by processes in which there is no apparent Si—Si bond formation or cleavage. Some of the more important or interesting compounds and their properties will also be discussed.

In § 2 it was pointed out that certain metal silicides contain Si—Si bonds. In the case of $CaSi_2$ it has been found that a solution of HCl in alcohol will 'remove' the calcium atoms to yield white, polymeric, solid, 'siloxene' of empirical formula $(Si_2H_2O)_x$, which is believed to contain Si—Si bonds in addition to Si—H bonds.[72]

An additional convenient method to those already described, for the synthesis of Si_2H_6, involves the reduction of Si_2Cl_6 with $LiAlH_4$.[73] Both n-Si_4H_{10} and i-Si_4H_{10} have been isolated from mixtures of higher silanes prepared by methods given in the previous section, and they have been characterized, chiefly by their proton n.m.r. spectra.[30, 74] The silanes for which melting and boiling points have been determined are:[41-43, 74-76] SiH_4 (m.p. $-185 \cdot 0$ °C; b.p. $-111 \cdot 9$ °C), Si_2H_6 (m.p. $-132 \cdot 5$ °C; b.p. $-14 \cdot 5$ °C), Si_3H_8 (m.p. $-117 \cdot 4$ °C; b.p. $52 \cdot 8$ °C), n-Si_4H_{10} (m.p. $-89 \cdot 9$ °C; b.p. $108 \cdot 1$ °C) and i-Si_4H_{10} (m.p. $-99 \cdot 1$ °C; b.p. $101 \cdot 4$ °C).

Disilane, Si_2H_6, reacts readily with HX (X = Cl, Br, I) in the presence of the corresponding aluminum halide catalyst to give compounds in which one or more of the hydrogen atoms are replaced by halogen,[77-80] namely

$$Si_2H_6 + HX \rightarrow Si_2H_5X + H_2$$

The greater the proportion of HX in the reaction mixture, the greater are the number of halogen atoms introduced into the Si_2H_6. Although this type of reaction also occurs with SiH_4, it proceeds at a reasonably rapid rate only on heating to about 100 °C during approximately 15 h.[81-84] However, Si_2H_6 reacts very much more rapidly, the reaction with HI being complete after 2–3 h at room temperature,[80] while with HBr, reaction proceeds to completion after a few minutes even at $-78°$.[78] No cleavage of the Si—Si bond has ever been observed in any of these reactions. Unfortunately no definite studies have yet been carried out on the mechanism of the reaction of HX with Si—H bonds, although it is possible to postulate mechanisms[80] to account for the difference in rates observed.

Disilanyl chloride and Si_2H_5Br may also be synthesized in good yields by passing the vapour of Si_2H_5I over AgCl or AgBr respectively at room temperature.[77, 79] Disilanyl chloride (also certain halides of monosilane) has recently been conveniently prepared in an analogous manner to that employed for the synthesis of GeH_3X and Ge_2H_5X (X = Cl, Br) by passing Si_2H_6 over AgCl at 90 °C. The reaction products apparently include hydrogen, silver and HCl.[85]

It has been found that a gaseous mixture of Si_2H_6 and BCl_3, containing excess Si_2H_6, reacts within a few hours at 0 °C to give quantitative yields of B_2H_6,[86] namely

$$6Si_2H_6 + 2BCl_3 \rightarrow 6Si_2H_5Cl + B_2H_6$$

Some polychlorodisilanes, e.g. $Si_3H_3Cl_3$, are also formed, particularly when greater proportions of BCl_3 are employed. It is interesting to note that no reaction occurred between Si_2H_6 and BF_3, or between SiH_4 or SiH_3Cl and BCl_3 under identical experimental conditions; however, with BBr_3 both SiH_4 and Si_2H_6 underwent reaction to give bromo-silanes and -disilanes respectively.[87]

No attempt has ever been made to fluorinate Si_2H_6 directly but Si_2H_5F may be prepared in good yield by the reaction of $(Si_2H_5)_3N$ with BF_3.[88] An adduct $(Si_2H_5)_3N.BF_3$ is formed at -134 °C. This dissociates partly and reversibly at low temperatures but decomposes on warming to room temperature, namely

$$(Si_2H_5)_3N + BF_3 \rightarrow Si_2H_5F + (Si_2H_5)_2NBF_2$$

No evidence for any Si—Si bond cleavage was obtained.

All the disilanyl halides are relatively stable thermally if *absolutely pure*,[77–80] but in the presence of small quantities of impurities—especially acids or bases—rapid randomization frequently occurs. Thus, Si_2H_5Br, either alone or when mixed with HBr, appears to undergo very little decomposition on standing for $2\frac{1}{2}$–3 days at 0 °C but, in the presence of Al_2Br_6, rapid decomposition occurs at 0 °C.[79] Disilanyl chloride, however, randomizes rapidly even at -24 °C in the presence of either HCl or Al_2Cl_6.[77] When Si_2H_5Br containing a small amount of $(CH_3)_3N$ was held at o °C for 16 h, the following randomization reaction appeared to occur:[89]

$$2Si_2H_5Br \rightarrow Si_2H_6 + Si_2H_4Br_2$$

The synthesis of higher perhalopolysilanes—both volatile and highly polymerized species—has already been described, but in general very little is known about them.

No inorganic pseudohalides of polysilanes are known. Although SiH_3CN is relatively stable at room temperature, repeated attempts to synthesize pure Si_2H_5CN failed since the material decomposed during purification processes.[77] This is not altogether surprising since, although permethylated polysilane derivatives are usually more stable thermally than the parent hydride derivatives, $Si_2(CH_3)_5CN$ disproportionates readily at 175 °C[68] (see § 2). Although pseudohalides such as $Si(NCO)_4$[90] are stable and may readily be prepared from $SiCl_4$ and $AgNCO$, no $Si_2(NCO)_6$ could be isolated from the analogous reaction with Si_2Cl_6.[91]

Bisdisilanyl ether (1,2-disilyldisiloxane), $(Si_2H_5)_2O$, may be synthesized in good yields by the hydrolysis of Si_2H_5I with a large excess of pure H_2O,[80] whereas the chlorinated derivative, $(Si_2Cl_5)_2O$, may be synthesized by the controlled, partial hydrolysis of Si_2Cl_6.[92] Silyldisilanyl ether, (1-silyldisiloxane),

$$SiH_3OSi_2H_5$$

may be prepared by the co-hydrolysis of Si_2H_5Br and SiH_3I.[93] Bisdisilanyl ether decomposes only slightly on heating to 70 °C for several hours, but $SiH_3OSi_2H_5$ randomizes at o °C[93] namely

$$2SiH_3OSi_2H_5 \rightleftharpoons (SiH_3)_2O + (Si_2H_5)_2O$$

It is interesting to note that this reaction is reversible and that $SiH_3OSi_2H_5$ may be formed from $(SiH_3)_2O$ and $(Si_2H_5)_2O$ at o °C.

The ether $Si_2H_5OCH_3$ could be formed quantitatively by the instantaneous reaction of $(Si_2H_5)_2S$ with CH_3OH,[89] the $(Si_2H_5)_2S$ being formed from the reaction of Si_2H_5I vapour with HgS at room temperature.[94]

It is known that $(C_2H_5)_2O$ is a stronger base than $(CH_3)_2O$ when the base strength is measured by infrared hydrogen-bonding techniques, which minimize the effect on the base strength of the dif-

TABLE I. $\Delta\nu$ and methyl proton chemical shift values in selected carbon and silicon ethers

	$\Delta\nu^*$ ($\pm$3 cm^{-1})	Chemical shift of methyl protons† ($\pm$0·01 ppm)
$(CH_3)_2O$	130 cm^{-1}	$-1\cdot76$ ppm
$(C_2H_5)_2O$	144 cm^{-1}	—
SiH_3OCH_3	92 cm^{-1}	$-1\cdot98$ ppm
$Si_2H_5OCH_3$	95 cm^{-1}	$-1\cdot96$ ppm

* $\Delta\nu = \nu_{OH(CH_3OH)} - \nu_{OH(hydrogen-bonded)}$.
† With 25% solution in cyclohexane as internal standard. Chemical shifts reported downfield from cyclohexane.

ferent steric requirements of the methyl as compared to the ethyl group.[95] The greater base strength of $(C_2H_5)_2O$ is therefore consistent with an inductive release of electrons by the CH_3 in the C_2H_5-group. The inductive effect of the SiH_3 moiety in a Si_2H_5-group has been evaluated in a similar manner, but since the base strengths of $(SiH_3)_2O$ and $(Si_2H_5)_2O$ are immeasurably small by infrared hydrogen-bonding studies[89] (which is consistent with the known properties of Si—O bonds[2]) the species SiH_3OCH_3 and $Si_2H_5OCH_3$, which have greater base strengths, were used.[89, 96] The results are given in Table 1. It can be seen that, within experimental error, the base strengths of SiH_3OCH_3 and $Si_2H_5OCH_3$ are essentially identical. In this respect it is interesting to note that the proton chemical shifts of the methyl groups in these two compounds are also identical to within experimental error. If it should be assumed that the proton chemical shifts of the methyl groups are controlled primarily by the inductive effects of the H_3SiO- and Si_2H_5O-groups, then this also suggests that the over-all electron-withdrawing effects of the SiH_3O- and Si_2H_5O-groups are the same. If a Si_2H_5-group is regarded as being derived from a H_3Si-group by replacing a hydrogen atom with a SiH_3 species, it is difficult to predict accurately from simple electronegativity data whether the SiH_3 moiety in the Si_2H_5-group should have a $+I$ or $-I$ inductive effect relative to the hydrogen atom it replaced, although the relative electronegativities of H (2·1) and Si (1·8) might suggest that the SiH_3-group would be electron-releasing.

The above data might therefore suggest that (a) the SiH_3 moiety in the Si_2H_5-group has neither a $+I$ nor $-I$ effect or (b) the SiH_3-group has a $+I$ effect which is fortuitously exactly compensated by an increase in $(p \to d)\pi$ bonding between the oxygen and the silicon in the Si_2H_5O-group. Although such fortuitous compensation seems rather unlikely, an increase in $(p \to d)\pi$ bonding between the oxygen and the silicon in the Si_2H_5O-group might possibly be expected if the electron density placed in the empty $3d$ orbitals of the alpha silicon could be transferred in part to the beta silicon by means of $(d \to d)\pi$ bonding (see § 4).

The reaction of BCl_3 with certain inorganic silicon ethers is of interest since it illustrates the greater rate of reaction of bonds attached to a Si_2H_5-group as compared to a SiH_3-group. It has been known for some time that $(SiH_3)_2O$ reacts readily at -78 °C with BCl_3,[97] namely

$$(SiH_3)_2O + BCl_3 \to SiH_3Cl + SiH_3OBCl_2$$

Although SiH_3OBCl_2 can be isolated it decomposes slowly at -78 °C, namely

$$3SiH_3OBCl_2 \to 3SiH_3Cl + BCl_3 + B_2O_3$$

The over-all reaction between $(SiH_3)_2O$ and BCl_3 may be expressed as

$$3(SiH_3)_2O + 2BCl_3 \to 6SiH_3Cl + B_2O_3$$

The reaction of $(Si_2H_5)_2O$ with a deficit of BCl_3 at -78 °C is completely analogous to the over-all reaction of $(SiH_3)_2O$ with BCl_3. No cleavage of Si—Si bonds takes place.[93]

The compound $SiH_3OSi_2H_5$ was also found to react with a deficit of BCl_3 at -78 °C.[93] The chief over-all reaction which occurs is

$$6Si_2H_5OSiH_3 + 2BCl_3 \to 6Si_2H_5Cl + 3(SiH_3)_2O + B_2O_3$$

although some SiH_3Cl was also formed. The observation that considerably more Si_2H_5Cl than SiH_3Cl was formed shows that the Si_2H_5—O bond is cleaved more rapidly than the SiH_3—O bond. The greater ease of cleavage of the Si_2H_5—O bond as compared to the SiH_3—O bond is also evidenced by the fact that $(Si_2H_5)_2O$ was consumed in preference to $(SiH_3)_2O$ when a mixture of these disiloxanes was treated with a deficit of BCl_3.[93] Assuming, as has been postulated previously,[97] that reaction of a disiloxane with a boron halide might involve a four-centre transition state, and assuming that the different rates of cleavage are due only to kinetic factors, then the more important complex in the reaction involving $SiH_3OSi_2H_5$ would appear to be

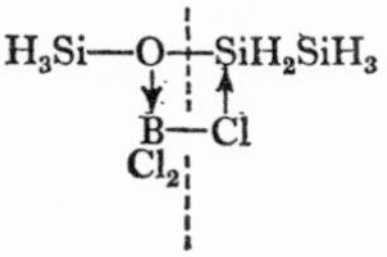

$$H_3Si-O-SiH_2SiH_3$$
$$B-Cl$$
$$Cl_2$$

rather than that in which the chlorine is attached to the SiH_3 group bonded directly to the oxygen. This is contrary to what would have been expected from steric considerations, since the presence of the large SiH_3-group would tend to make the silicon atom to which it was attached less vulnerable to attack. Regardless of the expected inductive effect of the SiH_3-group, it appears possible that this group, when present in a Si_2H_5-group, might be expected under certain circumstances to effectively withdraw electrons by means of $(d \to d)\pi$ bonding between adjacent silicon atoms. The actual transition complex, once formed, is assumed to involve a penta-coordinate silicon atom. The bonding at the silicon may actually approach that of a trigonal bipyramidal sp^3d hybrid.[2] A complex of the type postulated for a 'broadside' replacement reaction at silicon[2] is shown in the figure. The H_2SiO-group is regarded as lying in the xy-plane with the two hydrogen atoms and the oxygen atom being mutually separated from each other by angles of 120°.

If the d_{z^2} silicon orbital is involved in the hybridization, then the d_{xz} orbital would be so situated that it could overlap with a p_x orbital of the chlorine. This would give some measure of $(p \to d)\pi$ bonding in the complex in addition to the σ-bond formed by over-lap of the chlorine p_z orbital with the silicon $sp^3d_{z^2}$ orbital. Such π-bonding stabilization of the transition complex would be expected to be greater in the case of the Si_2H_5-group since the charge placed upon the alpha silicon by the π-bond might be transferred in part to the beta silicon by means of $(d \to d)\pi$ overlap[98] between the adjacent silicon atoms. Factors which stabilize the complex will also lower the energy of the highest energy transition state and will facilitate reaction.[2] Thus, in certain analogous reactions involving atoms having lone pairs of electrons, the Si_2H_5-group may be expected to react more rapidly than the SiH_3-group.

Ammonia, CH_3NH_2 and $(CH_3)_2NH$ all react instantly with Si_2H_5I or Si_2H_5Br at low temperatures to yield $(Si_2H_5)_3N$,[94] $(Si_2H_5)_2NCH_3$,[99] and $Si_2H_5N(CH_3)_2$,[88] respectively. In the case of $(CH_3)_2NH$ considerable quantities of $SiH_3SiH[NCH_3]_2$ are also obtained.[88] All of these compounds are fairly stable thermally; $(Si_2H_5)_3N$, which is the most stable, showed no decomposition on heating to 110 °C.[94]

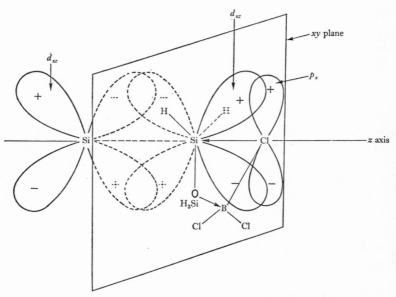

Fig. $(d \rightarrow d)\pi$ overlap in a silicon–silicon bond in the transition complex formed from $SiH_3OSiH_2SiH_3$ and BCl_3. The three hydrogen atoms attached to the SiH_3 group are omitted for simplicity. (Copyright, the American Chemical Society.)

Although the relative base strengths of the N-disilanyl N-methyl amines have not been studied, some evidence for their relative base strengths may be obtained from the proton chemical shifts of their methyl groups. It has been observed (see Table 2) that the proton chemical shifts of the methyl group in silylamines move to lower field as the SiH_3-groups are replaced by CH_3-groups.[100] This has been interpreted as indicating that the nitrogen atom has a greater effective electronegativity on proceeding from $(CH_3)_3N$ to

$SiH_3N(CH_3)_2$ to $(SiH_3)_2NCH_3$.[100] Since the relative base strengths have been shown by other methods to decrease on proceeding from $(CH_3)_3N$ to $(SiH_3)N(CH_3)_2$ to $(SiH_3)_2NCH_3$ to

$$(SiH_3)_3N,[5, 100, 101]$$

the n.m.r. data in Table 2 also suggest that the base strength of the disilanyl amines decreases in an exactly analogous manner.

TABLE 2. *Methyl proton chemical shifts of silicon amines**

Compound	Chemical shift	Compound	Chemical shift
$N(CH_3)_3$	$-0\cdot7$ ppm†	$N(CH_3)_3$	$-0\cdot7$ ppm†
$SiH_3N(CH_3)_2$	$-0\cdot9$ ppm†	$Si_2H_5N(CH_3)_2$	$-1\cdot07$ ppm‡
$(SiH_3)_2NCH_3$	$-1\cdot3$ ppm†	$(Si_2H_5)_2NCH_3$	$-1\cdot17$ ppm‡

* Measured in cyclohexane solution using cyclohexane as internal standard.
† Reference no. 100.
‡ Reference no. 99.

Relatively little is yet known concerning the Si_2H_5—P bond although $Si_2H_5PH_2$ has been synthesized together with the isomeric $(SiH_3)_2PH$ by the action of an ozonizer discharge on a mixture of PH_3 and SiH_4.[34] If Si_2H_6 and PH_3 are used, $Si_2H_5PH_2$ free of $(SiH_3)_2PH$ is obtained; no phosphorus insertion into the Si—Si bond appeared to occur.[35] A species of empirical composition Si_2AsH_7 has been isolated by treating a mixture of SiH_4 and AsH_3 in a similar manner.[33] The Si—P bond in $Si_2H_5PH_2$ is cleaved by water and by hydrogen chloride without rupture of the Si—Si bond to yield $(Si_2H_5)_2O$ and Si_2H_5Cl respectively, together with PH_3.[33]

4. Physical properties associated with the silicon–silicon bond

The Si—Si bond length does not appear to vary significantly in those compounds in which it has been measured and since the radius of Si ($1\cdot173$ Å) is about $1\cdot5$ times that of carbon ($0\cdot771$ Å), substituent atoms or groups about the Si atoms in a Si—Si bond should be substantially less sterically crowded than those about a C—C bond. This may indeed be the most important factor responsible for the relatively greater thermal stability of Si—Si chains as compared to C—C chains when the substituents are large (see § 5).

It will be recalled that the determination of bond energies, e.g. E_{M-M}, and bond dissociation energies, e.g. D_{M-M} (M=C, Si, etc.), are based on the principles expressed in the equations

$$MX_{4(gas)} + a \text{ kcal/mole} \rightarrow M_{(gas)} + 4X_{(gas)}$$

where X = H, halogen, etc., and hence

$$E_{M-X} = \tfrac{1}{4}a \text{ kcal/mole}$$

$$X_3M—MX_{3(gas)} + b \text{ kcal/mole} \rightarrow 2M_{(gas)} + 6X_{(gas)}$$

and hence $\qquad E_{M-M} = (b - 6E_{M-X}) \text{ kcal/mole}$

if it is assumed that E_{M-X} is identical in both MX_4 and M_2X_6. Also

$$X_3M—MX_{3(gas)} + c \text{ kcal/mole} \rightarrow 2X_3M_{(gas)}$$

and $\qquad\qquad D_{M-M} = c \text{ kcal/mole}$

Within experimental limits of accuracy, both mass spectroscopic appearance potential studies and thermochemical studies give identical values of E_{Si-Si} in Si_2H_6 (40 ± 10 kcal/mole[102] and $46 \cdot 4$ kcal/mole[103] respectively). However, for Si_2H_6 and $Si_2(CH_3)_6$, D_{Si-Si} values in the range 81–86 kcal/mole have been obtained,[104–106] although a value of 49 kcal/mole,[107] which has subsequently been criticized,[108] has been reported for $Si_2(CH_3)_6$.* In C_2H_6 the D_{C-C} value is essentially identical to the E_{C-C} value generally accepted for C compounds.[109] If, in the case of Si compounds, the neutral fragments in the mass spectroscopic studies have been correctly assumed, then the large difference between the E_{Si-Si} and D_{Si-Si} values may be related to the fact that d^3s as well as sp^3 hybridization is tetrahedral and that some d character may be present in the σ-bonding system of silicon.

The electronic configuration of silicon $[1s^2; 2s^2 2p^6; 3s^2 3p^2 (3d°)]$ differs from that of carbon $(1s^2; 2s^2 2p^2)$ in that silicon has relatively low-lying empty $3d$ orbitals. A very important observation which strongly suggests that the $3d$ orbitals of adjacent Si atoms may interact with each other is that at -90 °C cyclic $[Si(CH_3)_2]_x$

* Since this manuscript was submitted for publication a D_{Si-Si} value of 67 kcal/mole has also been reported for $Si_2(CH_3)_6$. [S. J. Band et. al, *Chem. Comm.* (1967) p. 723].

$(x = 5, 6, 7)$ are converted to $[Si(CH_3)_2]_x^-$ by Na/K alloy.[110] The e.s.r. spectra of the anion radicals show that the odd electron interacts equally with all the protons, all the silicon atoms and with all the carbon atoms. The presence of the electron in an overlapping $3d$ π-orbital system would be consistent with these observations.

In view of the above conclusions it would seem that some $(d \to d)\pi$ double-bond character might perhaps be expected in a Si—Si bond if one or more potentially $(p \to p)\pi$ bonding substituents were attached to one of the silicon atoms as in, for example, H_3Si—SiH_2—F. Some electron density would be fed into the silicon d orbital system by such an interaction and this electron density might tend to be delocalized by $(d \to d)\pi$ bonding between the adjacent silicon atoms. Although certain of the proton n.m.r. data for some Si_2H_5-compounds might suggest the presence of such a $(d \to d)\pi$ interaction,[111] a microwave study of Si_2H_5F shows no shortening of the Si—Si bond.[112] If such an interaction were only very small it would, of course, not give any significant shortening of the Si—Si bond and it might be detectable only by n.m.r. techniques.

A remarkable property of certain organopolysilanes is that their u.v. absorption maxima moves towards higher wavelengths with increasing chain length; furthermore, it seems that the position of the absorption maxima in the homologous series of linear polysilanes $X[Si(CH_3)_2]_yX$ (X = H, Cl, OCH_3, CH_3) depends only on the number of silicon atoms in the chain and not on the nature of the substituent. It seems that the Si—Si bond may act as a chromophore, probably through the use of the vacant $3d$-orbitals of silicon.[1, 113, 114]

Unlike carbon, no silicon compound (such as $R_2Si{=}SiR_2$, $R_2Si{=}CR_2$, $R_2Si{=}O$, etc.), has yet been isolated in which there is a p_π–p_π double bond. It has been suggested that with the larger silicon atoms the greater bond distance makes the overlap of p_π–p_π orbitals on adjacent silicon atoms small,[115] but this explanation has been criticized.[116, 117] Other explanations involving repulsions of bonding and non-bonding electrons have also been offered.[115] However, it seems likely that even if a p_π–p_π bond between two adjacent silicon atoms were just as strong as that between two adjacent carbon atoms, a species such as $H_2Si{=}SiH_2$ would still polymerize spontaneously to $(SiH_2)_x$. Although the polymerization of $CH_2{=}CH_2$ to the more thermodynamically stable polyethylene only occurs spontaneously in the presence of a catalyst it seems likely, in view of the fact that silicon compounds generally undergo a given type of reaction more rapidly than their carbon analogues,[2] that a species such as $SiH_2{=}SiH_2$ would polymerize rapidly and spontaneously even in the absence of a catalyst.

The proton n.m.r. spectra of a number of derivatives of the polysilanes, particularly those species containing an Si_2H_5-group, have been investigated.[111] There are certain interesting features of the n.m.r. spectra of compounds such as Si_2H_6, Si_3H_8, Si_2F_6, etc.,[30, 118, 119] a number of which are related to the presence of ^{29}Si satellite signals.

The proton n.m.r. spectra of Si_2H_5X species are qualitatively similar to those of their ethyl analogues with the exception of Si_2H_5I.[111] Where R is magnetically inactive, first-order A_3X_2 spectra are obtained with a 60 Mc instrument; where X is —F or —PH_2 the spectra are more complex but they can be analysed by a first-order interpretation,[34, 111] The effective symmetry of the molecules is increased by rapid rotation about the Si—Si bond.

The trend in the relative values of the proton chemical shifts of the SiH_3 and SiH_2 groups in Si_2H_5X in traversing the series Si_2H_5X (X = F, Cl, Br, I) is analogous to that observed in the ethyl halide series in that the SiH_2 resonance moves to higher field and the SiH_3 resonance moves to lower field as the electronegativity of the attached halogen decreases. Similar observations have been noted for the three known digermanyl halides, Ge_2H_5X (X = Cl, Br, I).[120] However, the spectrum of Si_2H_5I (and also Ge_2H_5I) differs from that of C_2H_5I in that the MH_2 (M = C, Si, Ge) resonance in Si_2H_5I (and also in Ge_2H_5I) falls at a higher field than the MH_3 resonance.

Although the internal chemical shift (δMH_3—δMH_2) in C_2H_5X compounds shows a reasonably good linear relationship to the electronegativity of the substituent X, no such simple relationship is found for Si_2H_5X compounds.[111] This is not necessarily surprising if the MH_2 chemical shifts should be controlled primarily by inductive effects in C_2H_5- and Si_2H_5-compounds and if the effective electronegativity of substituents in Si_2H_5X compounds are modified to varying extents by $(p \rightarrow d)\pi$ bonding between X and the alpha silicon atom. Other completely different effects may, of course, be responsible for these observations.

The β proton chemical shifts in C_2H_5-, i-C_3H_7-, t-C_4H_9-, cyclo-C_6H_{11}-, CH_3SiH_2-, $(CH_3)_2SiH$-, and $(CH_3)_3Si$-compounds decrease along the series X = H, N, O, F (which is consistent with increasing inductive deshielding);[121] however, in Si_2H_5X compounds the SiH_3 chemical shifts for these substituents vary much less (they are actually identical, to within experimental error, for X = H, N, and F). This has been interpreted as indicating an interaction between F, N or O and the SiH_3 group, which acts in a direction to oppose the inductive effects of these substituents and to make the net effect of a given substituent approximately equivalent to that of hydrogen. It is perhaps coincidental that F, N and O are believed to undergo the strongest $(p \rightarrow d)\pi$ bonding to silicon. The type of $(p \rightarrow d)\pi$ and $(d \rightarrow d)\pi$ bonding suggested earlier could supply the necessary effect opposing the inductive influence of a substituent.[111] A direct 'across space' sigma or pi interaction between the substituent, X, and the $3d$ orbitals of the SiH_3 group in a Si_2H_5X species could also possibly occur.

The non-additivity of certain ^{29}SiH coupling constant relationships[122] and the relationship between the 'silicon–silicon bond shift' and β proton chemical shift values have also been interpreted as indicating some type of interaction between X and the SiH_3 group in certain SiH_3SiH_2X compounds.[111]

5. Cleavage of the silicon–silicon bond

One of the more interesting properties of the Si—Si bond is that it does not readily undergo cleavage reactions even with many different types of reagents. This is particularly well illustrated by the fact that compounds containing the Si_2H_5-group may be readily interconverted without breaking the Si—Si linkage. However, several examples of Si—Si bond cleavage reactions in other species have been described in § 2. Further examples will be given in this section, together with certain other pertinent properties of inorganic polysilanes; however, as can be judged from the information in this section, very little is known about the cleavage of Si—Si bonds in inorganic compounds.

Although it is extremely difficult to evaluate the relative thermal stability of compounds, particularly when the studies are conducted by different investigators using different techniques, it appears that the parent polysilanes and perfluoropolysilanes are in general somewhat less thermally stable than the analogous carbon compounds, while the per-chloro, -bromo, -iodo and -phenyl polysilanes are more thermally stable than their carbon analogues.

Thus Si_2H_6 undergoes thermal decomposition at 311 °C,[123] while C_2H_6 decomposes at 485 °C.[124] Although thermal decomposition of Si_5H_{12} is almost complete after several months at room temperature,[41] later studies on Si_4H_{10} suggest that the thermal instability may be due, at least in part, to catalytic amounts of impurities.[43] Polymeric $(SiH_2)_x$ and $(SiF_2)_x$ decompose thermally in the range 250–380 °C.[58, 60] Although compounds such as $Si_{10}Cl_{22}$ (which distils at 215–220 °C *in vacuo*[12]) are known, no high molecular weight per-chloro, -bromo or -iodo polyalkanes have been prepared and even C_3Cl_8 decomposes (to $Cl_2C=CCl_2$ and CCl_4) at 200–260 °C.[125] Also, Si_2Br_6 may be distilled at 265 °C[54] while C_2Br_6 decomposes to $Br_2C=CBr_2$ and bromine at 200–210 °C.[126] The compounds Si_3Br_8 and Si_4Br_{10} are also moderately stable thermally.[39] Hexaiodosilane, Si_2I_6 is actually synthesized by heating SiI_4 with silver at 280 °C for 6 h.[18, 19] Hexaphenyldisilane, $Si_2(C_6H_5)_6$, melts without decomposition at 384 °C[127] and shows no evidence whatsoever of dissociating to $(C_6H_5)_3Si$, whereas $C_2(C_6H_5)_6$ readily gives free triphenylmethyl, $(C_6H_5)_3C$. The greater tendency of silicon, as compared to carbon, to form long, thermally stable, catenated chains in the presence of bulky substituents is probably related principally to the greater radius of silicon, as mentioned in § 4.

The greater reactivity of compounds containing a Si—Si bond as compared to the analogous monosilane species is exemplified by the fact that many compounds containing Si—Si bonds differ from their monosilane anologues in that they are spontaneously inflammable in air. For example, Si_2H_5I[80] and $(SiF_2)_x$[58] are spontaneously inflammable in air, whereas SiH_3I and SiF_4 are not.

Although the Si—Si bond is rapidly cleaved by aqueous alkali in the parent hydrides[4, 5] and perhalopolysilanes, it is relatively inert to this reagent and to most other reagents if a number of organic groups are attached to the molecule.[1, 2] However, in Si_2H_6 it is rapidly cleaved by potassium and also by KH in 1, 2-dimethoxy-ethane at room temperature to give $KSiH_3$,[128] namely

$$Si_2H_6 + KH \rightarrow KSiH_3 + SiH_4$$

Certain phenyl-containing disilanes are also readily cleaved by alkali metals under analogous conditions to give R_3SiM (M = Li, Na, K).[2]

Although Si_2H_6 reacts with iodine at room temperature to give polyiododisilanes, no cleavage of the Si—Si bond occurs.[4] However, when Si_2F_6 is mixed with chlorine or bromine and heated momentarily, the Si—Si bond is cleaved with explosive violence to yield SiF_3X (X = Cl, Br).[129, 130]

There is some evidence to suggest that if silicon atoms in a Si—Si bond are made more susceptible to nucleophilic attack by the presence of substituent chlorine or fluorine, then cleavage of the Si—Si bond occurs very much more readily. Hydrogen chloride does not cleave the Si—Si bond in Si_2H_6 or in Si_2Cl_6 at room temperature, but Si_5Cl_{12} in the presence of a mercury-derived catalyst, suffers stepwise cleavage of its Si—Si bonds, to give, in the first step, $HSiCl_3$ and Si_4Cl_{10}.[65] Mercuric cyanide breaks the Si—Si bond in Si_2Cl_6 at 100 °C,[131] namely

$$Si_2Cl_6 + Hg(CN)_2 \rightarrow 2SiCl_3CN + Hg$$

Also, Si_2F_6 reacts with $(CH_3)_2AsCl$ at room temperature, to give $(CH_3)_2As—As(CH_3)_2$,[132] namely

$$2(CH_3)_2AsCl + Si_2F_6 \rightarrow (CH_3)_2As—As(CH_3)_2 + 2SiF_3Cl$$

The chemistry of the Si—Si bond in inorganic compounds is still virtually unknown and much work needs to be done in this area in order to gain a fundamental understanding of the nature of polysilanes and their derivatives.

REFERENCES

1 M. Kumada and K. Tamao, in *Advances in Organometallic Chemistry* (ed. F. G. A. Stone and R. West), vol. VI. New York: Academic Press, 1967.

2 C. Eaborn, *Organosilicon Compounds*. London: Butterworths, 1960. L. H. Sommer, *Stereochemistry, Mechanism and Silicon*. New York: McGraw-Hill, 1965.

3 C. Eaborn and R. W. Bott, in *Organometallic Compounds of the Group IV Elements* (ed. A. G. MacDiarmid), vol. I. New York: Dekker, 1968.

4 E. A. V. Ebsworth, *Volatile Silicon Compounds*. Pergammon Press, 1963. E. A. V. Ebsworth, in *Organometallic Compounds of the Group IV Elements* (ed. A. G. MacDiarmid), vol. I. New York: Dekker, 1968.

5 A. G. MacDiarmid, in *Advances in Inorganic Chemistry and Radiochemistry* (ed. H. J. Emeléus and A. G. Sharpe), vol. III, p. 207. New York: Academic Press, 1961.

6 A. G. MacDiarmid, *Q. Rev. chem. Soc.* 1956, **10**, 208.

7 F. G. A. Stone, *Hydrogen Compounds of the Group IV Elements*. Englewood Cliffs, N.J.: Prentice-Hall, 1962.

8 A. G. MacDiarmid, in *Preparative Inorganic Reactions* (ed. W. L. Jolly), vol. I, p. 165. New York: Interscience, 1964.

9 R. Fessenden and J. S. Fessenden, *Chem. Rev.* 1961, **61**, 361.

10 B. J. Aylett, in *Preparative Inorganic Reactions* (ed. W. L. Jolly), vol. II, p. 93. New York: Interscience, 1965.

11 H. J. Emeléus and J. S. Anderson, *Modern Aspects of Inorganic Chemistry*, pp. 524–5. Routledge and Kegan Paul, 1960.

12 R. Schwarz and H. Meckbach, *Z. anorg. Chem.* 1937, **232**, 241.

13 R. Schwarz and C. Danders, *Chem. Ber.* 1947, **80**, 444.

14 R. Schwarz and R. Thiel, *Z. anorg. Allgem. Chem.* 1938, **235**, 247.

15 R. Schwarz, *Angew. Chem.* 1938, **51**, 328.

16 W. C. Schumb and E. L. Gamble, in *Inorganic Syntheses* (ed. H. S. Booth), vol. I, p. 42. McGraw-Hill, 1939.

17 A. Stock, A. Brandt and H. Fischer, *Chem. Ber.* 1925, **58**, 643.

18 W. C. Schumb, *Chem. Rev.* 1942, **31**, 587.

19 R. Schwarz and A. Pflugmacher, *Chem. Ber.* 1942, **75**B, 1062.

20 G. Schott and W. Herrmann, *Z. anorg. Chem.* 1956, **288**, 1. G. Schott and E. Hirschmann, *Z. anorg. Chem.* 1956, **288**, 9.

21 M. Schmeisser and M. Schwarzmann, *Z. Naturforsch.* 1956, **11**b, 278.

22 A. D. Craig and A. G. MacDiarmid, *J. inorg. nucl. Chem.* 1962, **24**, 161.

23 G. Fritz, *Z. Naturforsch.* 1952, **7**b, 507.

24 D. G. White and E. G. Rochow, *J. Am. chem. Soc.* 1954, **76**, 3897.

25 P. L. Timms, C. C. Simpson and C. S. G. Phillips, *J. chem. Soc. (A)*, 1967, p. 1467.

26 H. J. Emeléus and K. Stewart, *Trans. Faraday Soc.* 1936, **32**, 1577.

27 H. Niki and G. J. Mains, *J. Phys. Chem.* 1964, **68**, 304. M. A. Nay,

G. N. C. Woodall, O. P. Strausz and H. E. Gunning, *J. Am. chem. Soc.* 1965, **87**, 179.

28 G. A. Gibbon, Y. Rousseau, C. H. Van Dyke and G. J. Mains, *Inorg. Chem.* 1966, **5**, 114.

29 E. J. Spanier and A. G. MacDiarmid, *Inorg. Chem.* 1962, **1**, 432.

30 W. L. Jolly and S. D. Gokhale, *Inorg. Chem.* 1964, **3**, 946.

31 E. J. Spanier and A. G. MacDiarmid, *Inorg. Chem.* 1963, **2**, 215.

32 T. D. Andrews and C. S. G. Phillips, *J. chem. Soc. (A)*, 1966 p. 46.

33 W. L. Jolly and J. E. Drake, *Chemy Ind.* 1962, p. 1470.

34 S. D. Gokhale and W. L. Jolly, *Inorg. Chem.* 1964, **3**, 1141.

35 W. L. Jolly and S. D. Gokhale, *Inorg. Chem.* 1965, **4**, 596.

36 M. Abedini and A. G. MacDiarmid, *Inorg. Chem.* 1966, **5**, 2040.

37 K. A. Hertwig and E. Wiberg, *Z. Naturforsch.* 1951, **6**b, 336.

38 R. Schwarz and G. Pietsch, *Z. anorg. allgem. Chem.* 1937, **232**, 249.

39 A. Besson and L. Fournier, *C. r. hebd. Séanc. Acad. Sci., Paris*, 1910, **151**, 1055.

40 M. A. Ring, R. C. Kennedy and L. P. Freeman, *J. inorg. nucl. Chem.* 1966, **28**, 1373.

41 A. Stock, *Hydrides of Boron and Silicon*. Ithaca, N.Y.: Cornell University Press, 1933.

42 A. Stock and C. Somieski, *Ber. dtsch. chem. Ges.* 1916, **49**, 111.

43 H. J. Emeléus and A. G. Maddock, *J. chem. Soc.* 1946, p. 1131.

44 W. C. Johnson and S. Isenberg, *J. Am. chem. Soc.* 1935, **57**, 1349.

45 F. Fehér and W. Tromm, *Z. anorg. Chem.* 1955, **282**, 29.

46 K. Borer and C. S. G. Phillips, *Proc. chem. Soc.* 1959, p. 189.

47 C. S. G. Phillips and P. L. Timms, *Analyt. Chem.* 1963, **35**, 505.

48 C. S. G. Phillips, P. Powell, J. A. Semlyen and P. L. Timms, *Z. analyt. Chem.* 1963, **197**, 202.

49 P. L. Timms, C. C. Simpson and C. S. G. Phillips, *J. chem. Soc.* 1967, p. 1467.

50 P. Royen and C. Rocktäschel, *Angew. Chem.* 1964, **76**, 302.

51 F. Fehér and H. Strack, *Naturwissenschaften*, 1963, **50**, 570.

52 P. F. Antipin and V. V. Sergeev, *Zh. Prik. Khim.* 1954, **27**, 784; *Chem. Abstr.* 1955, **49**, 765.

53 W. C. Schumb and C. H. Klein, *J. Am. chem. Soc.* 1937, **59**, 261.

54 W. C. Schumb, in *Inorganic Syntheses* (ed. W. C. Fernelius), vol. II, p. 98. McGraw-Hill, 1946.

55 E. G. Rochow and R. Didtschenko, *J. Am. chem. Soc.* 1952, **74**, 5545.

56 P. L. Timms, personal communication, 1966.

57 P. L. Timms, R. A. Kent, T. C. Ehlert and J. L. Margrave, *J. Am. chem. Soc.* 1965, **87**, 2824.

58 J. C. Thompson and J. L. Margrave, *Science, N.Y.*, 1967, **155**, 669.

59 A. G. MacDiarmid, Y. L. Baay and J. F. Bald, Jr., unpublished observations, 1966; Y. L. Baay, Ph.D. dissertation, University of Pennsylvania, 1967.

60 R. Schwarz and F. Heinrich, *Z. anorg. Chem.* 1935, **221**, 277.

61 G. D. Cooper and A. R. Gilbert, *J. Am. chem. Soc.* 1960, **82**, 5042.

62 M. A. Ring, L. P. Freeman and A. P. Fox, *Inorg. Chem.* 1964, **3**, 1200.

63 J. A. Morrison and M. A. Ring, *Inorg. Chem.* 1967, **6**, 100.

64 G. Urry, *J. inorg. nucl. Chem.* 1964, **26**, 409.
65 A. Kaczmarczyk and G. Urry, *J. inorg. nucl. Chem.* 1964, **26**, 415.
66 A. Kaczmarczyk, J. W. Nuss and G. Urry, *J. inorg. nucl. Chem.* 1964, **26**, 427.
67 J. Urenovitch, R. Pejic and A. G. MacDiarmid, *J. chem. Soc.* 1963, p. 5563.
68 M. Schmeisser and K. Ehlers, *Angew. Chem.* 1964, **76**, 781.
69 A. Pflugmacher and I. Rohrman, *Z. anorg. allgem. Chem.* 1957, **290**, 101.
70 J. Urenovitch and A. G. MacDiarmid, *J. Am. chem. Soc.* 1963, **85**, 3372.
71 W. H. Atwell and D. R. Weyenberg, *J. organomet. Chem.* 1966, **5**, 594.
72 H. Kautsky and G. Herzberg, *Ber. dtsch. chem. Ges.* 1924, **57** *B*, 1665; H. Kautsky and A. Hirsch, *Z. anorg. Chem.* 1928, **170**, 1; *Ber. dtsch. chem. Ges.* 1931, **64** *B*, 1610; H. Kautsky and H. P. Siebel, *Z. anorg. Chem.* 1953, **273**, 113, 729; H. Kautsky and T. Richter, *Z. Naturforsch.* 1956, **11** *b*, 365.
73 A. E. Finholt, A. C. Bond, Jr., K. E. Wilzbach and H. I. Schlesinger, *J. Am. chem. Soc.* 1947, **69**, 2692.
74 F. Fehér, H. Keller, G. Kuhlbörsch and H. Luhleich, *Angew. Chem.* 1958, **70**, 402.
75 A. Stock, P. Stiebeler and F. Zeidler, *Ber. dtsch. chem. Ges.* 1923, **56** *B*, 1695.
76 R. Wintgen, *Ber. dtsch. chem. Ges.* 1919, **52** *B*, 724.
77 A. D. Craig, J. Urenovitch and A. G. MacDiarmid, *J. chem. Soc.* 1962, p. 548.
78 M. Abedini, C. H. Van Dyke and A. G. MacDiarmid, *J. inorg. nucl. Chem.* 1963, **25**, 307.
79 L. G. L. Ward and A. G. MacDiarmid, *J. inorg. nucl. Chem.* 1961, **20**, 345.
80 L. G. L. Ward and A. G. MacDiarmid, *J. Am. chem. Soc.* 1960, **82**, 2151.
81 H. J. Eleléus, A. G. Maddock and C. Reid, *J. chem. Soc.* 1941, p. 353.
82 H. E. Opitz, J. S. Peake and N. H. Nebergall, *J. Am. chem. Soc.* 1956, **78**, 292.
83 A. Stock and C. Somieski, *Ber. dtsch. chem. Ges.* 1918, **51**, 989.
84 A. Stock and C. Somieski, *Ber. dtsch. chem. Ges.* 1919, **52** *B*, 695.
85 R. P. Hollandsworth, W. M. Ingle and M. A. Ring, *Inorg. Chem.* 1967, **6**, 844.
86 C. H. Van Dyke and A. G. MacDiarmid, *J. inorg. nucl. Chem.* 1963, **25**, 1503.
87 J. E. Drake and J. Simpson, *Inorg. nucl. chem. Letters*, 1966, **2**, 219.
88 M. Abedini and A. G. MacDiarmid, *Inorg. Chem.* 1963, **2**, 608.
89 A. G. MacDiarmid and C. H. Van Dyke, unpublished observations, 1963; C. H. Van Dyke, Ph.D. dissertation, University of Pennsylvania, 1964.
90 G. S. Forbes and H. H. Anderson, *J. Am. chem. Soc.* 1940, **62**, 761.
91 G. S. Forbes and H. H. Anderson, *J. Am. chem. Soc.* 1947, **69**, 3048.
92 W. C. Schumb and R. A. Lefever, *J. Am. chem. Soc.* 1954, **76**, 2091.

93 C. H. Van Dyke and A. G. MacDiarmid, *Inorg. Chem.* 1964, **3**, 747.

94 L. G. L. Ward and A. G. MacDiarmid, *J. inorg. nucl. Chem.* 1961, **21**, 287.

95 C. H. Van Dyke and A. G. MacDiarmid, *J. phys. Chem.* 1963, **67**, 1930.

96 B. Sternbach and A. G. MacDiarmid, *J. Am. chem. Soc.* 1961, **83**, 3384.

97 M. Onyszchuk, *Can. J. Chem.* 1961, **39**, 808.

98 H. H. Jaffé, *J. phys. Chem.* 1954, **58**, 185. D. P. Craig, A. Maccoll, R. S. Nyholm, L. E. Orgel and L. E. Sutton, *J. chem. Soc.* 1954, p. 332.

99 A. G. MacDiarmid and M. Abedini, unpublished observations, 1962; M. Abedini, Ph.D. dissertation, University of Pennsylvania, 1963.

100 E. A. V. Ebsworth and N. Sheppard, *J. inorg. nucl. Chem.* 1959, **9**, 95.

101 S. Sujishi and S. Witz, *J. Am. chem. Soc.* 1954, **76**, 4631.

102 F. E. Saalfeld and H. J. Svec, *Inorg. Chem.* 1963, **2**, 50; F. E. Saalfeld, personal communication, 1967.

103 S. R. Gunn and L. G. Green, *J. phys. Chem.* 1961, **65**, 779; 1964, **68**, 946.

104 F. E. Saalfeld and H. J. Svec, *Inorg. Chem.* 1964, **3**, 1442.

105 W. C. Steele, L. D. Nichols and F. G. A. Stone, *J. Am. chem. Soc.* 1962, **84**, 4441.

106 G. G. Hess, F. W. Lampe and L. H. Sommer, *J. Am. chem. Soc.* 1965, **87**, 5327.

107 J. A. Connor, G. Finney, G. J. Leigh, R. N. Haszeldine, P. J. Robinson, R. D. Sedgwick and R. F. Simmons, *Chem. Comm.* 1966, p. 178.

108 I. M. T. Davidson and I. L. Stephenson, *Chem. Comm.* 1966, p. 746.

109 J. L. Cottrell, *The Strengths of Chemical Bonds*, 2nd ed., p. 273. Butterworths, 1958.

110 G. R. Husk and R. West, *J. Am. chem. Soc.* 1965, **87**, 3993. E. Carberth and R. West, *J. Am. chem. Soc.* 1965 (in the Press).

111 C. H. Van Dyke and A. G. MacDiarmid, *Inorg. Chem.* 1964, **3**, 1071.

112 A. P. Cox and R. Varma, *J. phys. Chem.* 1966, **44**, 2619.

113 H. Gilman, W. H. Atwell and G. L. Schwebke, *J. organomet. Chem.* 1964, **2**, 369.

114 H. Sakurai and M. Kumada, *Bull. chem. Soc. Japan*, 1964, **37**, 1894.

115 K. S. Pitzer, *J. Am. chem. Soc.* 1948, **70**, 2140.

116 R. S. Mulliken, *J. Am. chem. Soc.* 1950, **72**, 4493.

117 R. S. Mulliken, C. A. Rieke, D. Orloff and H. Orloff, *J. chem. Phys.* 1948, **17**, 1248; H. H. Jaffé, *ibid.* 1953, **21**, 258.

118 E. A. V. Ebsworth and J. J. Turner, *Trans. Faraday Soc.* 1964, **60**, 256.

119 R. B. Johannesen, T. C. Farrar, F. E. Brinckman and T. D. Coyle, *J. chem. Phys.* 1966, **44**, 962.

120 K. M. Mackay, P. Robinson, E. J. Spanier and A. G. MacDiarmid, *J. inorg. nucl. Chem.* 1966, **28**, 1377.

121 E. A. V. Ebsworth and S. G. Frankiss, *J. Am. chem. Soc.* 1963, **85**, 3516.

122 T. Yoshioka and A. G. MacDiarmid, *J. molec. Spectrosc.* 1966, **21**, 104.

123 K. Stokland, *Trans. Faraday Soc.* 1948, **44**, 545.

124 L. F. Fieser and M. Fieser, *Organic Chemistry*, p. 97. Heath and Co., 1944.

125 E. T. McBee, H. B. Hass, T. H. Chao, Z. D. Welch and L. E. Thomas, *Ind. Engng Chem.* 1941, **33**, 176; H. Gerding and G. W. A. Rijnders, *Recl. Trav. chim. Pay-Bas Belge*, 1946, **65**, 143; N. E. Aubrey and J. R. Van Wazer, *J. Am. chem. Soc.* 1964, **86**, 4380.

126 M. Pussol, *Bull. Soc. Chim. biol.* 1925, **37**, 161; A. Mouneyrat, *C. r. hebd. Séanc. Acad. Sci., Paris*, 1898, **127**, 109.

127 W. Schlenk, J. Renning and G. Racky, *Ber. dtsch. chem. Ges.* 1911, **44**, 1178.

128 D. M. Ritter and M. A. Ring, *J. Am. chem. Soc.* 1961, **83**, 802.

129 W. C. Schumb and E. L. Gamble, *J. Am. chem. Soc.* 1932, **54**, 3943.

130 W. C. Schumb and H. H. Anderson, *J. Am. chem. Soc.* 1936, **58**, 994.

131 A. Kaczmarczyk and G. Urry, *J. Am. chem. Soc.* 1959, **81**, 4112.

132 A. G. MacDiarmid, Y. L. Baay and T. A. Banford, unpublished observations, 1967; Y. L. Baay, Ph.D. dissertation, University of Pennsylvania, 1967.

9

METAL–METAL INTERACTION IN PARAMAGNETIC CLUSTERS

R. L. MARTIN

1. Introduction

Many inorganic molecules are known which contain more than one metal atom. Such entities are commonly termed 'clusters', the polynuclear metal core being embedded in or supported by the

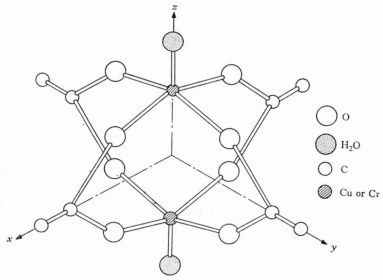

Fig. 1. Molecular structure of $[Cu_2(CH_3CO_2)_4(H_2O)_2]$.

ligand matrix. As a result, each cluster is effectively shielded from its near-neighbours and magnetic interactions between clusters are too small to be observed except at the very lowest temperatures.

The magnetic properties of a cluster will be determined by such factors as the electron configuration of the metal, the metal–metal separation, the electronic structures of the anion and associated ligands, and the geometrical disposition of the metal and ligand atoms. For example, binuclear chromium(II) acetate monohydrate (see Fig. 1) with the d^4-configuration is diamagnetic while the copper(II) analogue with the d^9-configuration is paramagnetic,[1] even though the two compounds are isostructural.[2] Subtle differences in the stereochemistries[3-5] of the dimeric anions $[Cr_2Cl_9]^{3-}$ and $[W_2Cl_9]^{3-}$ determine that the former is paramagnetic and the latter diamagnetic,[6] even though Cr^{III} and W^{III} possess identical d-shell populations (configuration d^3). On the other hand, the influence of the ligand may be discerned in the stronger Cu–Cu interaction ($\mu_{eff} = 1\cdot1$ B.M. at 300 °K) which occurs in the compound $[Cu_2(HCO_2)_4(pyridine)_2]$ compared with that ($\mu_{eff} = 1\cdot4$ B.M. at 300 °K) found[7,8] in

$$[Cu_2(CH_3CO_2)_4(pyridine)_2].$$

Likewise, bromo-bridged metal ions frequently interact more strongly than those bridged by chlorine atoms.[9,10]

These and many related observations during the past two decades have provoked a number of important physical questions concerning the manner in which two or more metal atoms interact within the confines of a magnetically isolated cluster. The mechanism by which unpaired electron spins couple and the factors which control the magnitude and sign of the spin–spin interaction continue to be imperfectly understood even though the underlying theoretical foundations were provided by Heisenberg, Dirac and Van Vleck nearly forty years ago.[11-14] However, the hypothesis of 'superexchange'—namely, the mechanism by which two paramagnetic atoms interact when shielded from each other by an intervening diamagnetic atom as in $[(NH_3)_5CrOCr(NH_3)_5]^{4+}$, has enabled important advances in the more profound understanding of the nature of magnetic interactions to be achieved.[15,17]

The measurement of paramagnetic susceptibility, and its temperature dependence, offers a particularly powerful probe for detecting the presence of metal–metal interactions. The sign of the spin–spin interaction (i.e. ferromagnetic or antiferromagnetic) can be established and the strength of the coupling between electrons can be estimated with considerable accuracy. Further, it is the most effective physical method for following experimentally the change with temperature of the relative populations of a manifold of spin states separated in energy by the order of kT (usually in the range 1–10^3 cm^{-1}). E.s.r. is limited in this regard in being confined usually to the lowest level of the spin manifold. However, in favourable cases, the temperature dependence of the intensity of an e.s.r. signal can be related to the magnitude and sign of the spin–spin interaction.[18] Refinements to current theories of electron-spin interactions must ultimately rest on the experimentally determined magnetic properties of both polynuclear systems and extended three-dimensional crystals.

In this article, some of the more qualitative concepts of metal–metal interaction are reviewed from the point of view of the inorganic chemist. Attention is focused on clusters, for in these systems the more complex aspects of cooperative magnetic phenomena (ferromagnetism, antiferromagnetism, ferrimagnetism, etc.) which are associated with the crystal lattice as a whole can be avoided. The theoretical discussion is intentionally limited to the HDVV model which, although oversimplified, continues to provide a sound basis for discussions of metal–metal interaction. No attempt will be made to review the field exhaustively and the discussion will be limited to a representative selection of clusters for which extended magnetic data are available. An excellent discussion of the evaluation of exchange interactions from experimental data has been given by Smart.[19]

2. The exchange effect and spin–spin coupling

Suppose we consider what happens when two atoms, such as copper(II) (configuration d^9), are brought together to form a cluster as in [Cu$_2$(CH$_3$CO$_2$)$_4$(H$_2$O)$_2$]. Interaction between the copper atoms will favour either a parallel or antiparallel alignment of the individual electron spins to give two states which differ in energy. Heisenberg[11, 12] was the first to show that this interatomic coupling has its origin in the quantum mechanical exchange effect. The reader will recall that for a system of two electrons moving under similar potential fields (e.g. as in the dimer H$_2$!) the Schrödinger equation takes the form

$$(H_0 + H')\Psi = E\Psi$$

where
$$H_0 = -\frac{h^2}{8\pi^2 m}(\nabla_1^2 + \nabla_2^2) - \frac{e^2}{r_{a1}} - \frac{e^2}{r_{b2}}$$

and
$$H' = \frac{e^2}{r_{ab}} + \frac{e^2}{r_{12}} - \frac{e^2}{r_{b1}} - \frac{e^2}{r_{a2}} \qquad [1]$$

where the subscripts a and b refer to the interacting atoms and the subscripts 1 and 2 refer to the two electrons. If the two electrons are virtually unconnected, the system is inherently degenerate since the electrons are indistinguishable. However, this twofold degeneracy of electron interchange is removed when the isotropic interaction described by the Hamiltonian H' in [1] is included, and the correct eigenfunctions are those which are 'symmetrical' [2] and 'antisymmetrical' [3] with respect to electron exchange, namely

$$\Psi_{sym}(1, 2) = (1/\sqrt{2})[\phi_a(1)\,\phi_b(2) + \phi_a(2)\,\phi_b(1)] \qquad [2]$$

$$\Psi_{anti}(1, 2) = (1/\sqrt{2})[\phi_a(1)\,\phi_b(2) - \phi_a(2)\,\phi_b(1)] \qquad [3]$$

If $2E_0$ is the energy of the dimeric system in the absence of the interaction potential, the appropriate eigenvalues are

$$E_{sym} = 2E_0 + \frac{K+J}{1+S_0^2} \qquad [4]$$

and
$$E_{anti} = 2E_0 + \frac{K-J}{1-S_0^2} \qquad [5]$$

Here the Coulomb integral K is given by

$$K = \iint \phi_a(1)\,\phi_b(2)\,[H']\phi_a(1)\,\phi_b(2)\,d\tau_1 d\tau_2$$

$$= \iint \phi_a(2)\,\phi_b(1)\,[H']\phi_a(2)\,\phi_b(1)\,d\tau_1 d\tau_2 \qquad [6]$$

and the exchange integral J by

$$J = \iint \phi_a(1)\,\phi_b(2)\,[H']\phi_a(2)\,\phi_b(1)\,d\tau_1 d\tau_2 \qquad [7]$$

where $\phi_i(k)$ are the usual one-electron wave functions when electron k is in an orbital i. Historically, the overlap integral

$$S_0 \quad (\equiv \int \phi_a \phi_b\, d\tau)$$

was frequently neglected in treatments of the exchange effect.[14] Despite the fact that the electrical interaction between the pair of atoms is independent of the electron spins, the energy of the system depends upon the total spin since the Pauli exclusion principle demands that the total wave functions must be antisymmetrical in every pair of electrons. Since the two-electron solutions to [1] may be written in the form of a product of spatial and spin coordinates, it follows that Ψ_{sym} must be combined with a singlet ($S = 0$) and Ψ_{anti} with a triplet ($S = 1$) spin wave function:

$$^1\Psi = \Psi_{sym}(1, 2)[\alpha(1)\,\beta(2) - \beta(1)\,\alpha(2)] \qquad [8]$$

$$^3\Psi = \Psi_{anti}(1, 2)\left\{\begin{array}{c} \alpha(1)\,\alpha(2) \\ \alpha(1)\,\beta(2) + \beta(1)\,\alpha(2) \\ \beta(1)\,\beta(2) \end{array}\right\} \qquad [9]$$

where α and β are functions of the spin coordinates only and the normalizing factors are omitted.

It can be shown[11, 20] that this correlation between orbital symmetry and spin alignment is equivalent to an apparent spin coupling between the orbitals ϕ_a and ϕ_b. Suppose $\mathbf{s}_1$ and $\mathbf{s}_2$ are the spin angular momentum operators of the two electrons measured in multiples of the quantum unit $h/2\pi$. If the vector sum of $\mathbf{s}_1$ and $\mathbf{s}_2$ is $\mathbf{s}_{12}$, then

$$\mathbf{s}_{12}^2 = (\mathbf{s}_1 + \mathbf{s}_2)^2 = \mathbf{s}_1^2 + \mathbf{s}_2^2 + 2\mathbf{s}_1\cdot\mathbf{s}_2 \qquad [10]$$

Since the eigenvalues (in multiples of $h/2\pi$) when $\mathbf{s}_{12}^2$, $\mathbf{s}_1^2$ and $\mathbf{s}_2^2$ operate on the wave functions [8] and [9] are $S(S+1)$, $s_1(s_1+1)$ and $s_2(s_2+1)$, respectively, the characteristic values of the scalar product $\mathbf{s}_1\cdot\mathbf{s}_2$ corresponding to $S = 0$ and $S = 1$ can readily be calculated. They are

$$S = 0: \quad \mathbf{s}_1\cdot\mathbf{s}_2 = \tfrac{1}{2}[0 - \tfrac{3}{4} - \tfrac{3}{4}] = -\tfrac{3}{4} \qquad [11]$$

$$S = 1: \quad \mathbf{s}_1\cdot\mathbf{s}_2 = \tfrac{1}{2}[2 - \tfrac{3}{4} - \tfrac{3}{4}] = +\tfrac{1}{4} \qquad [12]$$

where the state $S = 0$ is correlated with the symmetric and $S = 1$ the antisymmetric orbital solutions. From [4] and [5] we see that the characteristic values of the potential energy V of interaction between the electrons are

$$K + J \quad \text{when} \quad \mathbf{s}_1\cdot\mathbf{s}_2 = -\tfrac{3}{4} \qquad [13]$$

and
$$K - J \quad \text{when} \quad \mathbf{s}_1\cdot\mathbf{s}_2 = +\tfrac{1}{4} \qquad [14]$$

where we neglect the overlap integral S_0 and the term in E_0 since it is independent of the symmetry. Consequently, the total interaction between the atoms can be expressed as

$$V = K - \tfrac{1}{2}J - 2J_{12}\mathbf{s}_1 \cdot \mathbf{s}_2 \qquad [15]$$

where the effective spin coupling between the pair of atoms is equivalent to a potential energy of the form $-2J\mathbf{s}_1 \cdot \mathbf{s}_2$.

In the problems concerning magnetic interactions between paramagnetic centres, it is usually only the spin dependent term of [15] which is of interest. For two atoms, i and j, that have one electron each, the two-electron exchange Hamiltonian is written

$$\mathscr{H}_{ex} = -2J_{ij}\mathbf{s}_i \cdot \mathbf{s}_j \qquad [16]$$

where J_{ij} is the exchange integral for the two electrons. If the atoms have total spins of $\mathbf{S}_i = \Sigma\mathbf{s}_i$ and $\mathbf{S}_j = \Sigma\mathbf{s}_j$, then the exchange Hamiltonian is, $$\mathscr{H}_{ex} = -2J_{ij}\mathbf{S}_i \cdot \mathbf{S}_j \qquad [17]$$

providing all the electrons have the same exchange integral J_{ij}.

[16] shows that two electrons interact in a manner which suggests that there is strong coupling between their respective spins which, apart from the spin independent additive exchange term $-\tfrac{1}{2}J$, is proportional to the scalar product of these spin angular momenta. However, since the classical interaction energy between two magnetic dipoles $\boldsymbol{\mu}_1$ and $\boldsymbol{\mu}_2$ is

$$E = \frac{\boldsymbol{\mu}_1 \cdot \boldsymbol{\mu}_2}{\mathbf{r}^3} - \frac{3(\boldsymbol{\mu}_1 \cdot \mathbf{r})(\boldsymbol{\mu}_2 \cdot \mathbf{r})}{\mathbf{r}^5} \qquad [18]$$

it must be emphasized that the exchange effect resembles only the first term of this expression. Furthermore, actual magnetic dipole–dipole forces are very much weaker than the exchange effect occurring at the same value of the internuclear separation (see § 4.4). Accordingly, it will be seen that the exchange effect, although entirely orbital in nature, resembles a dipolar coupling between electron spins because the Pauli exclusion principle demands an orbital function of one symmetry for the triplet and another for the singlet state, i.e. the origin of J is electrostatic—not magnetic.

3. HDVV model of spin–spin interaction

We have seen that the Heisenberg–Dirac–Van Vleck model assumes that the effective spin coupling between orbits ϕ_a and ϕ_b with spin angular momenta S_a and S_b is equivalent to a potential energy of the form

$$V_{ab} = -2J_{ab}S_a \cdot S_b \qquad [19]$$

If orbital overlap is included, J_{ab} can be defined as one-half of the energy separation between the singlet and triplet states,

$$J_{ab} = \tfrac{1}{2}[E_{sym} - E_{anti}]$$
$$= (J - S_0^2 K)/(1 - S_0^4) \qquad [20]$$

If $J_{ab} > 0$, then the triplet state lies lowest (ferromagnetic coupling), whereas if $J_{ab} < 0$, the singlet state is more stable (antiferromagnetic coupling). Early workers neglected terms in S_0^2 and S_0^4 and assumed $J_{ab} = J$. Further comment on the sign of J_{ab} is reserved until § 4.

Before deducing the eigenvalues of [17], the assumptions implicit in the HDVV model should be considered.[14, 21] These are

(i) The interacting magnetic atoms are effectively in S-states, i.e. it is assumed that ligand fields are sufficiently strong to quench orbital angular momentum and leave only the spin angular momentum free.

(ii) The unpaired electrons are localized on the interacting magnetic centres, i.e. Heitler–London wave functions are used.

(iii) The paramagnetic centres are not necessarily composed of the same element although most published data refer to like atoms.

(iv) Ferromagnetic interactions are implied when $J > 0$ and antiferromagnetic interactions when $J < 0$, i.e. if the two electron spins are parallel the interaction energy is $-J$; if the spins are antiparallel an interaction of $+J$ is involved.

(v) It is customary, but not essential, to assume that the lowest excited electronic states of the metal atoms are not mixed with the ground state, i.e. the spectroscopic splitting factor is usually, but not necessarily, assumed to be 2·0.

(vi) Since the exchange integral is sensitive to orbital overlap, interactions on the same atom (intra-atomic) or between adjacent atoms (interatomic) alone are considered important (however, see § 5.2.4).

3.1. Binuclear clusters

For such compounds the spin–spin Hamiltonian reduces to

$$\mathcal{H}_{ex} = -2J_{12}\mathbf{S}_1 \cdot \mathbf{S}_2 \qquad [21]$$

and eigenvalues can be simply obtained by making use of the vector model.[21] If S is written for the spin quantum number $\mathbf{S}_i$ (i = 1, 2) the characteristic value of $(\mathbf{S}_1 \cdot \mathbf{S}_1) = S(S+1)$. Setting $\mathbf{S}' = \mathbf{S}_1 + \mathbf{S}_2$ enables us to write

$$(\mathbf{S}_1 + \mathbf{S}_2)^2 = (\mathbf{S}' \cdot \mathbf{S}') = 2S(S+1) + 2\mathbf{S}_1 \cdot \mathbf{S}_2$$

[21] can now be written

$$\mathcal{H}_{ex} = -J_{12}[(\mathbf{S}' \cdot \mathbf{S}') - 2S(S+1)] \qquad [22]$$

The energy levels under this Hamiltonian are simply

$$E(S') = -J_{12}[S'(S'+1) - 2S(S+1)] \qquad [23]$$

since $(\mathbf{S}' \cdot \mathbf{S}')$ can take eigenvalues $S'(S'+1)$ where the allowed values of S' are obtained by the addition rule for two spin vectors, namely $S' = (S_1 + S_2), (S_1 + S_2 - 1), \ldots, |S_1 - S_2|$. When the system is exposed to an external magnetic field (along the z-axis) the energy levels $E(S')$ are further split into $(2S'+1)$ levels, i.e. the first-order Zeeman term $g\mathrm{M}_{s'}\beta H$ must be added to the energy, where g is the Landé splitting factor and $\mathrm{M}_{s'}$ takes the values S', $(S'-1), \ldots, (-S'+1), -S'$. The effective magnetic moment, μ_{eff} (B.M.), of the binuclear cluster is obtained by summing the squares of the individual moments, $\mu^2(S')$, over all spin levels suitably weighted by the appropriate Boltzman factor. For a cluster containing more than two metal atoms, the spin degeneracy of the cluster is not removed entirely by the interaction V_{ij}. The number of states having a given value of S' may occur $\omega(S')$ times where $\omega(S') = \Omega(S') - \Omega(S'+1)$ with $\Omega(S')$ being the coefficient of $x^{s'}$ in the expansion of $(x^{s'} + x^{s'-1} + \ldots + x^{-s'+1} + x^{-s'})^n$ and n being the number of paramagnetic centres in the cluster. Then

$$\mu_{eff}^2 = \frac{g^2 \sum\limits_{s'} S'(S'+1)(2S'+1)\,\omega(S')\exp\left(-E(S')/kT\right)}{\sum\limits_{s'}(2S'+1)\,\omega(S')\exp\left(-E(S')/kT\right)} \qquad [24]$$

where $\omega(S') = 1$ for a dimer.

The above procedure can be illustrated by considering a binuclear complex such as the red $[Cr_2(CH_3CO_2)_4(H_2O)_2]$ in which the chromium atoms are separated by 2·64 Å. In this case $S = 2$ (Cr^{2+}, configuration $t_2^3 e^1$) so that the quantum number S' specifying the total spin of the system takes the values 0, 1, 2, 3 and 4. Spin–spin interaction between the pair of chromium atoms removes the 25-fold spin degeneracy of the ground state splitting it into the manifold of five levels shown in Fig. 2, the energies $E(S')$ being calculated from [23]. Figure 2 can be termed a spin state correlation diagram for it shows how the energy of each state $E(S')$

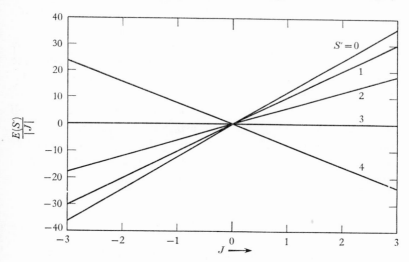

Fig. 2. Spin state correlation diagram for d^4–d^4 dimer.

varies with the sign and magnitude of the interaction, the slope of each line being given by $-[S'(S'+1)-2S(S+1)]$. If the Cr–Cr interaction is antiferromagnetic (i.e. J_{12} negative) the compound should be diamagnetic at 0 °K ($S' = 0$). If the interaction is ferromagnetic (i.e. J_{12} positive), a paramagnetic moment of 6·33 B.M. per Cr^{2+} is expected at this temperature. The temperature dependence of μ_{eff} per Cr^{2+} is obtained by summing $\mu^2(S')$ over these levels according to [24],

$$\mu_{\text{eff.}}^2 = 12 \frac{[e^{2x}+5e^{6x}+14e^{12x}+30e^{20x}]}{[1+3e^{2x}+5e^{6x}+7e^{12x}+9e^{20x}]} \qquad [25]$$

where $x = (J_{12}/kT)$. The corresponding expression for the molar magnetic susceptibility, χ_M, follows from the relation,

$$\chi_M = \frac{N\beta^2}{3kT}\mu_{\mathrm{eff}}^2 \qquad [26]$$

where $N\beta^2/3k = 1/8$.

The general features of the magnetic behaviour of binuclear complexes are best illustrated by plotting either the reduced magnetic susceptibility $[(|J_{12}|\chi)/N\beta^2]$ or its reciprocal as a function of the reduced temperature $kT/|J_{12}| = 1/|x|$ together with the dependence on reduced temperature of the effective magnetic moment. Typical curves are illustrated in Fig. 3a and b for $\frac{1}{2} \leqslant S \leqslant \frac{5}{2}$. The temperature T_c at which the maximum in susceptibility occurs for negative values of J_{12} is seen to increase as S increases. Providing T_c can be determined experimentally, the spin–spin coupling constant J_{12} can be determined from the relations summarized in Table 1. If the cluster is thermally stable, measurements of χ_M can be made in regions of temperature well above T_c where the $\chi_M^{-1}(T)$ curve becomes linear and obeys the Curie–Weiss law. The Weiss constant θ (which is best determined[22] by plotting $(\chi T)^{-1}$ against $(T)^{-1}$ and noting that $T^{-1} \to \theta^{-1}$ as $(\chi T)^{-1} \to 0$) is directly proportional to J_{12}, namely

$$\theta = \frac{2S(S+1)J_{12}}{3k} \qquad [27]$$

TABLE 1. *Relation between T_c, θ and J_{12} for binuclear clusters*

S	$J_{12} = -mkT_c$	$\chi_M \propto \dfrac{1}{T-\theta}$
1/2	$J_{12} = -1 \cdot 247 kT_c*$	$\theta = 1/2 J_{12}/k$
1	$J_{12} = -2 \cdot 048 kT_c$	$\theta = 4/3 J_{12}/k$
3/2	$J_{12} = -3 \cdot 087 kT_c$	$\theta = 5/2 J_{12}/k$
2	$J_{12} = -4 \cdot 328 kT_c$	$\theta = 4 J_{12}/k$
5/2	$J_{12} = -5 \cdot 761 kT_c$	$\theta = 35/6 J_{12}/k$

* The factors of $-1 \cdot 6$ and $-1 \cdot 193$ originally given in reference nos. 23 and 24 are slightly in error.

and can therefore be used to estimate the magnitude of either ferromagnetic or antiferromagnetic coupling. Unfortunately,

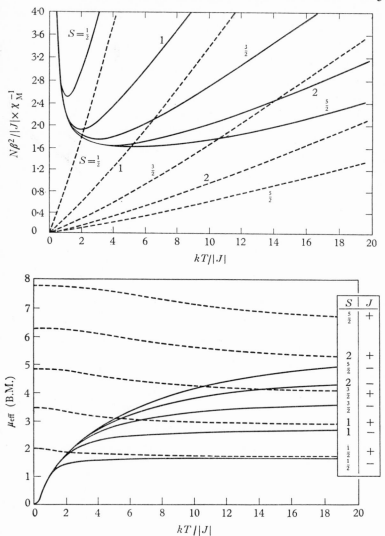

Fig. 3. (a) Reciprocal magnetic susceptibility per metal atom versus temperature. Full curves, J negative; broken curves, J positive. (b) Magnetic moment versus temperature per metal atom. Full curves, J negative; broken curves, J positive.

thermal decomposition of inorganic clusters in the temperature range $T \gg T_c$ often precludes the desired experimental determination of θ and thence J_{12}.

3.2. Trinuclear clusters

Clusters which contain three paramagnetic centres can be divided into three geometric classes. Extreme situations occur when three metal ions form either an equilateral triangle (D_{3h}) or a linear array ($C_{\infty v}$). The intermediate geometrical configuration is an isosceles triangle (C_{2v}). The magnetic properties of such systems can be deduced by extending the procedure described for binuclear compounds. The effects of nearest-neighbour isotropic spin–spin interaction may be obtained by generalizing the Hamiltonian [21],

$$\mathscr{H}_{ex} = -2[J_{12}(\mathbf{S}_1 \cdot \mathbf{S}_2) + J_{23}(\mathbf{S}_2 \cdot \mathbf{S}_3) + J_{31}(\mathbf{S}_3 \cdot \mathbf{S}_1)] \qquad [28]$$

Frequently, at least two of the interactions will be the same, so that we can write $J_{12} = J_{23} = J$ and

$$\mathscr{H}_{ex} = -2J[(\mathbf{S}_1 \cdot \mathbf{S}_2) + (\mathbf{S}_2 \cdot \mathbf{S}_3)] - 2J_{31}(\mathbf{S}_3 \cdot \mathbf{S}_1) \qquad [29]$$

By writing $\mathbf{S}' = \mathbf{S}_1 + \mathbf{S}_2 + \mathbf{S}_3$ and $\mathbf{S}^* = \mathbf{S}_3 + \mathbf{S}_1$ we have

$$(\mathbf{S}_1 + \mathbf{S}_2 + \mathbf{S}_3)^2 = (\mathbf{S}' \cdot \mathbf{S}') = 3S(S+1) + 2(\mathbf{S}_1 \cdot \mathbf{S}_2) \\ + 2(\mathbf{S}_2 \cdot \mathbf{S}_3) + 2(\mathbf{S}_3 \cdot \mathbf{S}_1)$$

and
$$(\mathbf{S}_3 + \mathbf{S}_1)^2 = (\mathbf{S}^* \cdot \mathbf{S}^*) = 2S(S+1) + 2(\mathbf{S}_3 \cdot \mathbf{S}_1)$$

and the Hamiltonian [29] can be rewritten,

$$\mathscr{H}_{ex} = -J[(\mathbf{S}' \cdot \mathbf{S}') - (\mathbf{S}^* \cdot \mathbf{S}^*) - S(S+1)] \\ - J_{31}[(\mathbf{S}^* \cdot \mathbf{S}^*) - 2S(S+1)] \qquad [30]$$

The quantum number S^* for the spin operator $\mathbf{S}^*$ takes the values permitted by the addition rule for two spin vectors, namely

$$S^* = (S_3 + S_1), (S_3 + S_1 - 1), \ldots, |S_3 - S_1| \\ = 2S, 2S - 1, \ldots, 1, 0.$$

Likewise the allowed values of S' for a given value of S^* are

$$S' = (S^* + S), (S^* + S - 1), \ldots, |S^* - S|$$

The energy levels under the Hamiltonian [30] are

$$E(S', S^*) = -J[S'(S'+1) - S^*(S^*+1) - S(S+1)] \\ - J_{31}[S^*(S^*+1) - 2S(S+1)] \qquad [31]$$

since $(\mathbf{S}' \cdot \mathbf{S}')$ and $(\mathbf{S}^* \cdot \mathbf{S}^*)$ can take eigenvalues $S'(S'+1)$ and

$S^*(S^*+1)$. For a triangular cluster $J_{31} = J$ and the energy levels are given by

$$E(S') = -J[S'(S'+1)-3S(S+1)] \qquad [32]$$

For a linear triad, with no interaction between the end members ($J_{31} = 0$), the manifold of spin levels is given by

$$E(S', S^*) = -J[S'(S'+1)-S^*(S^*+1)-S(S+1)] \qquad [33]$$

The relative ordering of the spin levels under the Hamiltonian [30] can be readily visualized for any d-configuration by using a spin correlation diagram (see Fig. 4) in which the energies $E(S', S^*)$ in

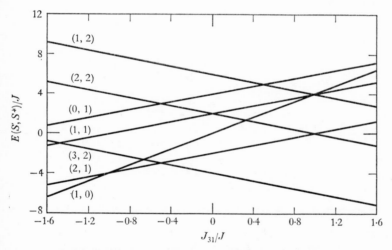

Fig. 4. Spin state correlation diagram for d^2–d^2–d^2 trimer.

units of J are plotted as a function of the ratio J_{31}/J (where J_{31} and J can, in principle, take positive and negative values). As an illustrative example, we can consider the trinuclear cluster[25]

$$[\text{Ni}_3(\text{acetylacetone})_6]$$

in which the d^8 configuration of Ni^{2+} leaves each metal atom with two unpaired e_g-electrons ($S = 1$). The total spin degeneracy of the triad is then $(2S+1)^3 = 27$. If the metal atoms form an equilateral triangle with $J_{31}/J = 1\cdot0$ (see Fig. 4), the metal–metal interaction partially lifts this degeneracy to give four new levels $E(S')$ characterized by the quantum number $S' = 0$, 1, 2 and 3,

TABLE 2. *Permitted values of the quantum numbers S, S* and S' for trinuclear clusters*

S	S*	S'								
1/2	1	—	1/2	3/2	—	—	—	—	—	—
	o	—	1/2	—	—	—	—	—	—	—
1	2	—	1	2	3	—	—	—	—	—
	1	o	1	2	—	—	—	—	—	—
	o	—	1	—	—	—	—	—	—	—
3/2	3	—	—	3/2	5/2	7/2	9/2	—	—	—
	2	—	1/2	3/2	5/2	7/2	—	—	—	—
	1	—	1/2	3/2	5/2	—	—	—	—	—
	o	—	—	3/2	—	—	—	—	—	—
2	4	—	—	2	3	4	5	6	—	—
	3	—	1	2	3	4	5	—	—	—
	2	o	1	2	3	4	—	—	—	—
	1	—	1	2	3	—	—	—	—	—
	o	—	—	2	—	—	—	—	—	—
5/2	5	—	—	—	5/2	7/2	9/2	11/2	13/2	15/2
	4	—	—	3/2	5/2	7/2	9/2	11/2	13/2	—
	3	—	1/2	3/2	5/2	7/2	9/2	11/2	—	—
	2	—	1/2	3/2	5/2	7/2	9/2	—	—	—
	1	—	—	3/2	5/2	7/2	—	—	—	—
	o	—	—	—	5/2	—	—	—	—	—

these levels being $\omega(S') =$ 1-, 3-, 2- and 1-fold degenerate, respectively (cf. Table 2). In fact, the X-ray structure analysis[25] shows that the triad of Ni atoms forms a linear array, and under the lower symmetry, the spin degeneracies are lifted further to leave six residual levels (if $J_{31}/J = 0$), each being characterized by the appropriate value of the quantum numbers S' and S^*. The variation of the energy of each $E(S', S^*)$ level with the ratio J_{31}/J is simply $-[S^*(S^*+1) - 2S(S+1)]$. Several other features of the spin correlation diagram for the Ni_3-system should be noted.

(i) The order of levels given in Fig. 4 corresponds to positive values of J; for negative values of J, the diagram should be inverted.

(ii) For a negative value of J with $J_{31} = 0$, the ground state of a linear trimer is not diamagnetic as has been implied by some authors.[26] The triplet spin state $(S', S^*) = (1, 2)$ lies lowest and, in fact, the diamagnetic level $(S', S^*) = (0, 1)$ can only become the ground state for values of $J_{31}/J > 0·5$. This result is not unexpected for linear molecules, and a consideration of simple spin coupled structures of the type

$$\uparrow \text{Ni} \uparrow\downarrow \text{Ni} \uparrow\downarrow \text{Ni} \uparrow$$

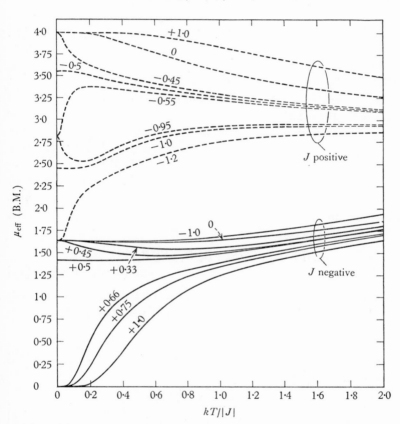

Fig. 5. Magnetic moment per metal atom versus temperature for d^2–d^2–d^2 trimer. Values of J_{31}/J are marked.

demonstrates that, if the end members are uncoupled, the ground state must be paramagnetic. When the (0, 1) and (1, 2) levels cross at $J_{31}/J = 0·5$, the limiting moment as $T \rightarrow \infty$ will be $\mu_{\text{eff}}^2 = 6(\text{B.M.})^2$ per Ni$_3$ or 1·41 B.M. per Ni atom.

(iii) For positive values of J, the spin septet (3, 2) lies lowest for $J_{31}/J > -0.5$. For $-1.0 < J_{31}/J < -0.5$ the quintet level (2, 1) is lowest, and for $J_{31}/J < -1.0$ a spin triplet (1, 0) is the ground state of the trimer.

The frequent crossing of spin levels as the parameter J_{31}/J is varied leads to a great variety of susceptibility and moment curves which often have an unusual dependence on temperature. The exceptional nature of these curves is illustrated in Fig. 5 in which μ_{eff}/Ni is plotted as a function of the reduced temperature $kT/|J|$.

3.3. Tetranuclear clusters

Recent structure determinations of several paramagnetic clusters have confirmed their tetranuclear nature. Among these are the tetrahedral compounds $[Cu_4OCl_6(Ph_3PO)_4]$[27] and possibly $[Co_4O(Me_3CCO_2)_6]$,[28] and linear $[Co_4(acac)_8]$.[29] For tetranuclear clusters involving a square (D_{4h}), tetrahedral (T_d) or intermediate configuration (D_{2d}) of the four metal atoms, the energies of the spin levels can be estimated by a further extension of the procedures outlined in the previous sections. For the square configuration, the Hamiltonian becomes

$$\mathscr{H}_{ex} = -2J[\mathbf{S}_1\cdot\mathbf{S}_2 + \mathbf{S}_2\cdot\mathbf{S}_3 + \mathbf{S}_3\cdot\mathbf{S}_4 + \mathbf{S}_4\cdot\mathbf{S}_1] \\ -2J_1[\mathbf{S}_2\cdot\mathbf{S}_4 + \mathbf{S}_1\cdot\mathbf{S}_3] \quad [34]$$

where J is the spin–spin interaction constant along the sides and J_1 that along the diagonals of the square. If we write

$$\mathbf{S}' = \mathbf{S}_1 + \mathbf{S}_2 + \mathbf{S}_3 + \mathbf{S}_4$$

$\mathbf{S}^* = \mathbf{S}_1 + \mathbf{S}_3$ and $\mathbf{S}^\dagger = \mathbf{S}_2 + \mathbf{S}_4$, it can be shown that the required eigenvalues for D_{4h} and D_{2d} are given by

$$E(S', S^*, S^\dagger) = -J[S'(S'+1) - S^*(S^*+1) - S^\dagger(S^\dagger+1)] \\ -J_1[S^*(S^*+1) + S^\dagger(S^\dagger+1) - 4S(S+1)] \quad [35]$$

For the tetrahedral configuration, $J_1 = J$ and

$$E(S') = -J[S'(S'+1) - 4S(S+1)] \quad [36]$$

The allowed values of the quantum numbers S', S^* and $S^\dagger$ are determined by the addition rule for the spin vectors. Spin correlation diagrams for tetramers (the case of $S = \frac{1}{2}$ is illustrated in

Fig. 6) enable the general pattern of magnetic behaviour to be determined by inspection. The linear arrangement of four metal ions is rather less tractable since four coupling constants are now required to describe properly the metal–metal interactions.

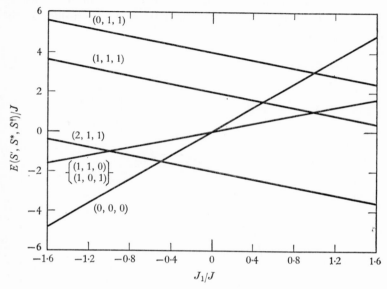

Fig. 6. Spin-state correlation diagram for d^1–d^1–d^1–d^1 tetramer.

4. Mechanisms of spin–spin coupling

For a system involving two interacting electrons we have seen that the combined action of the Pauli antisymmetry requirement and electrostatic interaction differentiates the energies of the singlet and triplet states, the relative stability of the two states being determined by the sign of J. In the previous section we have commented that the simple Hamiltonian [17] usually provides a good phenomenological description of metal–metal interactions in isolated clusters. However, it offers no information concerning the mechanism by which the spins couple. Before outlining some of the more recent ideas of exchange interaction, we comment briefly on historical speculations about the sign of J.

4.1. Historical speculations

Heisenberg considered that although J was negative for the H_2 molecule, it should change sign beyond some critical ratio of the internuclear separation to the mean radial extension of the atomic orbitals. That this possibly exists may be seen by writing J in the more explicit form

$$J = \frac{e^2}{r_{ab}} S_0^2 - 2S_0 \int \frac{e^2}{r_{b1}} \phi_a(1) \, \phi_b(1) \, d\tau_1$$
$$+ \iint \phi_a(1) \, \phi_b(2) \frac{e^2}{r_{12}} \phi_b(1) \, \phi_a(2) \, d\tau_1 d\tau_2 \qquad [37]$$

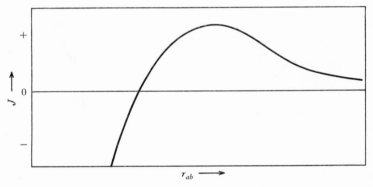

Fig. 7. Bethe's[30] curve relating exchange parameter J to the distance r_{ab}, between magnetic centres.

Clearly if the wave functions Φ_a and Φ_b have no nodes in the region of appreciable overlap so that the integrands are positive, then J will be positive if the contributions from the first and last terms of [37] exceed that from the middle term. The term in e^2/r_{12} will be dominant if the amplitudes of the wave functions are concentrated midway between the nuclei, since only in this region is r_{12} small. The terms $-e^2/r_{b1}$ (and $-e^2/r_{a2}$) are minimized for wave functions that have small radial extension compared to the internuclear spacing r_{ab}.

These conditions, first formulated by Sommerfeld and Bethe,[30] imply that the sign of the exchange integral should vary with inter-nuclear separation as shown schematically in Fig. 7. The $3d$ and $4f$

wave functions for the atoms of the transition and rare earth elements comply best with these requirements, and indeed cobalt, nickel, and gadolinium are ferromagnetic, which corresponds to positive values of J. More recently, the validity of Bethe's arguments has been questioned and the use of more refined wave functions in explicit calculations suggest that the curve of Fig. 7 is much oversimplified and may even be incorrect.[31]

The magnitude of the exchange integral will be numerically greatest when the integrand of [37] is large; that is, when the angular lobes of ϕ_a and ϕ_b are directed towards and overlapping one another so that the product $\phi_a \phi_b$ is large.

For orbitals which are truly orthogonal, the overlap integral $S_0 = 0$ and J reduces to the term in e^2/r_{12}, which is the 'true' positive exchange integral of Heisenberg, which forms the basis of the theory of ferromagnetism. The ferromagnetic sign of intra-ionic coupling arises in the same way and is the physical origin of Hund's rule. However, once the orbitals are no longer orthogonal, the terms in e^2/r_{b1} and e^2/r_{a2} become finite and can lead to negative value of J for situations in which the two electrons experience an attractive potential. In fact, the Heitler–London[32] scheme of chemical binding is based on just this attractive interaction between two electrons spinning antiparallel in non-orthogonal wave functions.

Since J is so sensitive to orbital overlap, it is safe to infer that in clusters, metal–metal interactions will be confined normally to neighbouring metal atoms, except when the ligands are polyfunctional and span other metal atoms which are more remote (e.g. $[Ni_3(acac)_6]$ in § 5.2.4).

4.2. Origin of superexchange

Metal–metal interactions in transition metal compounds frequently occur when the separation between the paramagnetic centres is much greater than the sum of their covalent radii. Because of the relatively large distances involved (> 4 Å) this type of interaction is universally known as 'superexchange' although Van Vleck[33] considered that the term 'indirect exchange' might be more appropriate. In these situations the interacting metal atoms are invariably screened from each other by anions, radicals, or molecules which are diamagnetic in their ground states. The problem to be answered is by what mechanism does the intervening ligand transmit the interaction between spins centred on neighbouring cations? The diamagnetic shielding effect of the closed shell structure can be removed by invoking an excited paramagnetic state of the anion and, in fact, Kramers's[15] original spin-coupling mechanism assigned a weight in the total ground-state wave function of the system to configurations in which the anions had some paramagnetism. The many recent observations of hyperfine interaction between ligand nuclear spin and electron spin of the magnetic ion amply confirm that ligand wave functions can acquire magnetic character. A more recent interpretation of Kramers's idea, which differs qualitatively from his original concept, views superexchange as arising from direct overlap of cation orbitals which have been 'expanded' by the intervening anion. In other words, the role of the latter is to provide a common orbital, which when suitably admixed with simple d orbitals, gives new antibonding cation orbitals which can then interact directly.

It is not the purpose of the present article to attempt to review the formal mathematical machinery currently employed to describe the various processes which give rise to superexchange; a comprehensive account has been given recently by P. W. Anderson.[34] On the other hand the system of semi-empirical rules which has evolved in the period since 1950 has proved to be generally applicable to many systems of interest to chemists, and the present discussion is restricted to a presentation of these qualitative ideas.[35-38]

The underlying mechanism[15, 16] behind the superexchange process can be illustrated by considering a simple perovskite crystal such as $KNiF_3$ in which Ni^{2+} cations (configuration d^8) are separated by a diamagnetic fluoride anion which lies on the Ni–Ni axis.[39, 40] For simplicity, we suppose that only three orbitals and four electrons are important: these are d_{z^2} (1 electron) centred on each Ni^{2+} and $2p_z$ (2 electrons) centred on F^-. The 'ground' orbital state (ionic configuration) is a spin singlet (cf. Fig. 8) if the Ni^{2+} spins are antiparallel (a) or a spin triplet if parallel (b). We now look for effects produced by mixing small amounts of excited states into the ground states (a) and (b). Anderson[16] originally considered the simplest such modification, namely the transfer of one electron from $2p_z$ on F^- into the half-filled d_{z^2} orbital on one Ni^{2+} to obtain excited orbital singlets as in (c) or triplets as in (d). Essentially this amounts to a configurational mixing of states like $Ni^+ F Ni^{2+}$, etc. In oxide systems like NiO, both electrons might also be transferred simultaneously to give the state $Ni^+ O Ni^+$ represented by (e), although it is considered unlikely[34] that the 'double' electron transfer process would be the dominant mechanism. The lone electron residing on F in (c) and (d) can now ex-

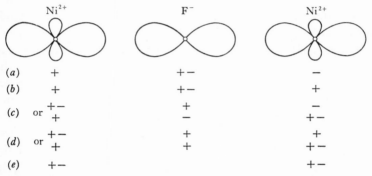

Fig. 8. Typical orbitals participating in a 180° superexchange pathway between two Ni^{2+} ions (d_{z^2}) via a F^- ion (p_z) as in $KNiF_3$.

change couple with the lone $3d_{z^2}$ electron on an adjoining Ni^{2+}. The preferred sign for this coupling is antiferromagnetic, since the orbitals are not orthogonal so that the singlet states (c) are lower in energy than the triplet states (d). Accordingly, since,

$$\Psi_{total}(S_{total}) = a\Psi_{ionic}(S_{total}) + b\Psi_{excited}(S_{total})$$

this configurational mixing leads to stabilization of the singlet state (a) with respect to the triplet (b). Quantitatively, Anderson has demonstrated that (a) and (b) become differentiated in energy by a quantity of the order of magnitude $b_{ij}^2 J/U^2$, where b_{ij} is the one-electron transfer or 'hopping' integral connecting (a) to (c), J is the exchange interaction between a $2p_z$ electron remaining on the anion and a $3d_{z^2}$ electron on Ni^{2+}, and U is the energy of (c) relative to (a). Although it is not immediately obvious, this coupling, although not quite the same as the usual exchange, is equivalent to an $\mathbf{S_i} \cdot \mathbf{S_j}$ coupling.

Even though the above discussion is qualitative, several pertinent observations can be made. The magnitude of the interaction will be closely related to the size of the orbital overlap since the transfer of an electron between two centres will be facilitated under these conditions. Since the one-electron transfer process couples spins belonging to $3d$ orbitals which overlap strongly with a single p-orbital, the superexchange process might be expected to reach a maximum when the metal–anion–metal configuration is $180°$. It turns out that this is not always so, for appreciable admixture of metal d_{xy} with anion p_π orbitals can lead to $90°$ interactions of a comparable order of magnitude. Admixture with anion s-orbitals must also be taken into account.[34]

The sign of the superexchange mechanism must (as we will see in § 4.3) be intimately connected both with orbital occupancy and the symmetry properties of the electron orbitals centred on the metal atoms and the intervening anion. For example, if one of the states to which electron transfer takes place is completely empty the interaction vanishes.[17] However, if there are other partially occupied orbitals on the metal atom, intra-atomic coupling with the transferred electron gives a ferromagnetic effect (see § 4.3, case (b)).

Another interesting consequence of the theory is that the transfer integral and hence the coupling should become larger as the electronegativity of the anion decreases, namely

$$J_{\mathrm{F}^-} < J_{\mathrm{O}^{2-}} < J_{\mathrm{S}^{2-}} < J_{\mathrm{Se}^{2-}}$$

In agreement with this deduction, it is usually found that exchange coupling between metal atoms bridged by Br^- (e.g. the linear antiferromagnet, $CuBr_2$) is greater[10] than that in the corresponding chloride (e.g. $CuCl_2$).

More recently, Anderson[17] has reformulated and refined the above model for the superexchange process. Bearing in mind that the magnetic wave functions originally localized on metal atoms are expanded over rather long distances through mixing with anion functions, he has emphasized that there is really no distinction in principle between the classical 'direct exchange' (i.e. direct overlap of magnetic ion wave functions) and the traditional 'superexchange' (i.e. exchange through nominally non-magnetic groups) as outlined above. In other words, this newer viewpoint no longer considers the direct transfer of an electron from the diamagnetic to the paramagnetic ion as a necessary precursor to exchange coupling, but adopts the view that the transfer is direct, all the way from cation to cation via expanded d (or f) wave functions (cf. Fig. 8e). After a careful evaluation of the relative magnitudes of the many mechanisms postulated for exchange coupling, Anderson concludes that only three spin-dependent mechanisms are of much quantitative significance (although it should be added that not all authors may agree with his conclusions). These are:

A. *Superexchange (or 'kinetic' exchange)*. The kinetic energy, or desire of electrons to delocalize themselves, promotes electron transfer between magnetic ions provided their d-orbitals overlap; that is 'incipient' chemical bond formation. Since the spins must be aligned antiparallel by the Pauli principle, the effect is always antiferromagnetic.

B. *Direct exchange (Heisenberg or 'potential' exchange)*. This is the e^2/r_{12} term of equation [37], which is always positive and represents the repulsive interelectronic potential energy. This ferromagnetic term is generally small compared with superexchange when the latter is present. However, it can become the major effect between two orbitals which are orthogonal by symmetry (e.g. coupling between a d_{z^2} orbital on one cation and a d_{xz} on its neighbour).

C. *Spin polarization ('indirect' exchange)*. This indirect effect arises because the solution of the unrestricted Hartree–Fock equation for ligand electrons of α-spin differs from the solution for those of β-spin, when in the presence of unpaired electrons on the cations. The effect is antiferromagnetic since, in a physical sense, each cation attracts ('polarizes') those ligand electrons which spin parallel towards its own spin. Anderson has estimated that the resulting exchange effect must be quite small; in fact, even when superexchange vanishes by orthogonality, spin polarization exchange will be swamped by direct exchange. Accordingly, it can be ignored in the qualitative discussions which follow.

From this more recent viewpoint, the three situations which are most commonly encountered in transition metal clusters, may be described as follows:

(*a*) *Overlap of two half-filled orbitals*. In the simplest model, the transfer of an electron from one cation to its neighbour gives an excited state (spins antiparallel) for the system of energy U above the ground state. The so-called transfer integral b_{ij} is proportional to the d-orbital overlap and must operate twice to return the system to its ground state. Since it carries the electron without change of spin, the ground state (spins antiparallel) is depressed relative to the ground state (spins parallel), which determines that the effect is always antiferromagnetic. The difference in energy between the singlet and triplet states of the system is

$$\Delta E \text{ (triplet–singlet)} = \frac{4b_{ij}^2}{U} \qquad [38]$$

which is equivalent to an antiferromagnetic exchange interaction between the two electrons $(s_1 = s_2 = \frac{1}{2})$ of

$$-2J_{12}\mathbf{s}_1\cdot\mathbf{s}_2 = \frac{4b_{ij}^2}{U}\,\mathbf{s}_1\cdot\mathbf{s}_2 \qquad [39]$$

so that the exchange coupling constant for the cation–cation interaction is

$$J_{12} = -2\frac{b_{ij}^2}{U} \qquad [40]$$

For a cluster with more than one unpaired d-electron per cation, the exchange parameter becomes

$$-2J\mathbf{S}_1\cdot\mathbf{S}_2 = \frac{s^2}{S^2}\frac{4b_{ij}^2}{U}\,\mathbf{S}_1\cdot\mathbf{S}_2 \qquad [41]$$

Since the transfer integral is proportional to the orbital overlap, it must increase exponentially with decreasing cation–cation separation. Anderson estimates that this spin-dependent effect should be associated with J_{ij} values of 200–1000 °K.

(b) *Overlap of a half-filled and an empty orbital.* The simple electron transfer between neighbouring cations is spin-independent in this situation. However, if there is a non-overlapping or orthogonal, partly occupied orbital centred on the acceptor cation a non-vanishing interaction results. For example, we can consider the coupling between $Ni^{2+}(t_{2g}^6 e_g^2)$ and $V^{2+}(t_{2g}^3 e_g^0)$ pairs. If the transfer integral carries an e_g electron from Ni^{2+} to the vacant e_g orbital of V^{2+} with its spin parallel to those of the t_{2g} electrons of V^{2+}, the system is stabilized by ferromagnetic intra-ionic exchange coupling within the V^{2+} ion, namely

$$J = +\frac{2b_{ij}^2}{4S^2U} \times \frac{J_{intra}}{U} \qquad [42]$$

This ferromagnetic superexchange coupling is weaker than the previous antiferromagnetic effect by the factor J_{intra}/U or by a factor of about 5–10. Hence, we might expect it to be discernible only when the latter is absent and under conditions where b_{ij} is large and U is small.

(c) *Overlap of a half-filled and a full orbital.* Suppose we now consider the transfer of a t_{2g} electron from Ni^{2+} to the half-filled t_{2g} subset of V^{2+}. The spin of the transferred electron must be anti-parallel to the t_{2g} spins on V^{2+}. Intra-ionic exchange coupling between the e_g and remaining t_{2g} electrons on Ni^{2+} ensures that the atomic moments of V^{2+} and Ni^{2+} are coupled ferromagnetically as in case (b).

We conclude this section by noting that several workers have proposed that a substantial contribution to superexchange can arise from electron correlation effects between the bridging ligand and its neighbouring cations.[41–44] This mechanism assumes that each of the electrons in a single ligand orbital is transferred simultaneously, one to each cation, as in Fig. 8e. This correlation effect (simultaneous incipient bond formation on each side of the ligand) is antiferromagnetic and closely resembles the more recent super-exchange mechanism outlined above. Anderson[34] has claimed that the magnitude of the correlation effect has been greatly over-estimated and that it contributes less to exchange than even spin polarization. The empirical evidence for the mechanism is also contentious. Accordingly, while correct in principle, the status of this effect remains in doubt at present.

It will be apparent to the reader at this stage that the net exchange coupling in any system will be the sum of a multiplicity of superexchange and direct exchange terms appropriate for the compound under study.

4.3. Metal–metal interactions

We have seen that the orthogonality relationship between d-orbitals of a cation and the occupied orbitals of the intervening anion are of prime importance for determining, in the first place, whether transfer of an electron between the metal and anion can be

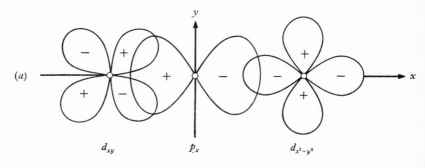

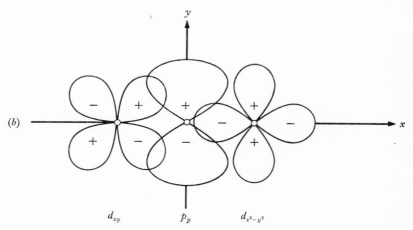

Fig. 9. Symmetry relations between (a) t_{2g}, e_g and p_σ; (b) t_{2g}, e_g and p_π orbitals orientated at 180°.

effected, and secondly, the sign of the exchange integral between the d-orbital and the anion orbital. The more important symmetry relationships are those illustrated in Fig. 9.

The p_σ (e.g. p_x) and the t_{2g} (e.g. d_{xy}) orbitals are orthogonal (cf. Fig. 9a) so that the net overlap is zero and the electron transfer from $p_\sigma \to t_{2g}$ cannot occur. On the other hand, the metal e_g (e.g. $d_{x^2-y^2}$) and anion p_σ (e.g. p_x) orbitals are not orthogonal and a $p_\sigma \to e_g$ pathway is available for electron transfer. Inspection of Fig. 9b reveals that the anion p_π orbital is orthogonal to the metal e_g orbitals but non-orthogonal to one member of each t_{2g} subset. Usually, electron transfer from anion to metal via the $p_\pi \to t_{2g}$ pathway should be less favourable energetically than $p_\sigma \to e_g$ owing to the smaller orbital overlap.

The s-orbital on the anion can also take part in the superexchange interaction. Thus the s and e_g orbitals are non-orthogonal, while the s and t_{2g} orbitals are orthogonal. However, since the s orbitals are lower in energy than the p_σ, electron transfer from the anion to the cation via the $s \to e_g$ pathway is likely to be less favourable.

Hybridization of s and p orbitals on the anion, and s, p and d orbitals on the cation will not affect the orthogonality relationships described above although the magnitude of the overlap integrals will be sensitive to these mixings.

To obtain a schematic representation of orbital and spin configurations, it is convenient[36] to subdivide an orbital set of quantum number l into $(2l+1)$ levels of α or positive spin and $(2l+1)$ levels of β or negative spin, each set being separated by the mean pairing energy π. For example, the ten spin-orbital levels of the $3d$ set in O_h symmetry can be represented schematically as in Fig. 10. Using this representation, the $^6A_1$ state of high-spin $[Mn(H_2O)_6]^{2+}$ has the configuration $(t_{2g}^-)^3(e_g^-)^2$ since the crystal field parameter $\Delta < \pi$ for the H_2O ligand (cf. Fig. 10c). Conversely, the $^2T_{2g}$ state of low-spin $[Mn(CN)_6]^{4-}$ can be written $(t_{2g}^-)^3(t_{2g}^+)^2$ since $\Delta > \pi$ for the CN^- anion (cf. Fig. 10d), the $+$ or $-$ superscript referring to the sign of the spin in the d-orbital. It will be recalled that the main configuration-dependent contribution to π is the intra-atomic exchange energy which favours a parallel alignment of spins within the atom.

We will now discuss several examples of metal–metal interaction to illustrate the dominant mechanisms which determine the sign and size of J. The discussion is phrased in terms of the original Kramers–Anderson concept in order to stress the relevant symmetry relationships between anions and cation orbitals. No attempt will be made to itemize every interaction which is possible for a particular system and the weaker effects such as polarization,[45–47] correlation,[41–44] anisotropic superexchange,[48] cation–anion–anion–cation interactions,[31] etc., are ignored.[34] It is customary and convenient to distinguish between two limiting situations: (*a*) exchange interaction between octahedrally coordinated cations with corner-sharing of the octahedra: so-called the 180° case; (*b*) exchange interaction between octahedrally coordinated cations with edge-sharing of the octahedra: so-called the 90° case.

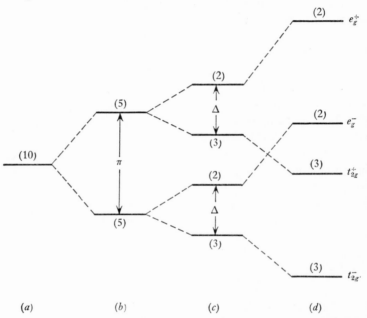

Fig. 10. Schematic energy level diagram for 3*d* orbitals in a cubic ligand field.

4.3.1. 180° Superexchange

(*a*) *Configuration* d^3/d^3. Consider the case of the polynuclear cation $[(NH_3)_5CrOCr(NH_3)_5]^{4+}$. Each Cr^{3+} ion is octahedrally coordinated with the O^{2-} ion being situated at the centre of the line connecting them (say, the z-axis). The $p_\sigma – e_g$ pathway (i.e. $p_z – d_{z^2}$) enables an excited state of the system to be formed by transferring an electron from O^{2-} to say the Cr^{3+} on the left of Fig. 11(*a*). Since the exchange interaction within this Cr^{3+} (i.e.

intra-ionic exchange) favours a parallel alignment of the four spins, the unpaired electron remaining on the O^{2-} ion spins in the opposite sense. It can likewise be transferred to the right-hand Cr^{3+} ion by a $p_\sigma - e_g$ pathway and couple ferromagnetically with the three t_{2g} of electrons of this cation. The resulting superexchange interaction is antiferromagnetic, the ground state for the binuclear system being $(t_{2g}^-)^3(t_{2g}^+)^3$. Interestingly, Anderson's more recent

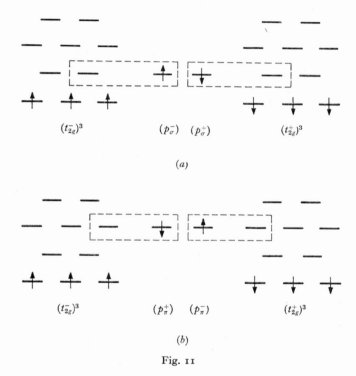

(a)

(b)

Fig. 11

interpretation of superexchange would give no σ-interaction in this case, since the e_g orbitals of both cations are empty.

In addition, electron transfer can be effected via the $p_\pi - t_{2g}$ pathway, although the smaller overlap here should lower the π contribution to superexchange. Mixing between two d_{yz} and a p_y or two d_{xz} and a p_x orbitals leads to the antiferromagnetic coupling shown in Fig. 11(b). At the same time there are direct and probably weaker ferromagnetic couplings on account of the orthogonality between some members of the t_{2g} sets (e.g. d_{xz} on one cation and d_{yz} on the other). The sum of these interactions should lead to a weak antiferromagnetic coupling between d^3 cations.

(b) *Configuration d^3/d^8*. Although no compounds of the type $[(NH_3)_5NiOCr(NH_3)_5]^{3+}$ have been reported, they should be of considerable interest since a moderate ferromagnetic interaction

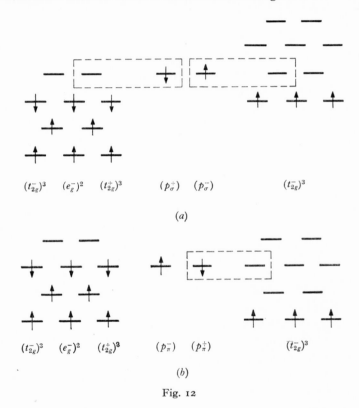

Fig. 12

between the metal ions is expected. The $p_\sigma - e_g$ pathways are shown diagrammatically in Fig. 12(a). Electron transfer from O^{2-} to Ni^{2+} is to an (e_g^+) orbital since the (e_g^-) orbitals are fully occupied. The (p_σ^-) electron remaining on O^{2-} couples ferromagnetically with the t_{2g} electrons on Cr^{3+} as described in (a) so that the resultant superexchange interaction between Ni^{2+} and Cr^{3+} is ferromagnetic.

Electrons cannot be transferred from O^{2-} to Ni^{2+} via a $p_\pi - t_{2g}$

pathway since the t_{2g} orbitals are fully occupied. However, transfer to Cr^{3+} can be effected as shown in Fig. 12(b). The spin remaining on the anion after one p_π electron has been transferred to Cr^{3+} couples ferromagnetically with the e_g electrons on Ni^{2+} since the p_π and e_g orbitals are orthogonal. The sum of the p_σ and p_π super-exchange interactions should couple the Ni^{2+} and Cr^{3+} ions ferromagnetically.

(c) *Configuration d^6/d^5.* The sign of the superexchange interaction will be sensitive to the ligand field strength for octahedrally co-ordinated cations with the configurations d^4, d^5, d^6, and d^7. For example, a linear high-spin d^6/d^5 system such as $Co^{3+}O^2-Co^{4+}$ should couple antiferromagnetically, the interaction via the σ-pathway being reinforced by that via the π-pathway as shown in Fig. 13(a). However, if $\Delta > \pi$ for the tetrapositive cation, the coupling in the system $Co^{3+}O^2-Co^{IV}$ can become ferromagnetic providing the σ-coupling (ferromagnetic) dominates the π-coupling (antiferromagnetic) as shown in Fig. 13(b) and (c).

The signs and qualitative estimates of the size of 180° cation–cation interactions can be derived in this fashion for all combina-tions of electron configurations. Extensive tabulations of 180° super-exchange couplings can be found in references nos. 31 and 37.

4.3.2. The 90° superexchange

The relation between the symmetries of the relevant orbitals when the lines connecting the interacting cations to the intervening anion make an angle of 90° is shown in Fig. 14. If the p_{σ_x} orbital of the intervening anion is orientated to the left-hand cation, it will be non-orthogonal to e_g and orthogonal to the t_{2g} subsets for this cation. However, p_{σ_x} will also be orthogonal to e_g and non-orthogonal to one of the t_{2g} orbitals (d_{xy}) of the cation located at 90° along the y-axis. The inverse situation holds for the p_{σ_y} orbital of the anion so that both the p_{σ_x} and the p_{σ_y} orbitals are simultaneously involved in the superexchange mechanism.

(a) *Configuration d^8/d^8.* The 90° interaction between Ni^{2+} cations is anticipated in compounds like the nickel(II) acetylacetonate trimer where NiO_6-octahedra are joined by sharing faces. Electron transfer can occur (cf. Fig. 15) from oxygen to nickel via the

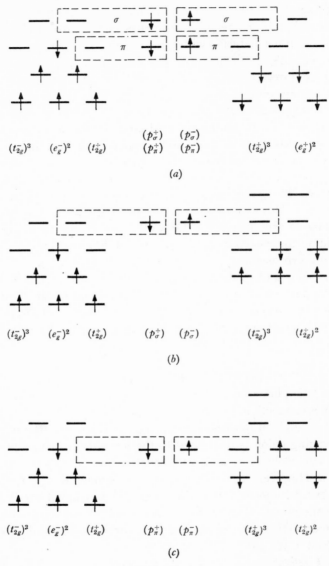

Fig. 13

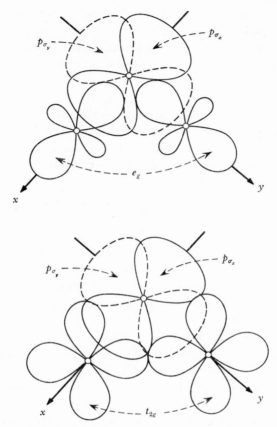

Fig. 14. Symmetry relations between (a) e_g and p_σ; (b) t_{2g} and p_σ orbitals orientated at 90°.

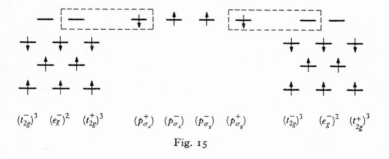

$(t_{2g}^-)^3$ $(e_g^-)^2$ $(t_{2g}^+)^3$ $(p_{\sigma_x}^+)$ $(p_{\sigma_x}^-)$ $(p_{\sigma_y}^-)$ $(p_{\sigma_y}^+)$ $(t_{2g}^-)^3$ $(e_g^-)^2$ $(t_{2g}^+)^3$

Fig. 15

$p_{\sigma_x} \to e_g(Ni_I)$ and $p_{\sigma_y} \to e_g(Ni_{II})$ pathways. Two factors then operate in concert to couple the two Ni atoms ferromagnetically. In the first place, since p_{σ_x} is orthogonal to $e_g(Ni_{II})$ and p_{σ_y} is orthogonal to $e_g(Ni_I)$, the exchange interaction between the remaining unpaired spins on the anion and the half-filled e_g orbitals of the cations will couple ferromagnetically, i.e. to give the configurations $(p_{\sigma_x}^-)(e_g^-)_{II}^2$ and $(p_{\sigma_y}^-)(e_g^-)_I^2$, respectively. Secondly, if electron transfer to each Ni atom is accomplished simultaneously, the unpaired spins produced on oxygen, $p_{\sigma_x}^-$ and $p_{\sigma_y}^-$, will align their spins parallel since these orbitals are orthogonal to one another (intra-ionic coupling). In this case, the filled t_{2g} orbitals play no part in the superexchange interaction.

(b) *Configuration* d^3/d^3. In a $90°$ $Cr^{3+}O^{2-}Cr^{3+}$ system, the $(p_\sigma - e_g)$ pathway is used for electron transfer as in case (a). Intra-ionic coupling on each Cr^{3+} ion will favour a parallel alignment of the spin of the transferred electron with that of the occupied t_{2g}^- set. Since p_{σ_x} is non-orthogonal to t_{2g} (Cr_{II}) and p_{σ_y} is non-orthogonal to t_{2g} (Cr_I), the exchange interaction between the electron remaining in $p_{\sigma_x}^+$ and the occupied t_{2g}^- (Cr_{II}) is negative. Intra-ionic ferromagnetic coupling between $p_{\sigma_x}^+$ and $p_{\sigma_y}^+$ on oxygen further favours a net ferromagnetism (cf. Fig. 16(a)).

The p_z orbital on O^{2-} is correctly orientated to give a π-overlap with d_{xz} on Cr_I and d_{yz} on Cr_{II} to give the antiferromagnetic coupling shown in Fig. 16(b). Inspection of Fig. 14(b) shows that antiferromagnetic coupling can be strengthened by direct overlap of d_{xy} orbitals if the Cr–Cr separation is not too great. Accordingly, the sign of the resultant exchange in d^3/d^3 systems is difficult to predict.

(c) *Configuration* d^8/d^3. The $90°$ interaction between a cation with a more-than-half-filled d-shell and a cation with a less-than-half-filled d-shell is expected to be antiferromagnetic contrary to the $180°$ case. Electron transfer via the σ-pathways results in anti-ferromagnetic superexchange as shown in Fig. 17. This interaction is reinforced by antiferromagnetic coupling between the remaining electron $(p_{\sigma_x}^-)$ and the occupied $(t_{2g}^+)^3$ set on Cr_{II}^{3+} as shown.

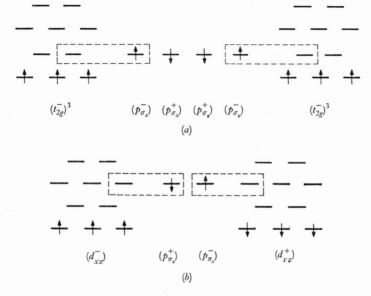

$$(t_{2g}^-)^3 \qquad (p_{\sigma_x}^-) \; (p_{\sigma_x}^+) \; (p_{\sigma_y}^+) \; (p_{\sigma_y}^-) \qquad (t_{2g}^-)^3$$

(a)

$$(d_{xz}^-) \qquad (p_{\pi_z}^+) \qquad (p_{\pi_z}^-) \qquad (d_{yz}^+)$$

(b)

Fig. 16

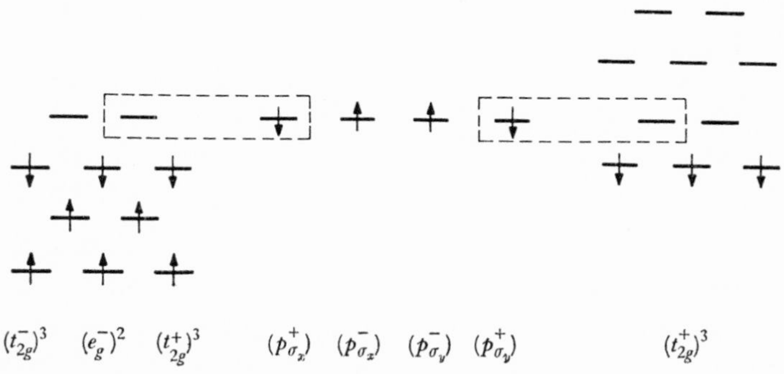

$$(t_{2g}^-)^3 \; (e_g^-)^2 \; (t_{2g}^+)^3 \qquad (p_{\sigma_x}^+) \; (p_{\sigma_x}^-) \; (p_{\sigma_y}^-) \; (p_{\sigma_y}^+) \qquad (t_{2g}^+)^3$$

Fig. 17

4.3.3. Superexchange via polyatomic anions

The superexchange mechanism is not in principle confined to monatomic anions. Polyatomic anions with well-developed π-systems such as carboxylates, Schiff base anions, cyanides, etc., can provide suitable pathways for electron transfer between two or more paramagnetic centres. This mechanism was first postulated[49] to account for the antiferromagnetism of copper(II) formate tetrahydrate, $Cu(HCO_2)_24H_2O$, and has been subsequently proposed[9] for several Schiff base dimers which show metal–metal interaction. Again the sign of the superexchange interaction will be determined by the symmetry relations between the anion σ- and π-orbitals and those of the interacting cations (cf. § 5.1.1 and Figs. 20 and 32).

4.4. Magnetic dipolar coupling

Not infrequently, the exchange coupling between the ith and j^{th} paramagnetic centres in a cluster has been referred to in the literature[26, 50] as 'dipolar coupling'. It must be emphasized that the actual magnitude of dipolar coupling is very much weaker than the exchange coupling described by [37]. This can be demonstrated by considering the magnetic dipole coupling between two electrons situated 1 Å apart with their spins parallel

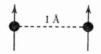

which from [18] is about 2 °K. Since metal–metal distances in clusters usually fall in the range 2–4 Å, dipolar couplings generally lie between a few tenths and a few hundredths of a degree. This estimate depends on the spin quantum number S and the relative orientation of the spin vectors. However, it can safely be assumed that in practice, dipolar couplings are 2–3 orders of magnitude smaller than exchange effects.

5. Experimental results

5.1. Dimers

5.1.1. Carboxylate–bridged dimers

Probably the simplest and most widely studied paramagnetic dimer is copper(II) acetate monohydrate[1] and its higher homologues[51-53] (cf. Fig. 1). Four bridging acetate groups support two $Cu(H_2O)$ groups so that the Cu–Cu separation (2·64 Å) is only

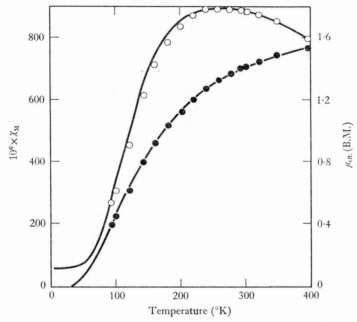

Fig. 18. Calculated (full curves) and experimental magnetic susceptibility (χ_M, empty circles) and moment ($\mu_{eff.}$, full circles) of copper(II) acetate monohydrate.

marginally greater than in the parent metal (2·56 Å). The temperature dependence of the magnetic susceptibility passes through a broad maximum at $T_c = 255$ °K (cf. Fig. 18) from which the strength of the metal–metal interaction is estimated to be $-2J_{12} = 284$ cm^{-1}. The $\chi_M(T)$ data are well reproduced by the Hamiltonian [17] with $S = \frac{1}{2}$ and S' taking the values $S' = 0$ and 1. The mechanism by which the spin coupling occurs has excited much interest and speculation[9, 54] for this molecule provides the simplest system yet located in which two isolated metal atoms interact. The original hypothesis[1] of a very weak Cu—Cu δ-bond which arises from a small lateral overlap of $3d_{x^2-y^2}$ orbitals centred on each copper atom (cf. Fig. 19) continues to be consistent with the available magnetic, e.s.r., and n.m.r. data. An alternative suggestion of a direct Cu—Cu σ-bond has now been discarded.[54]

The contribution to the interaction from the bridging π-system of the polyatomic carboxylate anions is difficult to estimate although

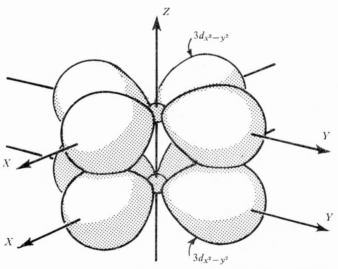

Fig. 19. Delta-bond formed by lateral overlap of two $d_{x^2-y^2}$ orbitals.

the closely related π-system of formate anion probably provides[49] the major pathway for superexchange in $Cu(HCO_2)_2,4H_2O$. In $[Cu_2(CH_3CO_2)_4(H_2O)_2]$ the π-pathway can only be made available by the mixing through spin-orbit coupling of a $^2B_2$ (hole in d_{xy}) with a $^2B_1$ (hole in $d_{x^2-y^2}$) state in D_{4h} symmetry, so that this contribution is expected to be small. In this connection the diamagnetism[55] of the isostructural $[Cr_2(CH_3CO_2)_4(H_2O)_2]$ is of special interest since the Cr–Cr separation[56] is identically 2·64 Å. It is the metal d_{xy} orbitals, which are fully occupied in the copper compound but only half-filled in the chromium analogue, which are correctly oriented to combine with the π-system of the acetate ligands.

Strong antiferromagnetic coupling (shown schematically in Fig. 20) via the π-pathway could separate the $S' = 0$ and $S' = 1$ levels of $[Cr_2(CH_3CO_2)_4(H_2O)_2]$ by 1000 cm^{-1} or more so that paramagnetism is not developed at room temperature. In sharp contrast, the diamagnetism[57] of the isoelectronic $[Mo_2(CH_3CO_2)_4]$, in which the Mo–Mo separation[58] is 2·11 Å (i.e. much less than that, 2·80 Å, in molybdenum metal), undoubtedly arises from a strongly covalent Mo—Mo multiple bond.[59] The green dia-

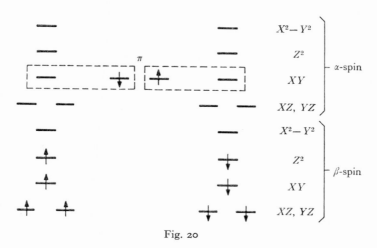

Fig. 20

magnetic[60] rhodium(II) acetate monohydrate (configuration d^7) and the related diamagnetic rhenium(III) compounds

$$[Re_2(CH_3CO_2)_4Cl_2]$$

(configuration d^4) provide further examples of covalent metal–metal bonding.[61]

The effects of varying the axial substituents[8, 53] as well as the carboxylate ligands[62, 63] in the copper acetate structure have been widely studied.[9] For example, formate ion[7, 49] and the anion of diazoaminobenzene[64, 65] produce stronger exchange couplings than acetate ion; axially situated pyridine type bases favour stronger interactions than aniline type bases.[8, 53]

5.1.2. Chlorine-bridged dimers

The closely related compounds $Cs_3Cr_2Cl_9$ and $K_3W_2Cl_9$ involve close-packed structures of cations and dimeric anions of composition $[M_2Cl_9]^{3-}$. The complex anion is formed by sharing a face between two adjacent MCl_6- octahedra.[3-5] In $K_3W_2Cl_9$ the W–W separation of $2\cdot41\text{Å}$ is much shorter than the sum of the metallic radii ($2\cdot82$ Å) and the near diamagnetism[6] ($\mu_{eff.} = 0\cdot47$ B.M. at 300 °K) is consistent with the presence of a strong W—W bond. On the other hand, the Cr–Cr separation of $3\cdot12$ Å in $Cs_3Cr_2Cl_9$ is considerably larger than the sum of the metallic radii ($2\cdot58$ Å) and $\mu_{eff.} = 3\cdot71$ B.M. (298 °K) is only slightly below the spin-only value of $3\cdot87$ B.M. for the t_{2g}^3 configuration.[66] The $\chi_M(T)$ curve is typical of a dimer with weak antiferromagnetic coupling ($T_c \approx 25$ °K so that $-2J_{12} = 43$ cm^{-1}). This result suggests that at $3\cdot12$ Å, coupling by direct overlap of t_{2g} orbitals and by the π-system dominates the ferromagnetic contributions from σ-orbitals centred on chlorine.

Although the available data are limited, there would appear to be a general and not unexpected rule that in any isoelectronic triad of transition metals (e.g. Cr, Mo, W) that a transition from weak exchange coupling (say, J's up to 1000 cm^{-1}) to strong covalent metal–metal bonding (say, bond energies up to 30 kcal.mole^{-1}) occurs as the atomic number of the metal increases due to the increased radial extension of the d-orbitals. This effect is counter-balanced by an increase in the oxidation state of the metal so that we find only a weak interaction[6] in Mo_2Cl_{10} ($\mu_{eff.} = 1\cdot64$ B.M. at 293 °K with $\theta = 15°$ over the range 90–293 °K) in contrast to the strong Mo—Mo bond in $[Mo_2(CH_3CO_2)_4]$. However, the isomorphous dimer W_2Cl_{10} appears to have a stronger W—W bond at $3\cdot30$ Å with a residual room temperature moment of

$$\mu_{eff.} = 1\cdot11 \text{ B.M.}$$

probably due to temperature independent paramagnetism.[67]

Binuclear di-π-cyclopentadienyl titanium(III) chloride, which is believed to involve pairs of chlorine-bridged titanium atoms at about $3\cdot5$ Å, provides an interesting example of antiferromagnetic d^1–d^1 interactions.[68] The maximum in susceptibility occurs at

$T_c = 170\,^\circ$K which corresponds to a singlet–triplet separation of $-2J_{12} = 192$ cm^{-1}.

5.1.3. Schiff base and related dimers

Ligands of the Schiff base type frequently give metal complexes with subnormal magnetic moments.[9] An excellent example is

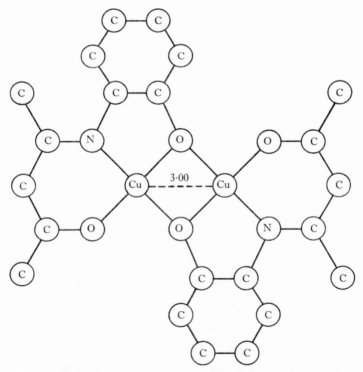

Fig. 21. Molecular structure of acetylacetone-mono(o-hydroxyanil)-copper(II) dimer.

provided by the dimeric compound acetylacetone-mono(o-hydroxyanil)copper(II), the structure[69] of which is shown in Fig. 21. The two copper atoms at $3\cdot00$ Å couple antiferromagnetically ($T_c = 270\,^\circ$K; $-2J_{12} = 298$ cm^{-1}) in this and the many related dimers which have been recently studied.[70–72]

The magnetic susceptibility of a series of closely related vanadyl derivatives down to $1\cdot4\,°K$ has demonstrated that the $d^1\text{-}d^1$ coupling is also antiferromagnetic.[73] Typical curves are illustrated in Fig. 22. From a study of the dependence of the interaction upon organic substituents, it has been suggested that a direct σ-overlap is involved in the V–V interaction but that superexchange via bridging oxygen atoms determines the Cu–Cu interaction in analogous complexes.[72, 73]

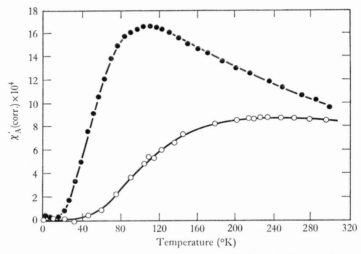

Fig. 22. Corrected magnetic susceptibility versus temperature curves for the 5-bromo-N-(2-hydroxyphenyl) salicylideneimine complex of VO²⁺ (full circles) and the 5-methyl-N-(2-hydroxyphenyl) salicylideneimine complex of Cu²⁺.

5.1.4. Oxo-bridged dimers

Several dimeric clusters are known in which the pair of metal atoms are shielded by an intervening oxide anion. The best charac-terized compounds of this type are $K_4[Cl_5RuORuCl_5]$, H_2O, $[(NH_3)_5CrOCr(NH_3)_5]Br_4$ and $[(phen)_2ClFeOFeCl(phen)_2]Cl_2$. The X-ray structure[74] of the first compound reveals that the Ru–O–Ru arrangement is linear and that the Ru—O bond distance ($1\cdot80$ Å) is considerably shorter than the sum of single bond radii ($1\cdot98$ Å). The long Ru–Ru distance of $3\cdot6$ Å (sum of metallic radii $= 2\cdot68$ Å) requires the involvement of oxygen orbitals to account for the observed diamagnetism of the $t_{2g}^4 - t_{2g}^4$ system. From the superexchange point of view, electron transfer via both the $p_\sigma - e_g$ and $p_\pi - t_{2g}$ pathways should couple the ruthenium atoms antiferromagnetically to give a stable diamagnetic ($S = 1$, $S' = 0$) ground state as shown in Fig. 23. An alternative approach has been proposed[75] in which the eight t_{2g}-electrons of the two ruthenium atoms together with four p_π-electrons of oxygen occupy bonding and non-bonding 3-centre molecular orbitals to give the closed configuration $(e_u)^4(b_{2g})^2(b_{1u})^2(e_g)^4$ under the D_{4h} symmetry of the dimer.

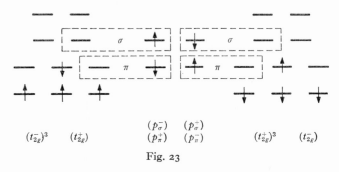

$$(t_{2g}^-)^3 \qquad (t_{2g}^+) \qquad \begin{matrix}(p_\sigma^-) & (p_\sigma^+)\\(p_\pi^+) & (p_\pi^-)\end{matrix} \qquad (t_{2g}^+)^3 \qquad (t_{2g}^-)$$

Fig. 23

The chromium(III) dimer[76] gives some evidence of exhibiting an underlying singlet–triplet magnetism consistent with either $S = \frac{3}{2}$ and $S' = 0, 1, 2, 3$ or $S = \frac{1}{2}$ and $S' = 0$ and 1. The moment at room temperature is $1\cdot3$ B.M. and unexpectedly the $\chi_M^{-1}(T)$ curve passes through a maximum at $T_c \sim 120\,°$K and the general antiferromagnetic behaviour is variable depending on the sample preparation.[77, 78] In fact, some workers[79] have even claimed that no maximum can be detected between $14°$ and $300\,°$K. The molecular orbital approach is not appropriate here, for the ground state predicted, namely $(e_u)^4(b_{2g})^2(b_{1u})^2(e_g)^2$ with $S = 1$, is not consistent with the experimental data.

The magnetic susceptibility of the compound formulated[80] as $[(phen)_2ClFeOFeCl(phen)_2]Cl_2$ which involves d^5–d^5 interactions confirms that the coupling between iron(III) atoms is again anti-ferromagnetic[78, 81] as would be expected if electron transfer pro-ceeds via the p_σ and p_π pathways.

5.1.5. Other bridged dimers

The magnetism of many other dimeric complexes have been reported although frequently the temperature dependence of the susceptibility has not been determined. A comprehensive survey of copper(II) complexes with subnormal magnetic moments is available[9] and this taken with the compilations of Figgis and Lewis,[26] Foex,[82] and König[83] are valuable sources of data for the interested reader.

5.2. Trimers

5.2.1. $d^3/d^3/d^3$

The best examples of a trinuclear molecule containing three metal atoms with the configuration d^3 are provided by the basic chromium(III) carboxylates formulated[84] as

$$[Cr_3O(RCO_2)_6(H_2O)_3]^+$$

The presence of Cr–Cr interaction in eighteen salts of this type was originally established by the magnetic investigations of Welo.[85] Each Cr^{3+} ion is nearly octahedrally coordinated by oxygen as shown in Fig. 24. The three metal atoms are arranged in an equilateral triangle about a central O^{2-} anion. The interaction between the metal atoms is antiferromagnetic the ground state of the $S = 3/2,3/2,3/2$ system corresponding to $S' = 1/2$

$$(\mu_{eff.} = 1 \cdot 05 \text{ B.M. per } Cr^{3+}).$$

The variation with temperature of the reciprocal of the magnetic susceptibility is reproduced[86, 87] by [24] and [26] when applied to the spin manifold shown in Fig. 25. The best value estimated for J_0 is $-15k$ ($= 10 \cdot 4$ cm^{-1}) although the calculated curve can be improved slightly by introducing[88] a subsidiary exchange integral $J_1 = -3 \cdot 25k$ ($= 2 \cdot 26$ cm^{-1}). The shape of the $\chi_M^{-1}(T)$ curve shown in Fig. 26 is characteristic for trinuclear clusters with $S' = 1/2$ ground states, being independent of the metal atom configuration. Since the lowest spin state is paramagnetic, one might anticipate that intercluster coupling might occur at sufficiently low temperatures. However, susceptibility measurements provide no evidence

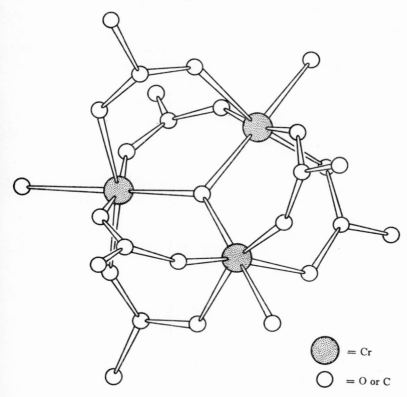

Fig. 24. Molecular structure of the trimeric cation $[Cr_3O(CH_3CO_2)_6]^+$.

for this down to $0·38$ °K, showing that the clusters are magnetically well insulated from one another. Because of the low molecular symmetry of this trimer,[84] the σ- and π-systems of both the trigonal oxide anion and the bridging carboxylate groups offer convenient pathways for superexchange. The Cr–Cr separation of $3·28$ Å is possibly too large for a direct interaction to account for the observed exchange coupling constant $J_0 = -15k$.

5.2.2. $d^3/d^3/d^5$

A trimer of special interest is the 'mixed' metal system

$$[Cr_2FeO(CH_3CO_2)_6(H_2O)_3]^+Cl^-,4H_2O$$

First prepared by Weinland and studied by Welo,[85] it provides a model compound with which to test the validity of the exchange Hamiltonian in [17] for a cluster which is comprised of different metal atoms. However, it must be remembered that while the compound analyses correctly for the above composition, there is no

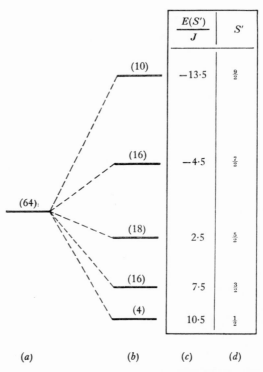

Fig. 25. Energy level scheme for a triangular triad of weakly interacting Cr^{3+} ions: (a) total spin degeneracy of ground state; (b) effect of spin–spin interaction of the form $-2J\mathbf{S_i \cdot S_j}$; (c) energy of spin levels $E(S')/J$; (d) values of S'.

evidence other than the susceptibility data with which to prove that each molecule is comprised of two chromium atoms and one iron atom. Between $1\cdot4$–$4\cdot2$ °K the compound follows a Curie–Weiss law corresponding again to an $S' = 1/2$ ground state for the $S = 3/2,3/2,5/2$ system. If it is assumed that the exchange integral J_0 is the same for Cr–Cr and Cr–Fe interactions, the experimental data between $1\cdot4°$ and 25 °K can be reproduced by [24] and [26] when applied to the spin manifold in Fig. 27. The mean value of J_0 is of the order of $-20k$, corresponding to an antiferromagnetic interaction. The simple model requires elaboration at more elevated temperatures; for example, by employing J_0 for the Cr–Cr interaction and J_1 for the two Fe–Cr interactions.[89] This has not been attempted.

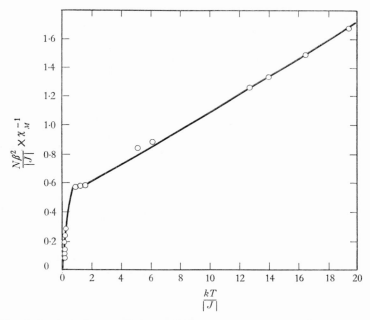

Fig. 26. Reciprocal magnetic susceptibility versus temperature of

$$[Cr_3O(CH_3CO_2)_6]Cl,9H_2O$$

O, Experimental data of Wucher and Gijsman[86] reduced for $J/k = -15$.
——, Theoretical curve for $S = 3/2$.

5.2.3. $d^5/d^5/d^5$

Basic carboxylates of iron(III) provide the best-known class of tri-nuclear complexes with the triad configuration $S = 5/2,5/2,5/2$. Welo[85] studied the magnetic properties of ten of these, six of which were associated with θ-values of about -600 °K and can probably be formulated[90] as $[Fe_3O(RCO_2)_6(H_2O)_3]^+$. Subsequent studies[87, 91–94] by other workers have confirmed that the iron atoms interact antiferromagnetically with a doublet $S' = 1/2$ being the ground state. The exchange integral is rather higher than that found for the isomorphous Cr^{3+} salts, being of the order of $-45k$.

More recently, it has been shown[95] that the temperature dependence of the susceptibility of trimeric iron(III) alkoxides,

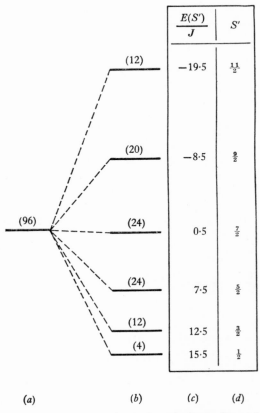

Fig. 27. Energy level scheme for a triangular 'mixed' metal cation
$[Cr_2FeO(CH_3CO_2)_6]^+$.

$[Fe_3(OR)_9]$, between $100°$ and $300 °K$, is consistent with anti-ferromagnetic Fe–Fe interactions and the spin manifold shown in Fig. 28. At $100 °K$, $\mu_{eff.} \sim 3.7$ B.M. per Fe atom and above this temperature the moment rises steadily, reaching a value of $\mu_{eff.} \sim 4.4$ B.M. at $300 °K$ in accord with an assumed D_{3h} model.

An extension of the measurements to liquid helium temperatures is now required to test this hypothesis more critically. It is noteworthy that the interaction ($J_0 \sim -15k$) in the iron(III) alkoxides, with no trigonal oxide anion, is reduced to about one-third

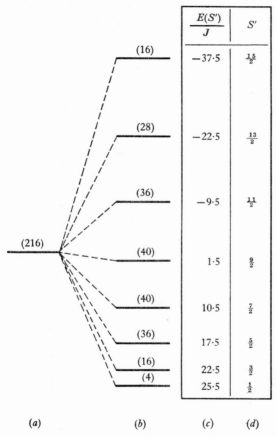

Fig. 28. Energy level scheme for a triangular triad of weakly interacting Fe^{3+} ions.

of that in the trimeric iron(III) carboxylates. However, the relative contributions of alkoxide versus carboxylate bridging groups, tetrahedral versus octahedral stereochemistry, and different Fe–Fe separations are impossible to ascertain.

5.2.4. $d^8/d^8/d^8$

The trimeric cluster $[\text{Ni}_3(\text{acac})_6]$ is noteworthy for it involves a linear (cf. Fig. 29) rather than triangular triad of Ni atoms.[25] Further, it provides the first unambiguous example of an isolated cluster in which the nearest neighbour metal–metal interactions are ferromagnetic.[96] The magnetic moment rises from 3·23 B.M. per

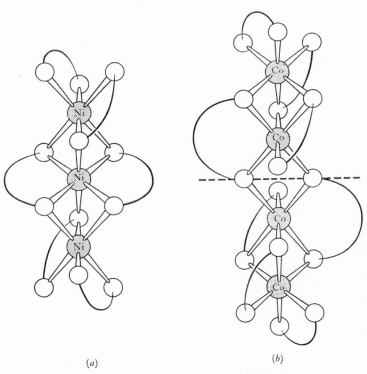

(a)   (b)

Fig. 29. Molecular structures of (a) $[\text{Ni}_3(\text{acac})_6]$; (b) $[\text{Co}_4(\text{acac})_8]$.

Ni^{2+} at 296 °K to 4·1 B.M. at 4·3 °K, corresponding to a ground molecular spin state $S' = 3$ (i.e. all six e_g-electron spins coupled parallel). The experimental data are consistent with the spin manifold shown in Fig. 30, where the ferromagnetic exchange integral for coupling between adjacent Ni atoms of the trimer is $J \sim +37k$ ($= +26$ cm^{-1}). In contrast the terminal Ni atoms are coupled

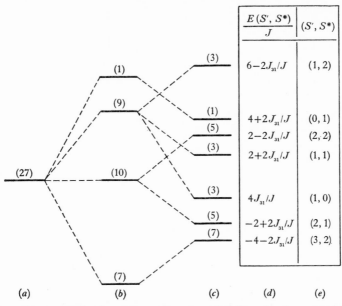

$\dfrac{E(S', S^*)}{J}$	(S', S^*)
$6 - 2J_{31}/J$	$(1, 2)$
$4 + 2J_{31}/J$	$(0, 1)$
$2 - 2J_{31}/J$	$(2, 2)$
$2 + 2J_{31}/J$	$(1, 1)$
$4J_{31}/J$	$(1, 0)$
$-2 + 2J_{31}/J$	$(2, 1)$
$-4 - 2J_{31}/J$	$(3, 2)$

Fig. 30. Energy level scheme for a triangular triad (b) and a linear triad (c) of Ni^{2+} ions with J positive and J_{31} negative in (c).

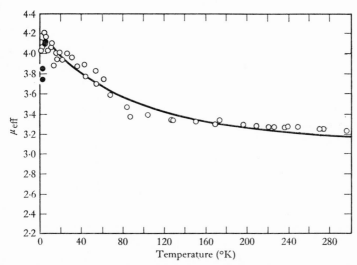

Fig. 31. Magnetic moment versus temperature for [Ni$_3$(acac)$_6$]. $\bigcirc$, Experimental data ($\theta = -0.2°$); $\bullet$, experimental data ($\theta = 0°$); full curve calculated with $g = 2.06$, $J = +37k$, $J_{31} = -10.3k$.

ENP

antiferromagnetically and more weakly with $J_{31} \sim -10k$ ($= -7$ cm^{-1}). The computed curve of best fit is shown in Fig. 31.

The ferromagnetic coupling between adjacent Ni atoms is consistent with the superexchange mechanism operating via the 90° Ni–O–Ni pathways as described in § 4.3.2(a).

For the exchange coupling between the two terminal Ni atoms, the only pathway evident is via the π-system of the two acetylacetone groups which bridge them. If an orientation for the p_π orbitals is chosen as shown in Fig. 32, then there is a finite overlap between this π-system and the half-filled $d_{x^2-y^2}$ orbitals of the

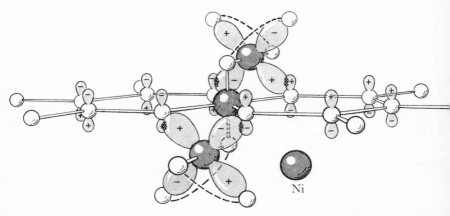

Fig. 32. Superexchange pathway provided by the π-system of two acetylacetone ligands for coupling between terminal Ni atoms of [Ni$_3$(acac)$_6$].

terminal Ni atoms. The resulting interactions must be antiferromagnetic superexchange.[97] It is also possible to orient the p_π orbitals to overlap the d_{xy} orbitals on the terminal Ni atoms. However, since these are both filled orbitals there is no exchange effect. There is no orientation of the π-system which will result in the simultaneous overlap of the $d_{x^2-y^2}$ orbital of one terminal Ni atom with the d_{xy} of the other. Accordingly, ferromagnetic coupling by electron transfer plus intra-atomic direct exchange is precluded.

Therefore the net interaction between the end members of the linear trimer is expected to be antiferromagnetic as is observed experimentally.

5.3. Tetramers

5.3.1. $(d^7/d^7/d^7/d^7)$

Three structural types of paramagnetic tetranuclear cluster have been discovered to date. These involve tetrahedral, trigonal bipyramidal, or octahedral coordination about the metal atom, the first two stereochemistries being associated with a tetrahedral M_4 cluster and the last with a linear M_4 quartet. The deep-blue tetrameric oxopivalate of cobalt (II) with the formula $[Co_4O(Me_3CCO_2)_6]$ appears[28] to be structurally analogous to the well-known basic beryllium (II) and zinc (II) acetates. At room temperature the effective moment of 3·86 B.M. per Co^{2+} is considerably below the values 4·4–4·8 B.M. usually observed for magnetically dilute tetrahedral $3d^7$ compounds. It is likely that antiferromagnetic coupling between cobalt atoms reduces the moment to this value although confirmation of both the sign and magnitude of J requires the extension of the magnetic measurements to much lower temperatures.

The crystal structure[29] of the violet cobalt(II) acetylacetonate reveals that the molecule is a centrosymmetric tetramer as shown in Fig. 29 b. Both terminal cobalt atoms are joined to the centre pair by sharing a common octahedral face with three bridge oxygen atoms. However, the central pair of cobalt atoms (Co–Co = 3·57 Å) are joined by sharing an edge with only two bridging oxygen atoms. Clearly, at least two different J-values, namely, J_{12} and J_{23}, should be required to describe magnetic interactions in this molecule. The magnetic moment has been reported[98] as $\mu_{\text{eff.}} = 4\cdot93$ B.M. at 297 °K and to follow a Curie–Weiss law between approximately 75° and 300 °K with $\theta = 15$ °K. More recent measurements[99] made on $Co(acac)_2$ recrystallized from n-hexane give a slightly higher moment at room temperature of 5·11 B.M. (295 °K). The $\chi_M^{-1}(T)$ curve follows a Curie–Weiss law to about 100 °K ($\theta = 20$ °K), but marked deviations occur between

$$1\cdot5 < T < 100 \ °K.$$

In fact, the moment reaches a maximum of 5·56 B.M. at 96 °K but then falls rapidly to 2·49 B.M. at 1·5 °K. Independent ferromagnetic and antiferromagnetic couplings appear to be competing in this molecule with the latter becoming dominant below about 75 °K.

5.3.2. $d^9/d^9/d^9/d^9$

The pattern of magnetic behaviour exhibited by the tetrahedral[27] cluster $[Cu_4OCl_6(OPPh_3)_4]$ shows certain similarities to that of $Co(acac)_2$. The effective moment rises from 1·87 B.M. at 295 °K to 2·12 B.M. at 54 °K but then decreases rapidly to 0·76 B.M. at 1·5 °K. A Curie–Weiss plot is obtained for temperatures between $60 < T < 300$ °K with $\theta \simeq 15$ °K. However, marked deviations are observed below 60 °K. Again it appears that antiferromagnetic coupling at lower temperatures is dominating the ferromagnetic interaction, which appears to be larger at higher temperatures.[99]

This review was written while the author was on study leave at the Bell Telephone Laboratories, Incorporated, Murray Hill, New Jersey, U.S.A. The author is indebted to A. P. Ginsberg and M. E. Lines for conversations which have clarified several aspects of the work reported here. The generous assistance given by Miss I. D. Bogdanski for the computations on which Table 1 and Figs. 3 and 5 are based is gratefully acknowledged.

REFERENCES

1 B. N. Figgis and R. L. Martin, *J. chem. Soc.* 1956, p. 3837.
2 J. N. Van Niekerk and F. K. L. Schoening, *Acta crystallogr.* 1953, 6, 227.
3 C. Brosset, *Ark. Kemi Miner. Geol.* 1935, 12A, no. 4.
4 G. J. Wessel and D. J. W. Ijdo, *Acta crystallogr.* 1957, 10, 466.
5 W. H. Watson and J. Waser, *Acta crystallogr.* 1958, 11, 689.
6 W. Klemm and H. Steinberg, *Z. anorg. Chem.* 1936, 227, 193.
7 R. L. Martin and H. Waterman, *J. chem. Soc.* 1959, p. 2960.
8 E. Kokot and R. L. Martin, *Inorg. Chem.* 1964, 3, 1306.
9 M. Kato, H. B. Jonassen and J. C. Fanning, *Chem. Revs.* 1964, 64, 99.
10 C. G. Barraclough and C. F. Ng, *Trans. Farad. Soc.* 1964, 60, 836.
11 W. Heisenberg, *Z. Phys.* 1926, 38, 411.
12 W. Heisenberg, *Z. Phys.* 1928, 49, 619.
13 P. A. M. Dirac, *Proc. Roy. Soc.* A, 1929, 123, 714.
14 J. H. Van Vleck, *The Theory of Electric and Magnetic Susceptibilities*, ch. XII. London and New York: Oxford University Press, 1932.
15 H. A. Kramers, *Physica*, 1934, 1, 182.

16 P. W. Anderson, *Phys. Rev.* 1950, **79**, 350.
17 P. W. Anderson, *Phys. Rev.* 1959, **115**, 2.
18 B. Bleaney and K. D. Bowers, *Proc. Roy. Soc.* A, 1952, **214**, 451.
19 J. S. Smart, *Magnetism* (ed. G. T. Rado and H. Suhl), vol. III, p. 63. New York: Academic Press, Inc., 1965.
20 P. A. M. Dirac, *Proc. Roy. Soc.* A, 1926, **112**, 661.
21 K. Kambe, *J. phys. Soc. Japan*, 1950, **5**, 48.
22 A. Danielian, *Proc. phys. Soc.* 1962, **80**, 981.
23 A. B. Lidiard, *Rep. Progr. Phys.* 1954, **17**, 201.
24 C. J. Ballhausen, *Introduction to Ligand Field Theory*, p. 147. New York: McGraw-Hill Book Co., Inc., 1962.
25 G. J. Bullen, R. Mason and P. Pauling, *Inorg. Chem.* 1965, **4**, 456.
26 B. N. Figgis and J. Lewis, in *Progress in Inorganic Chemistry*, vol. VI, p. 91. New York: Interscience Publishers, 1964.
27 J. A. Bertrand, *Inorg. Chem.* 1967, **6**, 495.
28 A. B. Blake, *Chem. Commun.* 1966, p. 569.
29 F. A. Cotton and R. C. Elder, *Inorg. Chem.* 1965, **4**, 1145.
30 A. Sommerfeld and H. A. Bethe, 'Ferromagnetism', in *Handbuch der Physik*, vol. XXIV, part II, p. 595. Berlin: Springer, 1933.
31 J. B. Goodenough, *Magnetism and the Chemical Bond*, pp. 79, 80. New York: Interscience Publishers, 1963.
32 W. Heitler and F. London, *Z. Phys.* 1927, **44**, 455.
33 J. H. Van Vleck, *J. Phys. Radium*, 1951, **12**, 262.
34 P. W. Anderson, *Magnetism* (ed. G. T. Rado and H. Suhl), vol. I, p. 25. New York: Academic Press, Inc., 1963.
35 J. B. Goodenough, *Phys. Rev.* 1955, **100**, 564.
36 J. B. Goodenough, *Phys. Chem. Solids*, 1958, **6**, 287.
37 J. Kanamori, *Phys. Chem. Solids*, 1959, **10**, 87.
38 E. O. Wollan, H. R. Child, W. C. Koehler and M. K. Wilkinson, *Phys. Rev.* 1958, **112**, 1132.
39 R. L. Martin, R. S. Nyholm and N. C. Stephenson, *Chemy Ind.* 1956, p. 83.
40 D. J. Machin, R. L. Martin and R. S. Nyholm, *J. chem. Soc.* 1963, p. 1490.
41 R. K. Nesbet, *Ann. Phys.* 1958, **4**, 87.
42 R. K. Nesbet, *Phys. Rev.* 1960, **119**, 658.
43 G. W. Pratt, *Phys. Rev.* 1955, **97**, 926.
44 J. B. Goodenough and A. L. Loeb, *Phys. Rev.* 1955, **98**, 391.
45 J. C. Slater, *Phys. Rev.* 1951, **82**, 538.
46 V. Heine, *Phys. Rev.* 1957, **107**, 1002.
47 J. H. Wood and G. W. Pratt, *Phys. Rev.* 1957, **107**, 995.
48 T. Moriya, *Phys. Rev.* 1960, **120**, 91.
49 R. L. Martin and H. Waterman, *J. chem. Soc.* 1959, p. 1359.
50 J. Lewis, F. E. Mabbs and A. Richards, *J. chem. Soc.* (*A*), 1967, p. 1014.
51 R. L. Martin and H. Waterman, *J. chem. Soc.* 1957, p. 2345.
52 R. L. Martin and A. Whitley, *J. chem. Soc.* 1958, p. 1394.
53 L. Dubicki, C. M. Harris, E. Kokot and R. L. Martin, *Inorg. Chem.* 1966, **5**, 93.

230   R. L. MARTIN

54   L. Dubicki and R. L. Martin, *Inorg. Chem.* 1966, **5**, 2203, and references therein.

55   W. R. King and C. S. Garner, *J. chem. Phys.* 1950, **18**, 689.

56   J. N. Van Niekerk, F. K. L. Schoening and J. F. deWet, *Acta crystallogr.* 1953, **6**, 501.

57   E. Bannister and G. Wilkinson, *Chemy Ind.* 1960, p. 319.

58   D. Lawton and R. Mason, *J. Am. chem. Soc.* 1965, **87**, 921.

59   F. A. Cotton, *Inorg. Chem.* 1965, **4**, 334.

60   D. J. Mackey, Ph.D. thesis, University of Melbourne, 1968.

61   F. A. Cotton, C. Oldham and W. R. Robinson, *Inorg. Chem.* 1966, **5**, 1798, and references therein.

62   E. Kokot, Ph.D. thesis, University of New South Wales, 1961.

63   J. Lewis, *Pure Appl. Chem.* 1965, **10**, 11.

64   C. M. Harris and R. L. Martin, *Proc. chem. Soc.* 1958, p. 259.

65   B. F. Hoskins, C. M. Harris and R. L. Martin, *J. chem. Soc.* 1959, p. 3728.

66   A. P. Ginsberg and M. B. Robin, unpublished results.

67   P. M. Boorman and N. N. Greenwood, unpublished results.

68   R. L. Martin and G. Winter, *J. chem. Soc.* 1965, p. 4709.

69   G. A. Barclay, C. M. Harris, B. F. Hoskins and E. Kokot, *Proc. chem. Soc.* 1961, p. 264.

70   M. Kishita, Y. Muto and M. Kubo, *Aust. J. Chem.* 1957, **10**, p. 386.

71   W. E. Hatfield and F. L. Bunger, *Inorg. Chem.* 1966, **5**, 1161.

72   A. P. Ginsberg, R. C. Sherwood and E. Koubek, *J. inorg. nucl. Chem.* 1967, **29**, 353.

73   A. P. Ginsberg, E. Koubek and H. J. Williams, *Inorg. Chem.* 1966, **5**, 1656.

74   A. McL. Mathieson, D. P. Mellor and N. C. Stephenson, *Acta crystallogr.* 1952, **5**, 185.

75   J. D. Dunitz and L. E. Orgel, *J. chem. Soc.* 1953, p. 2594.

76   W. K. Wilmarth, H. Graff and S. T. Gustin, *J. Am. chem. Soc.* 1956, **78**, 2683.

77   C. E. Schäffer, *J. inorg. nucl. Chem.* 1958, **8**, 149.

78   A. Earnshaw and J. Lewis, *J. chem. Soc.* 1961, p. 396.

79   H. Kobayashi, T. Haseda, E. Kanda and M. Mori, *J. phys. Soc. Japan*, 1960, **15**, 1646.

80   A. Earnshaw and J. Lewis, *Nature, Lond.* 1958, **181**, 1262.

81   N. Elliott, *J. chem. Phys.* 1961, **35**, 1273.

82   G. Foex, *Tables de constantes et donnés numériques.* Paris: Masson et Cie, 1957.

83   E. König, *Magnetische Eigenschaften der Koordinations und metall-organischen Verbindungen der 'Ubergangselemente'*, Neue Serie, Band 2, Landolt-Börnstein, 1966.

84   B. N. Figgis and G. B. Robertson, *Nature, Lond.* 1965, **205**, 694.

85   L. A. Welo, *Phil. Mag.* 1928, S7, **6**, 481.

86   J. Wucher and H. M. Gijsman, *Physica*, 1954, **20**, 361.

87   A. Earnshaw, B. N. Figgis and J. Lewis, *J. chem. Soc. (A)*, 1966, p. 1656.

88   J. T. Schriempf and S. A. Friedberg, *J. chem. Phys.* 1964, **40**, 296.

89 H. M.Gijsman, T. Karantassis and J. Wucher, *Physica*, 1954, **20**, 367.
90 L. E. Orgel, *Nature, Lond.* 1960, **187**, 504.
91 G. Foëx, B. Tsaë and J. Wucher, *C. r. hebd. Séanc. Acad. Sci., Paris*, 1951, **233**, 1432.
92 B. Tsaë and J. Wucher, *J. Phys. Radium*, 1952, **13**, 485.
93 A. Abragam, J. Horowitz and J. Yvon, *J. Phys. Radium*, 1952, **13**, 489.
94 J. Yvon, J. Horowitz and A. Abragam, *Rev. mod. Phys.* 1953, **25**, 165.
95 R. W. Adams, C. G. Barraclough, R. L. Martin and G. Winter, *Inorg. Chem.* 1966, **5**, 346.
96 A. P. Ginsberg, R. L. Martin and R. C. Sherwood, *Chem. Commun.* 1967, p. 856.
97 A. P. Ginsberg, R. L. Martin and R. C. Sherwood, *Inorg. Chem.* 1968, **7**, 932.
98 F. A. Cotton and R. H. Holm, *J. Am. chem. Soc.* 1960, **82**, 2979.
99 A. P. Ginsberg, R. L. Martin and R. C. Sherwood, unpublished results.

10

AMIDES AS NON-AQUEOUS SOLVENTS

R. C. PAUL

The excellent solvent properties of water have hindered the development of chemistry of non-aqueous solvents for quite some time. However, for certain reactions it has been recognized that the presence of water is actually harmful and for this reason many reactions are being carried out and studied in solvents other than water. Recent work has led to the development of chemistry of a number of non-aqueous solvents with a view to understanding their nature. But most of these solvents have low dielectric constants which make them less suitable for inorganic solutes. Apart from this, electrochemical work leading to a quantitative estimate of ionization has not been possible owing to the large degree of interionic attraction in these solvents.

Some compounds containing strong hydrogen bonds, such as sulphuric acid and hydrogen fluoride, have very high dielectric constants, even higher than that of water (100 at 25° and 84 at 0 °C respectively), and therefore they are expected to serve as good ionizing solvents. But they are strong acids and most of the solutes undergo solvolysis in them, forming sulphates or fluorides. Thus these solvents, in spite of their high dielectric constants, cannot be really classified as water-like.

Hydrogen bonds are also formed in some organic compounds containing the amide group. These compounds have high dielectric constants, some even higher than that of water, and they are only weakly basic. Thus there are greater chances for these compounds to act as water-like solvents and most electrolytes may dissolve in them and exist as ions in solutions just as in water.

A study of the chemistry of amides is quite fascinating in view of the similarity of their structure to that of the amino acids, peptides and proteins which form an essential part of our life. The amides contain the structural framework given above,

[233]

in which the positions 1, 2 and 3 are occupied by either hydrogen atoms or alkyl groups. In the present survey, three typical amides namely, formamide, acetamide and dimethylformamide, are being considered for discussion of the salient features of their solvent chemistry.

In Table 1 some of the physical constants of the amides have been given, along with those of water for the sake of comparison. As indicated by their melting and boiling points, these amides have a wide and convenient working range. These amides, except for NN'-dimethylformamide, have a high viscosity (higher than that of water), which may be an indication of their associated nature.[1, 2, 3] They have high dipole moments which also point to the possibility of their associated nature. The fairly high dielectric constants (in some cases even higher than that of water) are an indication of their good ionizing properties as solvents, and a low conductance points to their feebly ionic nature.

1. Solubility of substances in amides

A qualitative determination of the solubilities of various compounds has been reported in formamide,[4, 5] molten acetamide[6] and dimethylformamide[7] and the results have been summarized in Table 2(a) and (b).

Even strong electrolytes such as sodium chloride and potassium chloride are soluble in amides to a considerable extent. Ordinarily, inorganic compounds which dissolve in water and form hydrates have been found to exhibit higher solubilities in amides and form solvates. The solubility of these electrolytes may be attributed to the comparatively high dielectric constants of amides. Sodium bromide, sodium iodide, and potassium iodide have been found to form solvates with amides. Potassium iodide forms higher solvates (e.g. $KI.6CH_3CONH_2$)[8] and shows greater solubility which can also be attributed to the fact that iodides are more polarized and more covalent in character than the corresponding bromides and chlorides. In all the cases, bromides have been found to be more soluble than chlorides and this may also be due to the reason mentioned above. In the case of alkali metal halides particularly, where dissolution takes place resulting in the formation of ions, the

TABLE 1. *Physical constants of amides*

Property	Formamide	Acetamide	NN'-DMF	Water
Melting point (°C)	2·55	82·0	−61·0	0·0
Boiling point (°C/760 mm)	193·0	222·0	153·0	100·0
Specific gravity* (g.ml.$^{-1}$)	1·1296	0·983	0·9443	1·0 at 4°
Viscosity* (cP)	3·31	1·63	0·796	0·959
Dipole moment (debyes)	3·68	3·72	3·82	1·85
Dielectric constant* (ml.)	111·3	60·6	36·7	78·54
Specific conductance* (Ω^{-1}.cm^{-1})	$2-10 \times 10^{-6}$	$2-6.4 \times 10^{-6}$	$0.6-2.0 \times 10^{-7}$	6×10^{-8}
Surface tension* (dynes.cm^{-1})	58·2	37·98	—	—
Heat of fusion (kcal.mole^{-1})	—	3·9	—	1·35
Heat of vaporization (kcal.mole^{-1})	—	—	—	9·719
Trouton's constant* (cal.deg.$^{-1}$)	—	—	33·4	—
Cryoscopic constant (°C)	3·57	3·6	—	1·86

* At 25 °C. DMF = Dimethylformamide.

TABLE 2(a). *Solubilities of inorganic compounds in amides*

Compound	Solubility (g/100 g) in		
	Formamide	Acetamide	Dimethyl-formamide
LiCl	—	8·68	27·53
NaCl	9·32	3·49	0·05
KCl	6·30	2·11	0·05
LiBr	—	44·93	—
NaBr	35·84	14·75	3·23
KBr	21·38	11·12	12·79
NaI	56·62	30·44	3·72
NH_4Cl	11·02	10·07	0·05
KI	69·22	29·48	13·71
NH_4Br	36·11	32·04	—
NH_4I	104·20	43·20	—
$MgCl_2$	—	9·36	8·04
$ZnCl_2$	—	8·54	6·32
$CaCl_2$	22·2	21·68	1·98
$CaBr_2$	—	44·70	—
$SrCl_2$	15·90	17·45	—
$BaCl_2$	11·76	15·58	1·07
$BaBr_2$	29·85	40·50	—
$CdCl_2$	—	4·35	17·34
$CuCl_2$	—	14·64	—
$MnCl_2$	5·88	19·63	—
$MnBr_2$	—	38·64	—
$CoCl_2$	—	15·52	9·37
$NiCl_2$	—	14·21	7·49
$CoBr_2$	—	22·31	—
$NiBr_2$	—	15·16	—
$CrCl_3$	—	3·80	—
KNO_3	16·0	—	2·27
$NaNO_3$	41·3	—	13·07
$FeCl_3$	—	36·29	11·43
NH_4CNS	—	17·21	13·37
NaCNS	—	23·47	21·32
KCNS	65·9	11·94	15·97
NaCN	—	19·27	18·76
KCN	14·6	11·48	—
$KClO_3$	—	14·63	18·1
$KClO_4$	—	9·76	19·6
$NaClO_3$	—	21·92	23·4
$NaClO_4$	—	12·46	39·6

TABLE 2(a) (continued)

Compound	Solubility (g/100 g) in		
	Formamide	Acetamide	Dimethyl-formamide
CH₃COOLi	—	1·72	—
CH₃COONa	—	9·47	7·21
CH₃COOK	29·4	17·24	5·32
(CH₃COO)₂Pb	—	4·73	—
PbCl₂	5·62	11·10	—
Na₂S₂O₃	—	3·72	—
Na₂SO₄	1·41	5·79	—
K₂Cr₂O₇	—	4·31	—
KMnO₄	—	5·29	—
Na₂CO₃	1·45	3·78	—
CuSO₄	—	4·21	—
AgNO₃	—	18·83	—

TABLE 2(b). *Lewis acids, bases, protonic acids and quaternary ammonium salts soluble in amides*

Lewis acids	Bases	Protonic acids	Quaternary ammonium salts
AsCl₃	Aniline	H₂SO₄	R₄NX
SbCl₃	Benzylamine	HSO₃F	R₃NHX
SbCl₅	Pyridine	CH₃CO₂H	R = Me, Et
SnCl₄	α-, β-, γ-Picolines	CH₂ClCO₂H	X = Cl, Br
TiCl₄	Iso-quinoline	CHCl₂CO₂H	
	Piperidine	PhCO₂H	
	Morpholine		

solubilities depend not only on the dielectric constant of the solvent, but also on the lattice energies of the solutes. As the melting point and lattice energy rise, the solubility decreases. It is also evident that with an increase in the cation size, the solubility decreases while with an increase in the anion size the solubility increases. When the cation is smaller and the charge on it is higher, the solvation is also more pronounced. The data reveal that the solubility increases as the charge on the cation increases.

The high solubility of the protonic acids and miscibility in all proportions of a number of organic nitrogen bases in these amides indicates the formation of hydrogen bonds with amides. Thus it may be inferred that substances which form hydrogen bonds with amides are also soluble in it. This is in conformity with the associated nature of these solvents consequent upon hydrogen bonding.

With the above observations on the solubility of different substances in amides, a conclusion may be drawn that comparatively high solubility of electrolytes may be due to the fairly high dielectric constants, and solvation due to the hydrogen-bonded structure of these solvents.

2. Formation of solvates in amides

The formation of solvates in alcohols,[9, 10] carboxylic acids,[11] esters,[12] amides[13] and substituted amides[14, 15] are well known. The solvates are believed to be formed owing to the interaction of the solute and the solvent either by donating to or by accepting a pair of electrons from the solvent molecule or its ion. An examination of the structure of the amide group reveals that it has two donor atoms, oxygen and nitrogen. Both these atoms have a lone pair of electrons which can be donated. The evidence obtained from infrared spectral studies indicates the carbonyl oxygen to be the donor atom.[16] The n.m.r.[17, 18] studies also establish that protonation of amides takes place at the carbonyl oxygen.

TABLE 3. *Solvates of amides*

Composition	Formamide	Acetamide	Dimethylformamide
Monosolvates	BF_3, CH_2ClCO_2H	BF_3, BBr_3, SO_3, $SbCl_3$	BF_3, $SbCl_5$, SO_3, HCl, HBr, HSO_3F
Disolvates	$TiCl_4$, $SnCl_4$, $SbCl_3$, $SbCl_5$, $BiCl_3$, $AsCl_3$, SO_3, HCl, HBr	SO_3, $SnCl_4$, $TiCl_4$, $TeCl_4$, $SbCl_5$, $MgCl_2$, $ZnCl_2$, $MoCl_2$, $CuCl_2$	$SnCl_4$, $SnBr_4$, $SiCl_4$, $SiBr_4$, SiF_4, $TiCl_4$, $ZnCl_2$, $MgCl_2$, $CoCl_2$
Trisolvates	—	$AsCl_3$, $FeCl_3$	—
Tetrasolvates	$ZrCl_4$, $TiCl_4$, $FeCl_3$	$SnCl_4$, $SnBr_4$, $CaCl_2$	BI_3
Hexasolvates	$AlCl_3$	$AlCl_3$, $NiCl_2$, $CoCl_2$, $MgBr_2$, MgI_2	$AlCl_3$, $AlBr_3$, $FeCl_3$

The solvates isolated have been listed in Table 3. In those cases where one or more molecules of the amides are attached in excess over the number required according to the coordination number of the metallic atom of the Lewis acid, the additional amide molecules are probably attached through hydrogen bonding.[13] The fact that the metal atoms retain their normal coordination number has earlier been indicated in the case of the complex $ZrCl_4.4CH_3CONH_2$.[19] On the basis of the above, the structure of the solvates of amides with sulphur trioxide, antimony(V) chloride, tin(IV) chloride and tin(IV) bromide may be represented as:

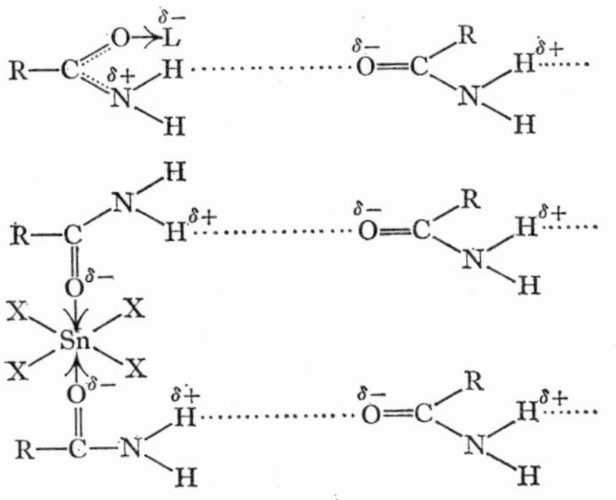

The Lewis acids are electron withdrawing groups. On coordination with the amides they would obviously increase negative charge on the oxygen end and positive charge on the nitrogen atoms. As a result of this reshuffling of charges, these solvates may ionize in solutions, as shown below:

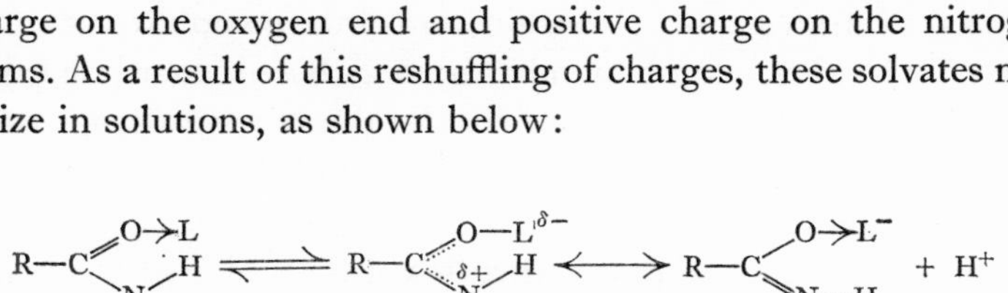

The disolvates of copper(II) chloride, manganese(II) chloride, zinc chloride, cadmium chloride and magnesium chloride with amides may have the following structures and the solvates $CoCl_2.6CH_3CONH_2$ and $NiCl_2.6CH_3CONH_2$ may have structures similar to those of their ammines:

$$M = Cu, Mn, Zn, Cd, Mg$$

The polar nature of the solvates of Lewis acids with amides is further supported by the high dipole moment (12·12 D) of $SO_3 . CH_3CONH_2$[16] and the comparatively high conductances of these solvates in the molten state, as given in Table 4.

TABLE 4. *Specific conductances of solvates of amides in the molten state*

Solvate	Specific conductance $(\Omega^{-1}.cm^{-1})$
$SnCl_4 . 4AcNH_2$	$1·23 \times 10^{-3}$ at 80°
$TiCl_4 . 2AcNH_2$	$4·78 \times 10^{-3}$ at 100°
$SbCl_5 . 2AcNH_2$	$5·88 \times 10^{-3}$ at 80°
$SnBr_4 . 4AcNH_2$	$4·03 \times 10^{-3}$ at 30°
$SO_3 . DMF$	$18·3 \times 10^{-4}$ at 140°
$SbCl_5 . DMF$	$4·9 \times 10^{-4}$ at 143°
$SnCl_4 . 2DMF$	$1·6 \times 10^{-4}$ at 235°
$SnBr_4 . 2DMF$	$8·2 \times 10^{-4}$ at 160°

3. Infrared spectral study of the complexes of amides with Lewis acids

From the infrared spectral examination of the complexes of amides with Lewis acids, information has been gathered regarding (a) the site of electron donation in amides, (b) the relative change in the structure of amide on complex formation, and (c) the total structure of the complex.

The following resonating structures of the amide group suggest that the amides exhibit donor properties, preferably through carbonyl oxygen because of its higher electronegativity than that of the nitrogen atom.

(i)   (ii)   (iii)

3.1. Dimethylformamide–Lewis acid systems

The complexes of dimethylformamide with sulphur trioxide, antimony(V) chloride and tin(IV) halides have been examined in chloroform solution.[16] The carbonyl stretching frequency in pure dimethylformamide appears at 1650 cm^{-1} and the absence of this band in antimony(V) chloride and sulphur trioxide complexes indicates a major change in the structure of the complexes as compared to that of dimethylformamide. This feature of the spectrum has been intrepreted as follows: (*a*) the site of complex formation is the carbonyl oxygen of dimethylformamide; (*b*) the carbonyl group is largely perturbed and its absorption frequency might have shifted to the absorption region for chloroform; (*c*) a medium intensity band at 1020 cm^{-1} in the case of the complex with antimony(V) chloride has been observed and this may be attributed to the vibration of $>$C—O-group; (*d*) the total structure of these complexes may be represented as:

(iv)   (v)

In the spectra of the complexes of tin(IV) halides with dimethylformamide, the band corresponding to the carbonyl stretching vibration shifts to lower frequencies by 25–30 cm^{-1} and can be assigned to the stretching frequency of the perturbed carbonyl group (νC $\cdots$ O $\to$ M). The perturbation causes a decrease in the bond order and the force constant of the $>$C$=$O-group.

Considering the relative perturbation of the carbonyl stretching frequency of dimethylformamide, the Lewis acids may be arranged in the order of their decreasing strength as:

$$SO_3 \approx SbCl_5 \gg SnBr_4 > SnCl_4$$

As depicted in structure (V), the positive charge centres on the nitrogen atom. Electron mobility from the (C—H) of the formyl group of dimethylformamide on complex formation is also

16

possible. The change in the absorption frequencies of the formyl C—H could not be studied due to the C—H vibrations of the solvent. But there is a clear indication that the carbon–nitrogen bond acquires a partial double-bond character on complex formation. The positive charge on the nitrogen atom may also cause a withdrawal of electrons from the methyl group attached to it, which evidently will produce strain on the methyl C—H bond. The chemical shift of the methyl proton as indicated from the proton resonance spectra of dimethylformamide complexes with Lewis acids, does not clearly support the above view, as it also indicates the broadening of the C—H band of the CHO group. The ionization of these complexes may therefore be due to the release of proton from a methyl or formyl group which is then solvated by the second solvent molecule.

3.2. Formamide–Lewis acid systems

The infrared spectra of the Lewis acid complexes of formamide indicate that on complex formation the $>C=O$, C—N and N—H stretching bands are affected the most. Pure formamide has two N—H stretching frequencies at 3515 and 3410 cm^{-1}. In its complexes with sulphur trioxide and antimony (V) chloride, one of the ν(N—H) disappears, possibly because of the perturbation caused to the N—H bond on complex formation. The structure of the complex may be represented as:

(vi)   (vii)

The amide I band, which is mainly due to carbonyl stretching frequency and appears at 1705 cm^{-1} in formamide, gets perturbed on complex formation with the Lewis acids. The significant feature of the spectra of sulphur trioxide complex with formamide is the absence of the carbonyl stretching band and the presence of a new weak band at 1100 cm^{-1}. The results are parallel to those observed in the complex of dimethylformamide with sulphur trioxide. In the complexes of the other Lewis acids like tin(IV) halides, there is a displacement of the carbonyl stretching band from 1705 to 1680 cm^{-1}.

3.3. Acetamide–Lewis acid systems

The infrared spectra of the adducts of acetamide with sulphur trioxide, antimony(V) chloride, tin(IV) chloride and bromide have been examined in chloroform while for some of them the nujolmull technique has been employed.

There are two N—H stretching frequencies at 3445 and 3430 cm^{-1} in acetamide molecule, while in the spectra of 1:1 complex of sulphur trioxide or the 1:2 complex of antimony(V) chloride there is only one band at 3415 cm^{-1}. In the complex of sulphur trioxide with acetamide, the carbonyl stretching frequency of acetamide at 1675 cm^{-1} is absent which suggests large structural changes in the amide molecule on coordination. These changes are similar to the changes which have already been proposed for analogous formamide complexes. The 1:1 complex of acetamide with antimony(V) chloride could not be isolated. In the infrared spectra of the 1:2 complex, the carbonyl stretching frequency is retained when the complex is examined in chloroform solution or in the solid state. The data are interpreted in terms of a formation which holds the second acetamide molecule through hydrogen bonding. The carbonyl stretching band observed is the outcome of the hydrogenbonded acetamide molecule. The Lewis-acid-bonded acetamide molecule experiences large structural changes as have been discussed for formamide and dimethylformamide–antimony(V) chloride systems. The structure of this 1:2 complex may be represented as:

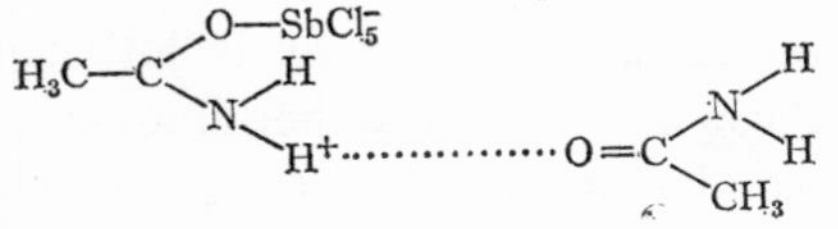

In confirmity with the above structure, the band at 1112 cm^{-1} may be assigned to the C—O-group.[20] The band corresponding to C⋯N might have shifted to the region of chloroform or nujol absorptions.

The solvent properties of acetamide could be explained on the basis of release of proton from the nitrogen atom with its simultaneous association with another acetamide molecule.

In the complexes of tin (IV) halides with two acetamide molecules there is a lowering of the carbonyl stretching frequency which again suggests that the complex formation is through carbonyl oxygen.

One definite conclusion which may be drawn from the above studies is that there is a large perturbation in the N—H group, apart from those observed in the $\diagup$C=O group. This suggests that N—H may contribute to the ionization of these complexes. In dimethylformamide, there is a competition between the formyl group (C—H) and the N—CH$_3$ group as already discussed.

4. Conductance studies in amides

4.1. Lewis acids

A large number of Lewis acids have been found to dissolve in amides and form stable solvates. Their specific conductance in the molten state indicates a fair degree of ionic character. For a more detailed study of their nature, the solvates of sulphur trioxide, antimony (V) chloride, tin (IV) chloride, tin (IV) bromide, titanium (IV) chloride and aluminium chloride have been selected.

In Fig. 1, the specific conductances of antimony (V) chloride, tin (IV) chloride, titanium (IV) chloride, arsenic (III) chloride and antimony (III) chloride in acetamide have been plotted against the concentration as represented by the molar ratios [Lewis acid: solvent]. Figures 2 and 3 represent similar curves for formamide and dimethylformamide as solvents.

The pure amides have comparatively low conductances. The Lewis acids are also known to be either non-conducting or to have only very low conductances.[21] The comparatively high conductances of the solutions of these Lewis acids in these amides suggest that the complexes formed are polar in nature. This view also gets support from the high values of the specific conductances of these solvates in the molten state and also from their high dipole moments. Considering the polar nature of the solvates formed, the following equilibria may be suggested to exist in their solutions:

$$R{-}C{=}\overset{+}{N}\diagdown\begin{smallmatrix}H\\[2pt]\\H\end{smallmatrix} + R{-}\overset{O}{\overset{\|}{C}}{-}NH_2 \rightleftharpoons [RCONH_2{\cdot}H]^+ + R{-}C{=}\overset{\cdot\cdot}{N}{-}H$$

$$MX_4 + 2RCONH_2 \rightarrow MX_4{\cdot}2RCONH_2$$

$$MX_4{\cdot}2RCONH_2 + 2RCONH_2$$
$$\rightleftharpoons 2[RCONH_2{\cdot}H]^+ + [MX_4{\cdot}2RCONH]^{2-}$$

$$(L = SO_3, SbCl_5;\ MX_4 = SnCl_4, TiCl_4)$$

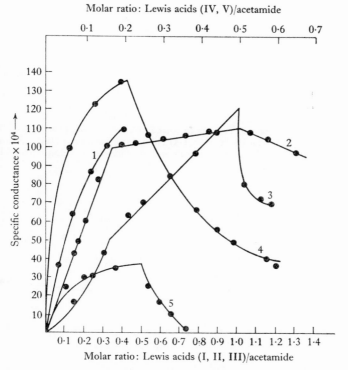

Fig. 1. Specific conductance of Lewis acids in molten acetamide at 100°. 1, Titanium(IV) chloride; 2, antimony(III) chloride; 3, arsenic(III) chloride; 4, antimony(V) chloride; 5, tin(IV) chloride.

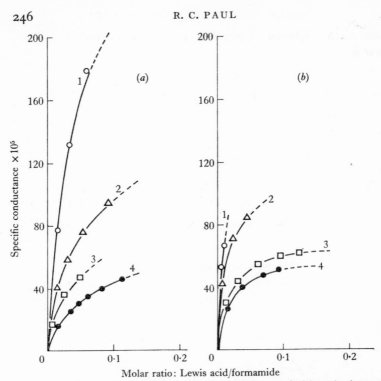

Fig. 2. Conductance of Lewis acids in formamide at 25°. (a) 1, Antimony(V) chloride; 2, tin(IV) chloride; 3, iron(III) chloride; 4, antimony(III) chloride. (b) 1, Zirconium(IV) chloride; 2, titanium(IV) chloride; 3, sulphur trioxide; 4, aluminium chloride.

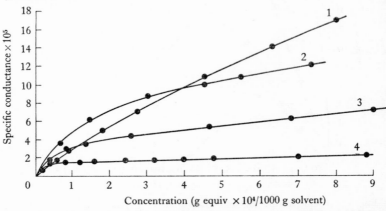

Fig. 3. Specific conductance of Lewis acid complexes in DMF. 1, Sulphur trioxide; 2, tin(IV) bromide; 3, antimony(V) chloride; 4, tin(IV) chloride.

The conductance curves also confirm the formation of the solvates actually isolated, as the breaks in the conductance–concentration curves indicate the composition of the solvates formed. In the case of acetamide the viscosities of the solutions of antimony(V) chloride and tin(IV) chloride increase tremendously beyond the molar ratio corresponding to the formation of solvates and this may be the cause of the sharp fall in the conductances of the solutions. No such phenomenon has been observed in the case of the solutions of Lewis acids in formamide and dimethylformamide.

Electrolysis of the solutions of sulphur trioxide, aluminium chloride, antimony(V) chloride, tin(IV) chloride and titanium(IV) chloride in acetamide and dimethylformamide indicates that hydrogen is evolved at the cathode which proves the existence of the hydrogen ions in these solutions.

The equivalent conductance of the solutions of Lewis acids, sulphur trioxide, aluminium chloride, antimony(V) chloride, tin(IV) chloride and bromide and titanium(IV) chloride has been determined in amides. The plots of equivalent conductance against square root of concentration have been extrapolated to zero concentration to get the values of the equivalent conductance at infinite dilution (λ_0) (Table 5). The fairly high values of (λ_0) may be explained as due to the formation of hydrogen ions and a charge transfer mechanism which is well known in other protonic solvents.[22] Thus, the autoprotolysis of the solvents suggested earlier seems to be quite probable. In comparison to dimethylformamide the values of (λ_0) are much lower in acetamide and formamide which indicates the possibility that formamide and acetamide are relatively weak proton donors. This could also be due to their much higher viscosity. The order of the strength of these Lewis acids at the concentration ranges studied appears to be as follows:

(a) In formamide:
$$SbCl_5 > TiCl_4 > AlCl_3 > SnCl_4$$

(b) In acetamide:
$$SbCl_5 > SnBr_4 > TiCl_4 > SnCl_4$$

(c) In dimethylformamide:
$$SO_3 > SbCl_5 > SnBr_4 > SnCl_4$$

TABLE 5. λ_0 values (Ω^{-1} cm^2) of Lewis acids, protonic acids and bases in amides

	Values in		
	Formamide	Acetamide	Dimethyl-formamide
Lewis acids			
SO$_3$	—	—	245
SbCl$_5$	78·6	170	223
SnCl$_4$	58·7	188·5	470
SnBr$_4$	—	140	480
TiCl$_4$	63·7	92	—
Protonic acids			
HCl	39·4	—	280
H$_2$SO$_4$	—	20·83	465
H$_2$S$_2$O$_7$	—	—	490
HSO$_3$F	—	59·52	267
AcOH	—	1·90	20
CH$_2$ClCO$_2$H	2·4	8·2	28·6
CHCl$_2$CO$_2$H	8·0	16·66	125
CCl$_3$CO$_2$H	23·0	10·90	—
EtCO$_2$H	—	—	16·7
p-C$_6$H$_4$CH$_3$SO$_3$H	—	—	242
iso-C$_3$H$_7$CO$_2$H	—	—	12·5
Bases			
Piperidine	—	48·0	0·21
Benzylamine	—	42·0	17·00
n-Butylamine	—	—	3·44
Triethylamine	—	—	2·20
Pyridine	—	13·0	—
Quinoline	—	18·0	—

4.2. Nitrogen bases

The organic nitrogen bases pyridine, α-, β- and γ-picolines, quinoline, benzylamine and piperidine, by virtue of their having a lone pair of electrons on nitrogen, behave as bases. Keeping in view the auto-ionization of amides, metal amides behave as an-solvo-bases. As the amide solvents are protonic in nature, they may donate protons to the bases. However, to check this aspect, the behaviour of the solutions of the above-mentioned nitrogen bases has been studied conductometrically.

In Fig. 4 the specific conductance of the solutions has been plotted against concentration as represented by the molar ratio of the base to the solvent. An examination of the plots reveals that there is no break in the curves for acetamide and dimethylformamide. Consequently no compound seems to be formed between these amides and bases. However, in the case of formamide, definite compounds with bases have been indicated (Figs. 5, 6).

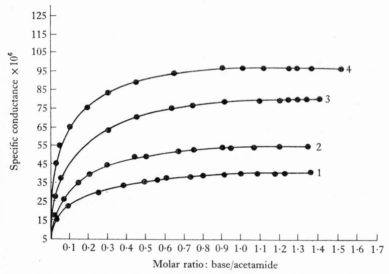

Fig. 4. Specific conductance of bases in molten acetamide at 100°.
1, Quinoline; 2, pyridine; 3, benzylamine; 4, piperidine.

A study of these curves, however, indicates that the conductances of the amides increase on the addition of these bases. This is an indication of the formation of the ions responsible for the conductance and constitutes evidence for the protonation of the bases by the solvent. The conductance studies suggest the following equilibria in the amides:

$$RCONH_2 + B \rightleftharpoons RCONH^- + BH^+$$

To understand further the nature of the solutions of these bases in amides, the equivalent conductance of the solutions of pyridine,

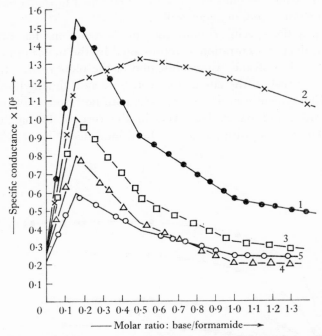

Fig. 5. Conductance of organic bases in formamide at 25°. 1, γ-Picoline;
2, pyridine; 3, α-picoline; 4, β-picoline; 5, aniline.

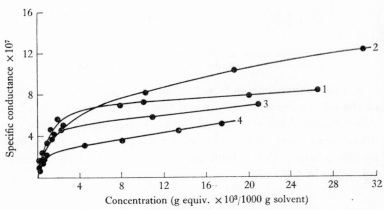

Fig. 6. Specific conductance of bases in DMF. 1, n-Butylamine; 2, piperidine;
3, benzylamine; 4, triethylamine.

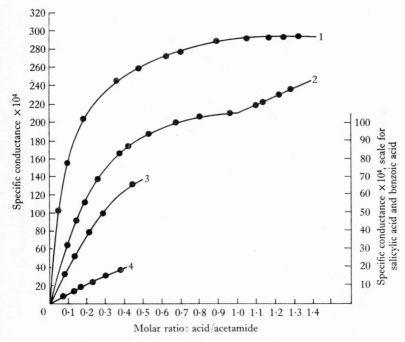

Fig. 7. Specific conductance of protonic acids in molten acetamide at 100°.
1, Fluorosulphuric acid; 2, sulphuric acid; 3, salicylic acid; 4, benzoic acid.

quinoline, benzylamine and piperidine have been determined. The values of the equivalent conductances at infinite dilution (λ_0) obtained as in the case of the Lewis acids are presented in Table 5.

For an electrolyte which undergoes ionization to furnish only two ions in a given solvent, the plots of log C versus log λ_c should be linear with a slope equal to $-\frac{1}{2}$,[23] so long as the degree of ionization is very small as compared to unity. It is certainly more difficult to account for the slope which is more negative than $-\frac{1}{2}$. A slope of $-\frac{3}{4}$ or (-0.75) requires that the ionization reaction produces four ions per molecule of the solute. Hence there might be a possibility that such bases which give a slope of $-\frac{3}{4}$ may exist as dimers. However, different possible mechanisms of dimerization point to the fact that protonation is necessary in the solutions of bases even in dimethylformamide.

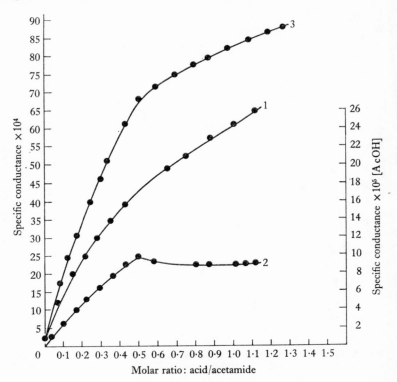

Fig. 8. Specific conductance of acetic acid and substituted acetic acids in molten acetamide at 100°. 1, Acetic acid; 2, monochloroacetic acid; 3, dichloroacetic acid.

4.3. Protonic acids

In the light of the auto-ionization of those amides which are protonic in nature, a study of the conductance of the solutions of protonic acids becomes all the more significant. The specific conductances of sulphuric acid, p-toluene sulphuric acid, acetic and substituted acetic acids indicate that the solutions are ionic in nature. In Figs. 7–12, plots of specific conductance of different protonic acids against the concentration have been presented. Trichloroacetic acid decomposes in molten acetamide and dimethylformamide, giving off CO_2. Mono- and dichloro-acetic acids, however, are apparently stable, indicating that three

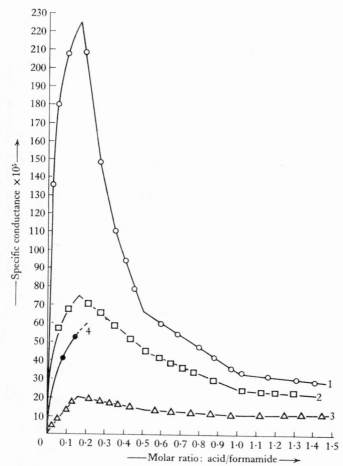

Fig. 9. Conductance of protonic acids in formamide at 25°. 1, Trichloroacetic acid; 2, dichloroacetic acid; 3, monochloroacetic acid; 4, *p*-toluenesulphonic acid.

chlorine atoms on the α-carbon atom weaken the carbon–carbon bond. The decomposition of trichloroacetic acid has already been indicated in water,[24] aniline[24] and N-methyl acetamide.[25]

The equivalent conductance of various protonic acids in amides at infinite dilution have been tabulated in Table 5. The comparatively low values of (λ_0) in acetamide and formamide as compared to dimethylformamide may be either due to the fact that the proton

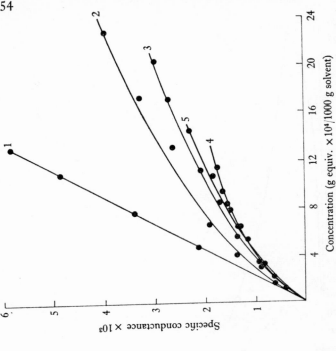

Fig. 11. Specific conductance of protonic acids against concentration in DMF. 1, Fluorosulphuric acid; 2, *p*-toluenesulphuric acid; 3, pyrosulphuric acid; 4, hydrochloric acid; 5, sulphuric acid.

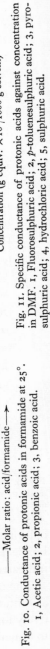

Fig. 10. Conductance of protonic acids in formamide at 25°. 1, Acetic acid; 2, propionic acid; 3, benzoic acid.

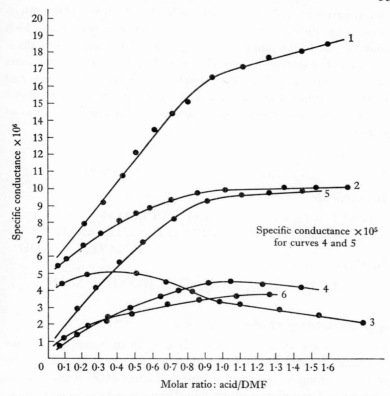

Fig. 12. Specific conductances of weak acids in dimethylformamide at different molar concentrations. 1, Acetic acid; 2, propionic acid; 3, isobutyric acid; 4, monochloroacetic acid; 5, dichloroacetic acid; 6, phenol.

gets highly solvated or the charge transfer mechanism is not as much pronounced in acetamide and formamide as in dimethylformamide. This may also be due to the high viscosity of formamide and acetamide.

The nature of the solutions of these protonic acids is further understood by comparing the experimental and the theoretical values of the Onsager slopes for the solutions of uni-univalent electrolytes using the following expression:

$$\lambda_c = \lambda_0 - \left[\frac{81 \cdot 66}{(DT)^{\frac{1}{2}} \eta} + \frac{8 \cdot 147 \times 10^5}{(DT)^{\frac{3}{2}}} \times \lambda_0 \right] \sqrt{c}$$

A comparison of the slopes in the cases of acetamide and dimethyl-formamide indicates a fair amount of interionic attraction which may be due to the comparatively low dielectric constant of these solvents. The degrees of dissociation and the dissociation constants of these protonic acids indicate that they are largely dissociated and behave as strong electrolytes. The order of the strength of protonic acids in different amides on the basis of their conductance is as follows:

In formamide

$$CCl_3CO_2H > CHCl_2CO_2H > CH_2ClCO_2H$$

In acetamide

$$HSO_3F > H_2SO_4 > \text{picric acid} > CCl_3CO_2H$$
$$> \text{salicylic acid} > CHCl_2CO_2H > CH_2ClCO_2H$$
$$> PhCO_2H > AcOH$$

In dimethylformamide

$$H_2S_2O_7 > HSO_3F > H_2SO_4 > HCO_2H > AcOH$$

4.4. Quaternary ammonium salts

Because of the close structural relationship of acetamide to ammonia, acetic acid and cyanides, the electro-chemical studies of the solutions of ammonium and substituted ammonium halides, alkali acetates and complex cyanides have been carried out in acetamide. Except for the complex cyanides, other electrolytes are fairly soluble in this solvent. The conductances of these solutions indicate that they are good conductors. The equivalent conductances at infinite dilution, degrees of dissociation and dissociation constants of these electrolytes have been tabulated in Table 6. The molar conductances of potassium ferricyanide and sodium ferrocyanide are also given in Table 7. The values of the degrees of dissociation and dissociation constants indicate that these electrolytes are largely ionized in acetamide. The λ_0 values decrease with the increase in size of the alkyl group or the cation. The behaviour thus seems to be consistent with the smaller ions having the greater tendency to associate and is in conformity with the findings in other protonic solvents such as N-methylacetamide,[26] nitromethane[27] and acetonitrile.[28]

TABLE 6. *The values of λ_0, degree of ionization and dissociation constants of various electrolytes in molten acetamide at different concentrations*

Electrolyte	Concentration	$\lambda_0(\Omega.\text{cm}^2)$	Degree of ionization	Dissociation constant $(K \times 10^{-2})$
NH_4Cl	0.286–0.409×10^{-2} N	32.24	0.8991	2.29
NH_4Br	0.342–0.422×10^{-2} N	41.70	0.8393	1.50
NH_4I	0.360–0.442×10^{-2} N	42.20	0.8258	1.40
Me_3NHBr	0.270–0.409×10^{-2} N	34.24	0.9637	6.90
Me_4NBr	0.348–0.483×10^{-2} N	38.76	0.9520	6.57
Et_3NHBr	0.366–0.448×10^{-2} N	33.33	0.8970	2.86
Et_4NBr	0.330–0.476×10^{-2} N	37.33	0.9329	4.28
CH_3COOK	0.193–0.302×10^{-2} N	36.36	0.7529	5.73
CH_3COONa	0.207–0.330×10^{-2} N	31.28	0.7341	4.18
CH_3COOLi	0.129–0.275×10^{-2} N	31.00	0.6451	1.50

TABLE 7. *Specific and molar conductances of sodium ferrocyanide and potassium ferricyanide in molten acetamide at $100°$*

	Concentration	Specific conductance	Molar conductance
$Na_4Fe(CN)_6$	6.02×10^{-3} M	$1.01 \times 10^{-4}\,\Omega^{-1}.\text{cm}^{-1}$	$16.76\,\Omega^{-1}.\text{cm}^2$
$K_3Fe(CN)_6$	4.94×10^{-3} M	$2.21 \times 10^{-4}\,\Omega^{-1}.\text{cm}^{-1}$	$44.44\,\Omega^{-1}.\text{cm}^2$

5. Solvolytic reactions in amides

The solvolytic reactions, besides providing superior methods of preparing new compounds, also establish the ionic or polar nature of the solvents. Amides are generally unstable in light and decompose at their boiling points. They are feebly ionic as is shown by their low specific conductance. It has been proposed that amides undergo auto-ionization as:

$$2RCONH_2 \rightleftharpoons RCONH_3^+ + RCONH^- \quad R{=}H,CH_3$$

Thus in solvolytic reactions the anions of the solute would be replaced by the anions characteristic of the amides.

The composition of these solvolysed products proves the existence of the anion $RCONH^-$ and consequently supports the auto-ionization of amides suggested earlier. Some of the solvolysed products have been listed in Table 8. It is worth while to mention that all the hydrides mentioned in Table 8 are strong reducing agents and reduce the amides to the corresponding amines if the components are heated together in the absence of an inert solvent.

6. Acid–base neutralization reactions in amides

Acid-base neutralization reactions have been most conveniently followed by physico-chemical methods. Conductometric titrations of acids and bases have been carried out in formamide, acetamide and dimethylformamide and have been explained on the basis of the auto-ionization of the solvent.

These studies clearly show that these substances which increase the concentration of the cations characteristic of the solvent behave as acids and those which increase the concentration of the anions characteristic of the solvent behave as bases.

In amides, protonic acids as well as Lewis acids have been shown to exist as ions and behave as ansolvo-acids and solvo-acids respectively, while the organic nitrogen bases behave as solvo-bases in amides. The conductance of the solutions of bases in amides, as discussed earlier, substantiates this fact. A typical acid-base neutralization reaction in an amide may be represented as shown below:

(*a*) formation of the solvo-acid:

$$MX_4 + 2RCONH_2 \rightarrow MX_4 . 2RCONH_2$$

(*b*) ionization of the solvo-acid:

$$MX_4 . 2RCONH_2 + 2RCONH_2$$
$$\rightleftharpoons MX_4(RCONH)_2{}^{2-} + 2(RCONH_2 . H)^+$$

(*c*) formation and ionization of the solvo-base:

$$RCONH_2 + B \rightarrow RCONH_2 . B \rightleftharpoons RCONH^- + BH^+$$

(*d*) neutralization reaction:

$$MX_4(RCONH)_2{}^{2-} + 2RCONH_2 . H^+ + 2RCONH^- + 2BH^+$$
$$\rightleftharpoons (BH)_2(MX_4 . 2RCONH) + 4RCONH_2$$

TABLE 8. *Products obtained through solvolytic reactions in amides*

Solute	Solvolysed product in the case of		
	Formamide	Acetamide	Dimethylformamide
Na	$HCONHNa.HCONH_2$	—	$NaN(CH_3)_2$
NaH	$HCONHNa.HCONH_2$	$CH_3CONHNa$	—
$NaNH_2$	$HCONHNa.HCONH_2$	$CH_3CONHNa$	$NaN(CH_3)_2$
$NaOCH_3$	—	$CH_3CONHNa$	$NaN(CH_3)_2$
$AlCl_3$	$AlCl(HCONH)_2.2HCONH_2$	$AlCl(CH_3CONH)_2$	—
$SnCl_4$	$SnCl_3(HCONH).2HCONH_2$	—	—
$SbCl_5$	$SbCl_3(HCONH)_2.HCONH_2$	$SbCl_3(CH_3CONH)_2.CH_3CONH_2$	—
$ZrCl_4$	$ZrCl_3(HCONH).2HCONH_2$	$ZrCl_2(CH_3CONH)_2$	—
$Cu(CH_3COO)_2$	—	—	$Cu(N(CH_3)_2)_2$
$Al(CH_3COO)_3$	—	—	$Al(N(CH_3)_2)_3$

Similarly the neutralization reaction of organic nitrogen bases with protonic acids in amides may be explained on the basis of the following equilibria:

$$HX + HCON(CH_3)_2 \rightleftharpoons H_2CON(CH_3)_2{}^+ + X^-$$
$$B + HCON(CH_3)_2 \rightleftharpoons (BH)^+ + CON(CH_3)_2{}^-$$
$$H_2CON(CH_3)_2{}^+ + X^- + BH^+ + CON(CH_3)_2{}^-$$
$$\rightleftharpoons BH \cdot HCON(CH_3)_2{}^+X^- + HCON(CH_3)_2$$
$$\Updownarrow$$
$$BH^+ + X^- + 2HCON(CH_3)_2$$

It is of interest that formamide has been used as a medium for carrying out conductometric, potentiometric and visual titrations of Lewis and protonic acids against organic nitrogen bases and alkali metal, copper and cadmium formamides. Copper and cadmium formamides have also been titrated conductometrically against alkali formamides and have been found to be amphoteric in character. The values of the equivalent conductance of Lewis acids, protonic acids and metal formamides indicate the comparatively low mobilities of H^+ and $HCONH^-$ ions in formamide.

REFERENCES

1   W. D. Kumler, *J. Am. chem. Soc.* 1935, **57**, 600.
2   A. M. Buswell, W. H. Rodebush and M. F. Roy, *J. Am. chem. Soc.* 1938, **60**, 2444.
3   M. J. Coppley, G. F. Zellhoefer and C. S. Marvel, *J. Am. chem. Soc.* 1938, **60**, 2666.
4   E. Colten and R. E. Brooker, *J. phys. Chem., Ithaca*, 1958, **62**, 1595.
5   R. Gopal and M. M. Husain, *J. Indian chem. Soc.* 1963, **40**, 272.
6   O. F. Stafford, *J. Am. chem. Soc.* 1933, **55**, 3987.
7   G. Jander, H. Spandau and C. C. Addison, *Chemistry in Non-aqueous Solvents*. Interscience Publishers, 1963.
8   L. Kahovec and K. Knollmüller, *Z. phys. Chem.* 1941, **51**B, 49.
9   A. A. Basushkim, *Izv. Akad. Nauk Fiz.* 1958, **22**, 1131.
10 P. Diehl and R. A. Ogg, *Nature, Lond.* 1957, **180**, 1114.
11 P. Pfieffer, *Liebigs Ann.* 1910, **376**, 285.
12 R. C. Paul and K. C. Malhotra, *Z. anorg. Chem.* 1963, **321**, 56.
13 R. C. Paul and R. Dev, *Ind. J. Chem.* 1965, **3**, 315.
14 R. C. Paul and B. R. Sreenathan, *Ind. J. Chem.* 1964, **2**, 97.
15 R. S. Drago, D. W. Meek, M. D. Joesten and L. LaRoche, *Inorg. Chem.* 1963, **2**, 124.
16 R. C. Paul, B. R. Sreenathan and S. L. Chadha, *J. inorg. nucl. Chem.* 1966, **28**, 1225.

17 R. J. Gillespie and T. Birchall, *J. Am. chem. Soc.* 1960, **82**, 4478.
18 G. Fraenkel and C. Franconi, *Can. J. Chem.* 1963, **41**, 148.
19 A. Clearfield and E. J. Malkiewich, *J. inorg. nucl. Chem.* 1963, **25**, 237.
20 P. J. Stone and H. W. Thompson, *Spectrochim. acta*, 1957, **10**, 17.
21 P. Walden, *Z. anorg. Chem.* 1900, **25**, 209.
22 B. E. Conway, J. M. Bockris and H. Linston, *J. chem. Phys.* 1956, **24**, 834.
23 W. S. Muney and J. F. Coetzee, *J. phys. Chem.* 1962, **66**, 89.
24 F. H. Verhoek, *J. Am. chem. Soc.* 1934, **56**, 571.
25 L. R. Dawson, J. W. Vaughn, M. E. Pruitt and H. C. Eckstrom, *J. phys. Chem.* 1962, **66**, 2684.
26 L. R. Dawson, E. D. Wilhoit and P. G. Sears, *J. Am. chem. Soc.* 1956, **78**, 1569.
27 A. K. R. Unni, L. Elias and H. I. Schiff, *J. phys. Chem.* 1963, **67**, 1216.
28 A. C. Harckness and H. M. Daggett, *Can. J. Chem.* 1965, **43**, 1215.

11

DEFECT AGGREGATION IN SOLID STATE CHEMISTRY

A. L. G. REES

Point defects have been known to interact with one another for almost as long as their existence in crystalline solids has been accepted. The preoccupation with the rather spectacular physical consequences of isolated point defects, particularly the spectroscopic and electronic properties of solids containing them, led to neglect of early study of the nature of their mutual interactions. It is clear that the chemical properties of crystalline solids arise from the interaction of defects—both like and unlike—to produce defect complexes and, through cooperative action, aggregates or ordered arrangements of the defects or their complexes. This article cannot be properly described as a review, but rather as a statement of a point of view. It will attempt to present physical models and outline general mechanisms consistent with the chemical behaviour and certain relevant physical properties of solids. Line defects or dislocations interact with point defects and with each other and certainly have some significance in the kinetics of chemical change. However, they are excluded from this discussion.

1. The interaction of point defects

Point defects, both intrinsic and non-stoicheiometric, have been catalogued adequately many times before,[1] so that it is relevant here to discuss only those of their characteristics that play a significant role in their interaction. Probably the first explicit use of defect interaction was that by Bragg and Williams[2] in 1934 in their classic paper on order–disorder transitions in certain binary alloys. It was used by Lacher[3] in his statistical thermodynamic treatment of the palladium–hydrogen system; Lacher's theory is analogous in this feature to the theories of regular solutions and localized monolayers, which were developed at about the same time largely by

R. H. Fowler and his collaborators.[4] Anderson[5] developed the basic theory of non-stoicheiometric oxides and sulphides on the same kind of assumption. In essence the theory postulates an attractive interaction between defects and predicts the separation of a new phase under certain conditions. Fundamentally nothing further is required as an additional characteristic of a solid to explain both the separation of new phases in a chemical reaction and perhaps certain phase changes.

There is a thermodynamic requirement for the existence of point defects of one kind or another in all crystalline solids at finite temperatures, simply because the completely ordered arrangement of atoms in a crystal is the configuration of lowest potential energy and any disturbance of this perfection requires the expenditure of energy. In low concentration (< 1 in 10^5) such defects presumably behave fairly independently of each other, although for some systems suggestions of clustering have been made. At higher concentrations interaction becomes important. The types of point defect that we are concerned with are (i) vacancy, (ii) interstitial and (iii) substitutional. Each of these introduces local strain and a local redistribution of charge into the lattice.

At this point it may be of value to examine in a little more detail the origin of the interactions. For like defects the elastic and Coulomb interactions will undoubtedly be repulsive, but these will be counteracted by an attractive interaction which is presumably associated with quantum–mechanical exchange, leading to a minimum in the interaction energy curve at some separation, the depth of the minimum being in general shallower the greater the separation at which it occurs. If the attractive component of the interaction is large then it would be normal for defects to associate as closely as possible; the stable configuration will consist of defects on nearest-neighbour sites. However, there is no reason why potential minima should not occur at next nearest-neighbour sites or at more remote sites. Provided the depth of a minimum is larger than the thermal energy (kT) it will have some significance in determining stable configurations of defects (e.g. superlattices).

Unlike defects will of course interact strongly because here the elastic or electrical interactions are attractive rather than repulsive; the point defects will tend to migrate towards one another so that the stresses and electrical interactions are optimized. The most obvious interactions of unlike defects are between those of opposite charge. Simple and familiar examples are:

(a) Electrons associating with vacant anion sites in an ionic crystal where the effective positive Coulomb field of the vacancy binds the electron in a hydrogen-like complex known as an F-centre, represented by the symbol $(e|\square^{-})$.

(b) Anion vacancies associating in simple complexes with anions of greater charge than those of the anions of the host crystal in which they are present substitutionally. This situation is exemplified by the association of sulphide ions and anion vacancies in silver bromide crystals containing small quantities of sulphide, represented by the symbol $(S^{2-}|\square^{-}\mathfrak{X}\square^{-})$.

Such defect complexes have reasonable stability. In some circumstances it has been established that the very act of association of two defects can introduce stresses which find relief in a rearrangement of the atoms or ions in the immediate neighbourhood. This is simply a statement of the fact that a local configuration different from that of the host lattice has a lower potential energy and that this new configuration or symmetry persists only within the small volume affected by the defect complex. The now classic example of this is the interaction of Fe^{3+}, ions in ferrous oxide of composition $Fe_{1-x}O$ with cation vacancies.[6] Here the defect complex

$$(Fe^{3+}|\square^{+}\mathfrak{X}\square^{+})$$

is not the most stable configuration. The Fe^{3+} ion on a normal cation site will be a centre of effective positive charge. The cation vacancy, which balances the deficiency in iron in the lattice, carries an effective negative charge so a strong interaction results. The Fe^{3+} ion, being a little smaller than the Fe^{2+}, can be accommodated in the interstitial position between its ideal site and the site of the vacancy to which it is attracted; it moves into this interstitial position, leaving behind another cation vacancy. The net result is a more symmetrical complex of a Fe^{3+} ion in an interstitial position between two cation vacancies; $(Fe^{3+}|\triangle\mathfrak{X}\square^{+})_2$. A similar rearrangement has been established for the oxygen-rich UO_{2+x} phases derived from the UO_2 fluorite-type structure,[7] in which on formal grounds one had assumed the excess oxygen atoms to be accommodated interstitially. Actually the interstitial oxygen repels a neighbouring lattice oxygen, pushing it into an interstitial site on the side of the lattice site remote from the excess interstitial oxygen. The complex so formed is then a vacant anion site with two interstitial oxygens associated with it. The complex may be symbolized $(\Box^- \mathfrak{X} O^{2-} \,|\, \triangle)_2$. The type of situation exemplified here is probably widespread; in fact, it is reasonable to infer that such rearrangement complexes are a necessary feature of any system from which a new phase of different basic structure separates out on appropriate change of one of the variables.[8]

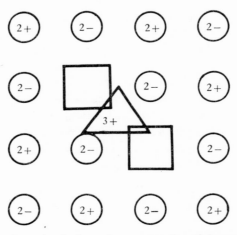

Fig. 1. Two-dimensional schematic representation of the rearrangement defect complex, $(Fe^{3+} \,|\, \triangle \mathfrak{X} \Box)_2$, in $Fe_{1-x}O$.

2. Statistics of defect populations

So far we have considered only the interaction of like and unlike point defects to form simple complexes which can originate at any point in the host crystal. The separation of a new phase undoubtedly requires the coalescence of these units to form ordered aggregates which ultimately become a new phase. Until recently there has been little explicit understanding of the implications of the simplest statistical mechanical models of interacting point defects in crystals. It would be profitable to examine the case of a simple solid containing one kind of simple defect, requiring the expenditure of energy E_a for its formation. If these defects are assumed to interact attractively—and there is ample evidence that this is correct in principle—one may define another energy parameter $\frac{2}{z} E_{aa}$, which denotes the average interaction energy per defect pair.

z is the coordination number (the number of adjacent defect sites) for a defect site in the crystal. If the system variables are (i) the composition, expressed here as the defect population, $\theta = N_a/N$, where N_a and N are the number of defects present and the number of potential defect sites respectively, (ii) the pressure of the volatile

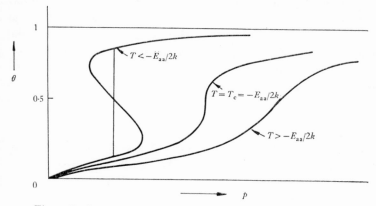

Fig. 2. Isotherms showing two-phase region, critical behaviour and homogeneous composition range.

component of the solid phase, p, and (iii) the temperature, T, then the statistical thermodynamic equation connecting these variables is of the form

$$f(p) = A \left(\frac{\theta}{1-\theta}\right) \exp\left[-E_a - 2\theta E_{aa}/kT\right]$$

where A is a composite factor which varies slowly with temperature.[9] For the isothermal dependence of θ on p we find that for temperatures below a critical temperature defined by $T_c = -E_{aa}/2k$ the θ–p relation has a familiar form shown in Fig. 2. This is simply a quantitative statement that at $T < T_c$ the thermal energy is not adequate to prevent the cooperative interaction of defects to produce a new phase and that for $T > T_c$ the thermal energy is sufficient to prevent the aggregation proceeding to a point where a new phase separates out, *but it does not mean that the defect population consists entirely of isolated defects.* It is obvious that the key to the separation of a new phase is the interaction energy parameter. If E_{aa} were zero no new phases would be possible. What we wish to explore is the nature of the defect population along the θ–p curve (i) up to the point of appearance of the second phase for isotherms for which $T < T_c$ and (ii) along the whole of the θ–p curves for $T > T_c$.

In setting up the statistical mechanical problem the contribution from defect interaction to the energy is $\bar{N}_{aa}E_{aa}$, where $\bar{N}_{aa}$ is the average number of defect pairs. Relating $\bar{N}_{aa}$ to the defect population N_a is the significant problem here. To make the mathematical expression useful for comparison with experiment one counts the average number of pairs on the assumption that the interaction energy is zero. Even on this crude assumption of random distribution one finds that pairs exist to the extent

$$\bar{N}_{aa} = \frac{\frac{1}{2}zN_a^2}{N},$$

which leads to the $2\theta E_{aa}$ term in our θ–p–T relation. The average number of atoms involved in pairs $\bar{N}_p$ can be shown by simple counting to be given by

$$\bar{N}_p = \frac{2}{z}\bar{N}_{aa} = \frac{N_a^2}{N}$$

The fraction of defects involved in pairs is then $\theta_p = \bar{N}_p/N = \theta^2$. It must be pointed out here that these pairs are not necessarily isolated pairs, but that defects having $1, 2, \ldots, z$ nearest-neighbour defects have been taken into account. Expressed another way, we can say that a distribution of aggregate sizes exists.

If the defect pairs are counted with due recognition of a finite attractive interaction then the relation is algebraically more complex, but for interaction energies much greater than kT the number of defects involved in pairs is equal to the total number of defects, as one might expect, i.e. $\theta_p \to \theta$. Figure 3 shows the range of involvement of defects in aggregates distributed in size from 2 to N_a for $T < T_c$ and $T > T_c$. These simple statistical results demonstrate that a significant proportion of defects is involved in aggregates or clusters under almost any circumstances of practical import. Although this discussion has been developed for non-stoicheiometric defects, the same arguments apply to other kinds of thermal defect (e.g. intrinsic).

The extension to defect complexes involves no difficulties in principle, but there are substantial mathematical complications. There is, however, a problem of mechanism which is not relevant to the statistical thermodynamic treatment. Our understanding of diffusion processes provides an easy picture of the way in which aggregates of simple defects grow. However, it is difficult to

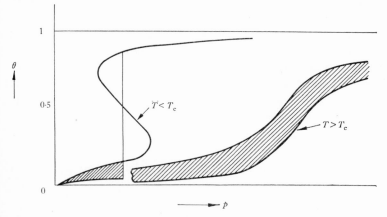

Fig. 3. Isotherms for temperatures below and above the critical temperature showing ranges of proportion of defects involved in aggregates. Lower boundaries of hatched areas—no interaction. Upper boundaries of hatched areas—large interaction.

imagine how a population of isolated defect complexes can aggregate by diffusion of individual complexes. In fact, it is physically improbable for complexes to move through a lattice as entities, except perhaps for those involving electrical defects, namely trapped electrons and positive holes. Even for aggregation of the simplest defect complex, the F-centre, a stepwise accretion of various vacancies and electrons has been established. For complexes such as that in $Fe_{1-x}O$ the impossibility of integral diffusion of the complex of two vacancies and an interstitial atom is obvious; such complexes must dissociate and the simple components diffuse and accrete separately at the growing cluster, the configurational readjustment occurring there. The system is dynamic; at any instant there are present in the population, clusters, clusters plus single defects, defect complexes and isolated simple defects of both kinds.

3. Premonitory effects

In the previous sections a connection between the defect aggregate existing in a homogeneous phase and the product phase that separates out when the variables are changed appropriately has been implied, but not examined in detail. The connection is easy to comprehend when the atoms of the host lattice maintain their structural relationship in the product lattice with dimensional changes only, such as occurs in alkali halides containing F-centres whose aggregates lead ultimately to alkali metal phase separation. This has been discussed many times.[10] The significant feature is that the macroscopically homogeneous reactant phase contains premonitory evidence of the as yet non-existent product phase.

The neutron diffraction studies of Roth[6] on $Fe_{1-x}O$ and Willis[7] on UO_{2+x} demonstrated for the first time that the structure of the defect complexes was related to the ideal stoicheiometric phases to which they were tending, namely Fe_3O_4 and U_4O_9 respectively. Less direct evidence for similar premonitory phenomena has been deduced for other systems.[11] The evidence and permissible conclusions have been presented very lucidly by J. S. Anderson in his 1963 Liversidge Lecture to the Chemical Society.[8] The defect aggregates have been referred to as 'microheterogeneoities' or 'microdomains'—they are properly ordered small volumes of the crystal having distinctly different structure from the host, but are not thermodynamically recognizable as a separate phase. The presence of microdomains raises difficulties in properly defining a phase. This problem was discussed at length at a conference on Defects in Solids held in Perth, Australia, in 1966, but no uniformly acceptable definition emerged. If one accepts Ariya's description of the TiO and VO phases as containing more than one type of microdomain and that the solid is composed of a 'single-phase' mixture of them, then the difficulties are even more acute. These proposals need closer study before acceptance, since in the author's view long-distance order of one of the phases bounding the homogeneous region is a necessary prerequisite of a single phase.

The structural and other evidence for microdomains is entirely consistent with the simplest statistical mechanical models; as pointed out earlier the very presence of an attractive interaction energy demands their existence as anticipatory steps. The precise description of systems involving the more elaborate defect complexes and sequences of phases is not possible at the moment, although some progress has been made in the latter problem.[9]

Shear structures which occur in many metal oxide systems displaying homogeneous series and which have been extensively studied and characterized by Magneli and Wadsley are of a further order of complexity and must await the unravelling of the problems of the simpler systems for adequate statistical thermodynamic description. It is not clear from the structural evidence that homogeneous ranges of variable composition of finite extent exist in these systems, so that no discussion of premonitory effects is possible. For information on these systems reference should be made to the excellent reviews by Wadsley.[12]

The recognition of premonitory interaction of defects is of importance to the understanding of reactions in the solid state and provides a reasonable basis for the defect aggregation and nucleation steps. Unless of course these are rate-determining steps their presence as part of the total reaction mechanism will be unobservable, but they remain necessary conceptual parts of the mechanism.

4. A premonitory model of melting

It is tempting to explore the applicability of cooperative defect interaction in a more general way, particularly to transformations of one kind or another in simple solids of fixed composition. Certainly transformations of the order–disorder type, magnetic and ferroelectric ordering, are precisely of this nature.[13] Another transition for which there is convincing evidence for premonitory effects in macroscopic properties, but for which there is no satisfying model, is that of melting.[14] The most significant physical features of melting are (i) the persistence of long-distance order right up to the melting temperature, and its discontinuous disappearance at that temperature, (ii) the sudden collapse of the crystal at the melting temperature, (iii) the existence of pre-melting and post-melting effects in various physical properties (e.g. specific heat, thermal expansion coefficient, etc.). The proposals about a mechanism of melting presented here are centred around the occurrence of vacancy defects. No one would doubt that the most important difference between a solid and liquid is the presence of long-distance order in the solid, but not in the liquid, and various models have been proposed in the past to simulate the discontinuous change in the entropy that certainly takes place on melting. Some time ago attempts were made to associate melting with the change in order brought about by the increase in the number of vacant lattice points as the melting point is approached. Several theories[15] based on this notion have been proposed, but they are subject to the objections (i) that the equilibrium concentration of vacant sites must be small right up to the melting point and (ii) that if a sufficiently high concentration of vacancies could be produced in the crystal such a high concentration is not consistent with the maintenance of long-range order.

Any model of the melting mechanism must provide an explanation of (i) the phenomena of pre- and post-melting, exemplified in such properties as specific heat, viscosity, etc., (ii) the persistence of long-distance order, as demonstrated by X-ray diffraction, right up to the melting point, (iii) supercooling, (iv) the effects of impurities on the melting point, and (v) a quantitative description of the change in the various thermodynamic parameters with temperature as the melting point is approached and at the melting point itself.

The model of melting described briefly here[16] is a defect aggregation mechanism. However, it does not assume that vacancies interact to form large aggregates or voids whose volume is many times that of a single vacancy. The model takes account of the further interaction between a small void (an aggregate of a small integral number of vacancies) and the atoms (or molecules or ions; referred to hereafter collectively as atoms) occupying proper lattice positions in the surface of such a void. There will be a tendency for atoms in the surface of a void, which are in special position energetically speaking, to collapse or 'evaporate' into the voids and, since the potential energy surface will be rather flat over the whole volume of the void (see Fig. 4), there will be little tendency for these atoms to take up ordered positions. They will, in fact, tend to form volumes of disorder, presumably of a random close-packed character in the Bernal sense.[17] The processes of creation and aggregation of vacancies and the collapse of atoms into the void so produced would not be expected to proceed independently, but rather according to a step-wise process of alternate accretion of vacancies

Fig. 4. Schematic representation of the flat-bottomed potential well presented by a square aggregate of four vacant sites to an additional atom entering the void.

and void-peripheral atoms according to a scheme of the following type:

$$\square + \square \rightarrow \square \mathfrak{X} \square$$
$$\square \mathfrak{X} \square + (A_v \mid \square) \rightarrow (A_v \mid \square_3)$$
$$(A_v \mid \square_3) + \square \rightarrow (A_v \mid \square_4)$$
$$(A_v \mid \square_4) + (A_v \mid \square) \rightarrow (A_{v_2} \mid \square_5)$$
$$\rightarrow (A_{v_n} \mid \square_m)$$

[A_v signifies a void-peripheral atom]

The general symbol $(A_{v_n} \mid \square_m)$ describing a disordered region denotes an aggregate of n void-peripheral atoms occupying a total volume equivalent to m lattice sites. The important point about this model is that a considerable volume of a crystal may be transformed into the disordered aggregate by the creation of relatively few intrinsic vacancies (see Fig. 5), simply because each void-peripheral atom collapsing into an aggregate brings with it the volume of one crystal lattice site. Clearly the number of additional vacancies required must be determined by the increase in molecular volume of the random close-packed state over that of the crystalline solid.

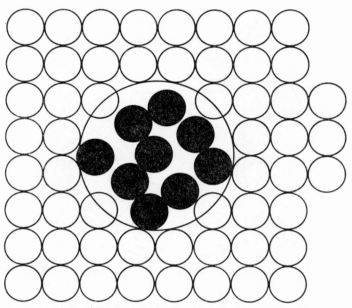

Fig. 5. Disordered aggregate of nine atoms occupying the equivalent of twelve lattice sites, but with the creation of three vacancies only.

A simple calculation establishes a rough relation between the number of atoms in random close-packed regions with the number of vacancies created, namely

$$N_r = \left(\frac{\rho_1}{\rho_s - \rho_1}\right) N_h,$$

where $\rho_{1,s}$ are the densities of the liquid, solid. Since we know that density changes on melting are usually of the order of 1%, we expect the 'amplification factor', as it were, to be of the order of

100. A few crystals, such as those of H_2O, Sb, Bi, Ga, decrease in volume on melting and simple calculations obviously do not apply.

This concept leads to a geometrical model of the melting process and use may be made of the results to determine the parameters of a simple statistical mechanical theory. At any temperature approaching the melting point we can imagine the initiation of small random close-packed regions throughout the relatively perfect

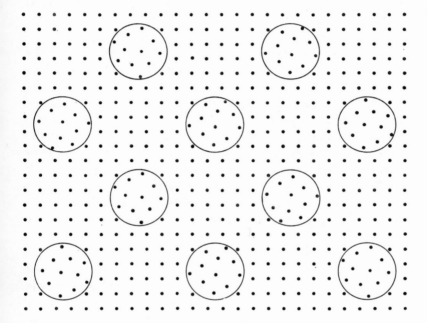

Fig. 6. Disordered regions nucleating in a regular array throughout an ordered structure.

crystal lattice. In the ideal isotropic case one can assume these volumes to be spherical and to grow at a constant rate. In the simplest and, of course, naïve and unrealistic model one can imagine the centres of these disordered regions to be distributed on a regular 'lattice' (Fig. 6). The implications of such a model are immediately obvious. The liquid-like phase is contained in isolated islands or clusters, but the long-distance order is still retained throughout the host crystal. The fraction of the volume required by the random close-packed regions is independent of the number of separate regions for the same total number of molecules involved in disordered regions. At some point in their growth the spherical regions will coalesce at the same instant. In fact, if the disordered regions are distributed on a cubic close-packed lattice, this occurs when the total volume of the disordered regions reaches $\pi/(3\sqrt{2})$ of the total volume of the solid. At this point continuity of the crystalline region is replaced by continuity of the disordered region with consequent disappearance of mechanical rigidity of the specimen; the temperatures at which this occurs are identified with the melting point.

The most obvious deficiency of this model is the assumption of a regular distribution of disordered regions. The disordered regions will nucleate at random points throughout the volume of the crystal. It is perhaps not unrealistic to assume that the regions each achieve the same size at equilibrium at a particular temperature, but that the initial distribution is important in determining the point at which all the disordered regions coalesce. It is possible to calculate the average distance between nearest neighbours from a distribution function for the nearest-neighbour distances in a stationary random distribution of points.[18] From this one may obtain the proportional volume of clusters at the point of coalescence. It is

$$\alpha = \{\Gamma(4/3)/2\}^3 = 0\cdot0893$$

We note here that the volume-ratio is independent of the number of disordered regions and that collapse of the long-distance order is predicted to take place when the disordered volume is less than 10 % of the total. It is relevant to point out that the liquid just above the melting point would contain ordered crystalline remnants dispersed through the disordered liquid.

A crude statistical mechanical treatment may be used on this model, starting from the transfer of lattice atoms to positions of disorder without concern for the steps in the mechanism. If we define for a crystal of N atoms

N_r the number of molecules in random close-packed (r.c.p.) volumes at temperature T,

E_r the average energy required to transfer a molecule from an ordered site to an r.c.p. site,

$2E_{rr}/z$ the average interaction energy per pair of molecules in r.c.p. sites,

z the coordination number of an atom in an r.c.p. assembly,

$a(T)$ the extra factor in the partition function associated with the changed location of an atom transferred to an r.c.p. site,

$\overline{N}_{rr}$ the average number of nearest-neighbour pairs of molecules in r.c.p. sites,

then we may write for the partition function factor

$$R(T) = \sum_{N_r} \frac{N!}{(N-N_r)!N_r!}\, [a(T)]^{N_r} e^{-[N_r E_r + (2\overline{N}_{rr}E_{rr}/z)]/kT}$$

By using the approximate expression to evaluate $\overline{N}_{rr}$ and by putting $[\partial \ln R(T)]/[\partial N_r] = 0$, we find the temperature dependence of N_r to be given by

$$\left(\frac{\theta}{1-\theta}\right) e^{\theta E_{rr}/kT} = a(T)e^{-E_r/kT}$$

where $\theta = N_r/N$. For simplicity write $A = 2E_{rr}/k$, $B = E_r/k$ and $C = a(T)$, assumed to be varying slowly with temperature. The relationship may now be written

$$\frac{1}{-(A\theta+B)} \ln\left[\frac{1}{C}\left(\frac{\theta}{1-\theta}\right)\right] = \frac{1}{T}$$

For certain values of the parameters this equation exhibits three roots. Actually, for it to exhibit this behaviour

$$\left(\frac{-2A}{A+2B}\ln C\right) \quad \text{must be} \quad > 4$$

and $A < 0$. When these conditions are satisfied the relation is of the general form shown in Fig. 7. Beyond $\theta_s = C/(C+1)$ the equation loses physical significance. However, this is usually outside the significant range of θ values. The statistical mechanical theory predicts both the occurrence of aggregates of atoms in disordered volumes at temperatures below the melting point and the survival of molecules in ordered positions above the melting point. On the assumption of strong interaction, which is really demanded by the model, almost all atoms in random close-packed sites are in aggregates. From the above relation one can derive, by putting $\theta = 0.5$, the following relation for the melting point, T_m,

$$T_m = \frac{A + 2B}{2 \ln C}$$

From the model we expect the onset of coalescence at $\theta = 0.1$, so we may identify this value with the smallest root θ_1 of this

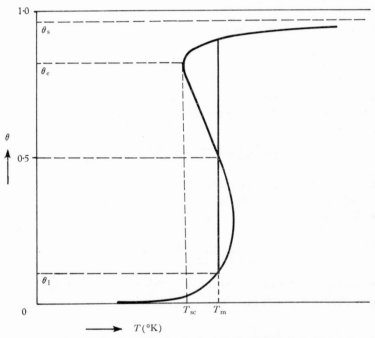

Fig. 7. Dependence of extent of disorder on temperature showing the discontinuous change on the disorder at the melting temperature. Specific values of θ referred to in the text are indicated on the diagram.

equation to obtain a further relationship between the parameters of the equation. We find

$$\left(\frac{A}{A + 2B}\right) \ln C = \frac{1}{1 - 2\theta_1} \ln \left(\frac{\theta_1}{1 - \theta_1}\right) = D$$

For $\theta_1 = 0.1$, $D = -2.75$, and $2D = A/T_m$ or $A = -5.50 T_m$.

One may identify the expenditure of energy NE_r with the production of 1 mole of non-interacting atoms in random locations (a highly compressed perfect gas) and $-NE_{rr}$ with the energy expended in interaction to produce a mole of liquid. Hence the latent heat of fusion is given by $\Lambda_m = N(E_{rr}+E_r) = A+2B$. The artificial intermediate state is related to vapour, so we might expect $-NE_{rr}$ to be related to Λ_v, the latent heat of vaporization of the liquid at T_m. There is some justification for writing

$$-A = F\Lambda_v$$

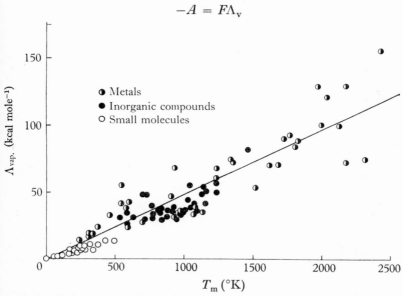

Fig. 8. Experimental values of the latent heat of vaporization plotted against melting temperature for solid elements, gaseous elements, simple molecules and salts.

where F is a temperature-independent factor, so that

$$T_m = \frac{F\Lambda_v}{5\cdot50}$$

Plots of Λ_v versus T_m for elements, simple molecules and salts[19] are given in Fig. 8. The excellent linear dependence for the rare gases and the halogen molecules (Fig. 9) suggests that there is some measure of correctness in the model.

From the expression for contributions to the Helmholtz free energy F one can derive expressions for the contributions to the entropy S and the specific heat C_v. One observes that these reproduce the premonitory behaviour found experimentally. Moreover, the entropy of fusion at the melting point, using the identification of $A + 2B$ with Λ_m, turns out to be

$$\Delta S_m = \frac{0 \cdot 8 \Lambda_m}{T_m}$$

as expected.

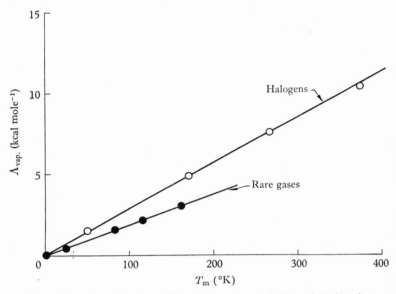

Fig. 9. Experimental values of latent heat of vaporization plotted against melting temperature for the rare gases and halogen molecules.

This model may also be used to derive a relationship between the temperature of maximum supercooling T_{sc} and the melting point. The relation is

$$\frac{T_{sc}}{T_m} = - 2D\theta_e(1 - \theta_e)$$

where θ_e is the value of θ at the lower limit of T on the upper loop of the θ–T curve. Now it is obvious that for any values of the parameters, θ_e must lie close to 0·80 so that

$$\frac{T_{sc}}{T_m} \simeq 0 \cdot 80$$

which is found experimentally.[20] Superheating of the solid is also possible on this model, but this is unlikely to be achieved experimentally.

The qualitative implications of this model are fairly obvious. The more important are:

(i) A physical picture of the sudden collapse of crystal rigidity at the melting point is provided by the model.

(ii) Premelting phenomena are explicitly predicted. The increase in the number of molecules in random close-packed regions with temperature below the melting point which the model demands allows one to calculate the anomalous increase in specific volume, specific heat, etc., up to the melting point.

(iii) Post-melting phenomena are predicted because of the existence of calculable amounts of crystalline remnants at temperature above the melting point.

(iv) These predictions are consistent with the X-ray evidence that X-ray reflections retain their sharpness up to the melting point, but lose intensity, and that diffuse scattering increases progressively up to the melting point.

(v) Highly anisometric molecules in highly anisotropic structures will tend to form cylindrical or prismatic rather than spherical r.c.p. regions. The crystalline remnants would tend to be cylindrical or lath-like, leading to a mesomorphic phase (liquid crystals).

For such a crude model to account qualitatively for the observed facts so satisfactorily is encouraging. Refinements of such a defect aggregation model of melting could lead to a quantitative description of the whole phenomenon of melting. Experimental tests of the model could be made by neutron or X-ray diffraction, since the fraction of the crystal in a totally disordered state is significant at temperatures well below the melting point. There is some evidence that this is so. The experimental comparison of X-ray and dilatometric densities for sodium metal[21] shows significant divergence at temperatures 75 °K below the melting point and a calculated disorder of $\sim 4\%$ of the volume at the melting point.

5. Conclusion

Defect aggregation and the attractive interaction energy associated with aggregation are undoubtedly of central importance in anticipating the production of a new phase. Whether this concept would

282   A. L. G. REES

be of value in discussing models of first-order transitions between two solid phases of the same composition is not clear, but the possibility would bear examination.

REFERENCES

1 F. Seitz, *J. appl. Phys.* 1945, **16**, 553. J. S. Anderson, *Rep. Prog. Chem.* 1947, **43**, 104. A. L. G. Rees, *Chemistry of the Defect Solid State*, London, Methuen, 1954. *Electrochemistry* (ed. J. A. Friend and F. Gutmann), p. 3, Pergamon, 1964.
2 W. L. Bragg and E. J. Williams, *Proc. Roy. Soc.* A, 1934, **145**, 699.
3 J. R. Lacher, *Proc. Roy. Soc.* A, 1937, **161**, 525.
4 R. H. Fowler and E. A. Guggenheim, *Statistical Thermodynamics*, chapters x and xiii. Cambridge University Press, 1939.
5 J. S. Anderson, *Proc. Roy. Soc.* A, 1946, **185**, 69.
6 W. L. Roth, *Acta crystallogr.* 1960, **13**, 140.
7 B. T. M. Willis, *Proc. Roy. Soc.* A, 1963, **274**, 134; *J. Phys., Paris*, 1964, **25** (5), 431.
8 J. S. Anderson, *Proc. chem. Soc.*, June 1964, p. 166.
9 J. S. Anderson, *Proc. Roy. Soc.* A, 1946, **185**, 69. A. L. G. Rees, *Trans. Faraday Soc.* 1954, **50**, 335.
10 J. W. Mitchell, *Fundamental Mechanisms of Photographic Sensitivity*, p. 242. London: Butterworth, 1951.
11 S. M. Ariya and M. P. Morozova, *Zhur. obshch. Khim.* 1958, **28**, 2617. S. M. Ariya and Y. G. Popov, *Zhur. obshch. Khim.* 1962, **32**, 2077.
12 A. D. Wadsley, *Adv. Chem.* no. 39, p. 23 (1963). *Non-stoichiometric Compounds* (ed. L. Mandelcorn), p. 98, Academic Press, 1963.
13 See, for example, J. M. Ziman, *Principles of the Theory of Solids*. Cambridge University Press, 1964.
14 A. R. Ubbelohde, *Q. Rev. chem. Soc.* 1950, **4**, 356; *Melting and Crystal Structure*, Oxford, 1965. G. Borelius, *Solid St. Phys.* 1963, **15**, 2.
15 F. C. Frank, *Proc. Roy. Soc.* A, 1939, **170**, 182. S. Bresler, *Acta phys.-chim. URSS*, 1939, **10**, 491.
16 A. L. G. Rees, unpublished.
17 J. D. Bernal, *Proc. Roy. Soc.* A, 1964, **280**, 5299. *Liquids: Structure, Properties, Solid Interactions* (ed. T. J. Hughel), p. 25. Amsterdam: Elsevier, 1965.
18 S. Chandrasekar, *Rev. mod. Phys.* 1943, **15**, 1.
19 Data from *Handbook of Chemistry and Physics*, 47th edn (ed. R. C. Weast), pp. D–33 *et seq.* Cleveland: Chemical Rubber Co., 1966.
20 D. Turnbull, *J. appl. Phys.* 1950, **21**, 1022.
21 R. Feder and H. P. Charbnau, *Phys. Rev.* 1966, **149**, 464.

12

TRANSITION METAL DERIVATIVES OF SILICON, GERMANIUM, TIN AND LEAD

F. G. A. STONE

1. Introduction

One of the most active areas for research in inorganic chemistry at the present time is the study of compounds possessing covalent metal–metal bonds. In one such class of compound a transition metal is bonded to silicon, germanium, tin, or lead and it is with these complexes that this chapter is concerned.

Most transition metal derivatives of the Group IV B elements have been characterized in the last five years, although Hein and Pobloth[1] and Hieber and Teller[2] reported the first examples over twenty-five years ago. The original formulations proposed for these complexes, e.g. $Sn[Co(CO)_4]_2$ or $Et_2Pb:Fe(CO)_4$,* must now be regarded as untenable, but it should be remembered that this work occurred prior to the renaissance of inorganic chemistry, at a time when separation techniques were less developed than they are today, and when such aids as infrared and nuclear magnetic resonance spectroscopy were unknown.

Development of this field awaited the stimulus provided by the discovery of several stable σ-bonded alkyl derivatives of the transition metals.[4,5] The recognition[4,6] that transition metals could be 'conditioned' by the presence of π acceptor ligands to form σ bonds with carbon atoms made it logical that these same metals should be capable of forming similar bonds with the other Group IV elements. Indeed, the silicon–iron complex $Me_3SiFe(CO)_2(\pi\text{-}C_5H_5)$ was characterized[7] at much the same time as was $H_3CFe(CO)_2(\pi\text{-}C_5H_5)$.[8] Gorsich's[9] study of the formation of complexes $R_{4-n}M[Mn(CO)_5]_n$ from the reaction between the anion $[Mn(CO)_5]^-$ and organo-tin and -lead halides provided much needed impetus to this field.

* In 1947 Hein and Heuser[3] reported that this type of complex was dimeric.

2. Synthesis

Table 1 provides the reader with a summary of the five methods that have been used to prepare the majority of the binuclear and polynuclear metal complexes, and the many examples given are intended to illustrate the extent of the field.

TABLE 1. *Methods of synthesis, and some examples of complexes of transition metals with Group IV atoms as ligands*

I. Metathetical reactions involving a carbonyl metal anion

$H_3SiI + Co(CO)_4^- \rightarrow H_3SiCo(CO)_4 + I^-$	10
$GeCl_4 + 2\pi\text{-}C_5H_5Fe(CO)_2^- \rightarrow Cl_2Ge[Fe(CO)_2(\pi\text{-}C_5H_5)]_2 + 2Cl^-$	11
$Ph_3GeCl + \pi\text{-}C_5H_5W(CO)_3^- \rightarrow Ph_3GeW(CO)_3(\pi\text{-}C_5H_5) + Cl^-$	12
$Bu_3SnCl + Co(CO)_4^- \rightarrow Bu_3SnCo(CO)_4 + Cl^-$	13
$2Ph_2SnCl_2 + 2Fe(CO)_4^{2-} \rightarrow [Ph_2SnFe(CO)_4]_2 + 4Cl^-$	14
$Me_2SnCl_2 + 2Re(CO)_5^- \rightarrow Me_2Sn[Re(CO)_5]_2 + 2Cl^-$	15
$Ph_3SnCl + Ir(CO)_3PPh_3^- \rightarrow Ph_3SnIr(CO)_3PPh_3 + Cl^-$	16
$Me_3SnCl + Rh(CO)_2(PPh_3)_2^- \rightarrow Me_3SnRh(CO)_2(PPh_3)_2 + Cl^-$	16
$2Et_3PbCl + Fe(CO)_4^{2-} \rightarrow (Et_3Pb)_2Fe(CO)_4 + 2Cl^-$	14, 17
$Ph_2PbCl_2 + 2Co(CO)_4^- \rightarrow Ph_2Pb[Co(CO)_4]_2 + 2Cl^-$	14
$Ph_3PbCl + \pi\text{-}C_5H_5Cr(CO)_3^- \rightarrow Ph_3PbCr(CO)_3(\pi\text{-}C_5H_5) + Cl^-$	12

II. Reactions involving elimination of a neutral molecule

$2Ph_3SiH + Mn_2(CO)_{10} \rightarrow 2Ph_3SiMn(CO)_5 + H_2$	15
$2Cl_3SiH + Co_2(CO)_8 \rightarrow 2Cl_3SiCo(CO)_4 + H_2$	18
$GeH_4 + 2HMn(CO)_5 \rightarrow H_2Ge[Mn(CO)_5]_2 + 2H_2$	19
$6Me_3SiH + 2Ru_3(CO)_{12} \rightarrow 3Me_3SiRu(CO)_4Ru(CO)_4SiMe_3 + 3H_2$	20
$HGeCl_3 + ClMn(CO)_5 \rightarrow Cl_3GeMn(CO)_5 + HCl$	21
$(Me_3Ge)_2Hg + cis\text{-}(Et_3P)_2PtCl_2$ $\rightarrow trans\text{-}Me_3GePt(Cl)(PEt_3)_2 + Hg + Me_3GeCl$	22
$SnCl_2 + \pi\text{-}C_5H_5Fe(CO)_2HgCl \rightarrow Cl_3SnFe(CO)_2(\pi\text{-}C_5H_5) + Hg$	23
$Me_3SnNMe_2 + trans\text{-}(Ph_3P)_2Pt(H)Cl$ $\rightarrow trans\text{-}Me_3SnPt(Cl)(PPh_3)_2 + Me_2NH$	24
$Me_3SnNMe_2 + \pi\text{-}C_5H_5W(CO)_3H \rightarrow Me_3SnW(CO)_3(\pi\text{-}C_5H_5) + Me_2NH$	24
$Ph_4AsGeCl_3 + Cr(CO)_6 \xrightarrow{h\nu} Ph_4As[Cr(CO)_5GeCl_3] + CO$	24 a

III. Oxidative-addition reactions*

$(EtO)_3SiH + trans\text{-}(Ph_3P)_2Ir(CO)Cl \rightarrow (EtO)_3SiIr(H)(Cl)(CO)(PPh_3)_2$	26
$SnCl_2 + trans\text{-}(Ph_3P)_2Ir(CO)Cl$ $\xrightarrow{\text{acetone}} Cl_3SnIr(H)(Cl)(CO)(PPh_3)_2, Me_2CO$	27
$Ph_3SnCl + (Ph_3P)_4Pt \rightarrow Ph_3SnPt(Cl)(PPh_3)_2 + 2Ph_3P$	28

* These are reactions (25) in which either 4 or 5 coordinate transition metal complexes with d^8 configurations add a molecule, forming hexacoordinate complexes with the metals in a d^6 configuration, or transition metal complexes with d^{10} configurations react to form d^8 complexes.

IV. Insertion of germanium (II) or tin (II) halides into metal–metal
and metal–halogen bonds

$GeI_2 + Co_2(CO)_8 \rightarrow I_2Ge[Co(CO)_4]_2$	29
$GeI_2 + [\pi\text{-}C_5H_5Fe(CO)_2]_2 \rightarrow I_2Ge[Fe(CO)_2(\pi\text{-}C_5H_5)]_2$	11
$SnCl_2 + [\pi\text{-}C_5H_5Fe(CO)_2]_2 \rightarrow Cl_2Sn[Fe(CO)_2(\pi\text{-}C_5H_5)]_2$	30
$SnCl_2 + [\pi\text{-}C_5H_5NiCO]_2 \rightarrow Cl_2Sn[Ni(CO)(\pi\text{-}C_5H_5)]_2$	29
$SnCl_2 + Co_2(CO)_8 \rightarrow Cl_2Sn[Co(CO)_4]_2$	23, 29
$SnBr_2 + (Bu_3P)_2Co_2(CO)_6 \rightarrow Br_2Sn[Co(CO)_3PBu_3]_2$	23
$SnCl_2 + \pi\text{-}C_5H_5Fe(CO)_2Cl \rightarrow Cl_3SnFe(CO)_2(\pi\text{-}C_5H_5)$	30
$SnCl_2 + \pi\text{-}C_5H_5Fe(CO)_2I \rightarrow Cl_2ISnFe(CO)_2(\pi\text{-}C_5H_5)$	31
$SnCl_2 + (Ph_3P)_2PtCl_2 \rightarrow Cl_3SnPt(Cl)(PPh_3)_2$	32
$SnCl_2 + trans\text{-}(Ph_3P)_2Pt(H)Cl \rightarrow Cl_3SnPt(H)(PPh_3)_2$	33
$GeI_2 + [W_2(CO)_{10}]^{2-} \rightarrow [(OC)_5WGeI_2W(CO)_5]^{2-}$	33a

V. Action of lithium derivatives of Group IV elements on transition
metal halide complexes

$2Ph_3GeLi + (Et_3P)_2PtCl_2 \rightarrow (Ph_3Ge)_2Pt(PEt_3)_2 + 2LiCl$	34
$2Ph_3GeLi + (Et_3P)_2PdCl_2 \rightarrow (Ph_3Ge)_2Pd(PEt_3)_2 + 2LiCl$	35
$2Ph_2MeSiLi + (PhPMe_2)_2PtCl_2$	
$\rightarrow (Ph_2MeSi)_2Pt(PMe_2Ph)_2 + 2LiCl$	36
$Ph_3SnLi + (Ph_3P)_2PtCl_2 \rightarrow Ph_3SnPt(Cl)(PPh_3)_2 + LiCl$	37

Some comments on the various syntheses are pertinent. Reaction I has been used to attach three, or even as many as four, transition metal atoms to a Group IV element; for example,

$$MeSnCl_3 + 3NaCo(CO)_4 \rightarrow MeSn[Co(CO)_4]_3 + 3NaCl \;^{38}$$

$$SnCl_4 + 4NaFe(CO)_2(\pi\text{-}C_5H_5)$$
$$\rightarrow Sn[Fe(CO)_2(\pi\text{-}C_5H_5)]_4 + 4NaCl \;^{39}$$

The carbonyl anion reaction can be controlled so as to effect a partial substitution of halogen atoms on tin. Thus treatment of Me_2SnCl_2 with an equimolar amount of $[\pi\text{-}C_5H_5Mo(CO)_3]^-$ affords $ClMe_2SnMo(CO)_3(\pi\text{-}C_5H_5)$, whereas an excess of the anion yields $Me_2Sn[Mo(CO)_3(\pi\text{-}C_5H_5)]_2$.[12] Stepwise halogen replacement has been used to obtain molecules with a sequence of three different metal atoms.[40]

$$Me_2SnCl_2 \xrightarrow{\quad Mn(CO)_5^-\quad} (OC)_5MnSnMe_2Cl$$

$$\xrightarrow{\quad \pi\text{-}C_5H_5Mo(CO)_3^-\quad} (OC)_5MnSn(Me_2)Mo(CO)_3(\pi\text{-}C_5H_5)$$

The complexes $(OC)_5MnSn(Ph_2)M(CO)_3(\pi\text{-}C_5H_5)$ [M = Mo or W], $(OC)_5MnSn(Ph_2)Re(CO)_5$ and $(OC)_5MnSn(Ph_2)Co(CO)_4$ have been similarly prepared.[41]

A combination of methods I and IV (Table 1) has been used to obtain a novel pentanuclear metal complex[42]

$$[\pi\text{-}C_5H_5Fe(CO)_2]_2 \xrightarrow{SnCl_2} [\pi\text{-}C_5H_5Fe(CO)_2]_2SnCl_2$$

$$\xrightarrow{\pi\text{-}C_5H_5Mo(CO)_3^-} [\pi\text{-}C_5H_5(OC)_2Fe]_2Sn[Mo(CO)_3(\pi\text{-}C_5H_5)]_2$$

Sometimes treatment of a halide with a carbonyl anion yields an unexpected product. Thus the di-tin compound $\{Sn[Re(CO)_5]_3\}_2$, with eight metal atoms per molecule, is obtained[43] from $NaRe(CO)_5$ and $Br_3SnRe(CO)_5$. Apparently the pentanuclear complex $Sn[Re(CO)_5]_4$ is not formed. Similarly, treatment of methyltin trichloride with $Fe(CO)_4^{2-}$ yields $Me_4Sn_3Fe_4(CO)_{16}$.[44]

The term 'insertion reaction' is used to describe several processes in metal carbonyl chemistry, such as the conversion of alkylmetal carbonyls into acylmetal carbonyls on treating the former with carbon monoxide, tertiary phosphines, or sulphur dioxide. Although some useful suggestions have been made, the mechanisms of these reactions are not well understood, and they may not be the same for all types of reactant.

For the insertion of an $SnCl_2$ group into a transition metal–halogen bond one can consider two possible mechanisms. The reaction might proceed by coordination of an $SnCl_2$ molecule, perhaps solvated, to the transition metal, followed by an intramolecular migration of the halide group from the transition metal to the tin atom. Alternatively, the active species might be $SnCl_3^-$, produced under the reaction conditions, which could displace halide ions in S_N1 or S_N2 reactions. Experiments suggest that the former mechanism operates in one such reaction.[31] The compound $\pi\text{-}C_5H_5Fe(CO)_2I$ reacts rapidly with $SnCl_2$ in methanol affording $Cl_2ISnFe(CO)_2(\pi\text{-}C_5H_5),MeOH$. Under similar conditions $\pi\text{-}C_5H_5Mo(CO)_3I$ is inert. These observations indicate that the iodine atom remains associated with the $\pi\text{-}C_5H_5Mo(CO)_3I$ molecule during the reaction. As one might expect, formation of the seven coordinate intermediate $Cl_2Sn,Fe(CO)_2I(\pi\text{-}C_5H_5)$ would be easier than formation of the eight-coordinate intermediate

$Cl_2Sn,Mo(CO)_3I(\pi\text{-}C_5H_5)$. Therefore, the inertness of π-cyclo-pentadienylmolybdenum tricarbonyl iodide is understandable.

Many complexes are known wherein SnX_3 groups are bonded to Pt, Pd, Rh, Ru or Ir through Sn (Table 2). The tin-containing species are often anionic, or sometimes cationic, so that the products are isolated as salts by addition of ions such as R_4E^+ (E = N, P or As) or Ph_4B^-.[32, 45–53] The method of preparation involves treating a halide complex of the transition metal with an alcoholic or acetone solution of the tin (II) halide. Hydrochloric or hydrobromic acids are sometimes added. It is therefore not clear in what form the tin reacts—for example, whether the species $SnCl_3^-$ or $SnCl_2L$ (L = H_2O,Me_2CO) are involved, or whether insertion or ligand displacement reactions are occurring. Because of the nature of the products, the reactions can conveniently be treated as being of the 'insertion' type (Table 1, method IV), although the processes are unlikely to be mechanistically the same in all cases.

The insertion of $SnCl_2$ or GeI_2 into the metal–metal bond of a metal carbonyl appears to depend on the structure of the carbonyl.[29] Reaction occurs readily when the metal–metal bond in the carbonyl is supported by bridging carbonyl groups [e.g. $Co_2(CO)_8$], but it appears that insertion either does not occur [e.g. $Mn_2(CO)_{10}$]* or occurs more slowly (e.g. $[Ph_3PCo(CO)_3]_2$†) when bridging carbonyl groups are absent.

A number of compounds have been obtained by methods which do not clearly fall within the classifications set out in Table 1, although in many cases the reactions are obviously related to those delineated. Thus $SnCl_2$ in MeOH reacts with an aqueous solution of $CoCl_2$ and KCN to form $K_6[(NC)_5CoSnCl_2Co(CN)_5]$.[54] Also related to the insertion of tin (II) halides into M–M bonds is the reaction:[55]

$$[\pi\text{-}C_5H_5Fe(CO)_2]_2 + SnX_4$$
$$\rightarrow \pi\text{-}C_5H_5Fe(CO)_2SnX_3 + \pi\text{-}C_5H_5Fe(CO)_2X$$

Similarly octacarbonyl dicobalt and tin (IV) halides afford $XSn[Co(CO)_4]_3$ (X = Cl, Br or I).[38]

* Very recently it has been shown [29a] that $SnCl_2$ can be made to insert into the Mn–Mn bond of $Mn_2(CO)_{10}$ at 190 °C in a hydrocarbon solvent.

† A bridged form of this molecule may exist in equilibrium with the non-bridged form in solution. As pointed out elsewhere,[29] reaction with $SnCl_2$ could occur via the bridged form.

TABLE 2. *Platinum–metal tin complexes*

Compound	Colour	M.p. (°C)	Reference
$(Me_4N)_2[Pt(SnCl_3)_2Cl_2]$	(a) Red	> 250 dec.	46
	(b) Yellow	195–200 dec.	46
$(Ph_3MeP)_2[Pt(SnCl_3)_2Cl_2]$	Yellow	50	32, 46
$(Et_4N)[HPt(SnCl_3)_2(PEt_3)_2]$	Yellow	—	47
$(Ph_3MeP)_3[Pt(SnCl_3)_5]$	Red	—	32
$(Me_4N)_3[HPt(SnCl_3)_4]$	Brownish yellow	—	47
$(Me_4N)_4[Pt_3Sn_8Cl_{20}]$	Dark red	—	49
$(Et_4N)_4[Pt_3Sn_8Cl_{20}],Me_2CO$	Red	> 250	49
$(Ph_3P)_2Pt(SnCl_3)_2Cl_2$	Orange-yellow	260–270 dec.	37
$(Ph_3P)_2Pt(SnCl_3)_2$	Orange	—	32
$(Ph_3P)_2Pt(SnCl_3)Cl$	Pale yellow	—	32, 46
	White	280–290 dec.	37
$(Ph_3P)_2Pt(SnMe_3)Cl$	Yellow-brown	234	24
$(Ph_3As)_2Pt(SnCl_3)Cl$	Orange	150	46
$(Ph_3P)_2Pt(SnPh_3)Cl$	White	> 278 dec.	37
	White	205	28
$(Ph_3P)_2Pt(SnCl_3)H$	Orange	172	33
	Orange-yellow	130–135	37
$(Et_3P)_2Pt(SnCl_3)H*$	White	100	48
$(Ph_3P)_2Pt(SnCl_3)OH$	Yellow	165–170	33
$(C_8H_{12})_3Pt_3Sn_2Cl_6†$	Red-orange	—	49
$(Ph_4As)_4[Pd_2Cl_2(SnCl_3)_4]$	Red	50 dec.	52
$(Me_4N)_4[Ir_2Cl_6(SnCl_3)_4]$	Yellow-orange	> 300	46
$(Me_4N)_2[IrCl_3(SnCl_3)_2(CO)]$	Pale yellow	> 250 dec.	27
$(Ph_3PH)_4[Ir_2Cl_6(SnCl_3)_4]$	Yellow	112–117	46
$(Ph_3As)_2(C_8H_{12})IrSnCl_3†$	Yellow	171	46
$(Ph_3P)_2(C_8H_{12})IrSnCl_3†$	Yellow	127 dec.	46
$(C_8H_{12})_2IrSnCl_3†$	Yellow	178 dec.	46
$(C_7H_8)_2IrSnCl_3‡$	Pale yellow	230 dec.	46
$(Ph_3P)_3(Cl)Ir(SnCl_3)(H)$	Yellow-orange	132	27
$(Ph_3P)_3(H)_2Ir(SnCl_3)$	White	206	27
$(Ph_3P)_2(Cl)Ir(SnCl_3)(H)(CO)$	Pale yellow	155	27
$(Ph_3P)_2(H)_2Ir(SnCl_3)(CO)$	Pale yellow	191	27
$Ph_3P(Cl)_2Ir(SnCl_3)(CO)$	Pale yellow	270	27
$Ph_3PIr(SnMe_3)(CO)_3$	—	—	16
$Ph_3PIr(SnPh_3)(CO)_3$	—	—	16
$[Ph_3PIr(CO)_3]_2SnMe_2$	—	—	16
$(Me_4N)_2[Rh(SnCl_3)_2(CO)Cl]$	Orange	> 200 dec.	46
$(Ph_3PH)_4[Rh_2(SnCl_3)_4Cl_2]$	Yellow-orange	137	46
$(Me_4N)_4[Rh_2(SnCl_3)_4Cl_2]$	Orange	> 250 dec.	46
$(Ph_3P)_3RhSnCl_3$	Red-brown	130	46
$(C_7H_8)_2RhSnCl_3‡$	Yellow	170 dec.	46

TABLE 2 (cont.)

Compound	Colour	M.p. (°C)	Reference
$(Ph_3P)_2(C_7H_8)RhSnCl_3$‡	Orange	134	46
$(Ph_3As)_2(C_7H_8)RhSnCl_3$‡	Orange	177 dec.	46
$(Ph_3Sb)_2(C_7H_8)RhSnCl_3$‡	Orange	200 dec.	46
$(Ph_3P)_2Rh(SnMe_3)(CO)_2$	—	—	16
$(Ph_3PH)_2[Ru(SnCl_3)_2Cl_2]$	Yellow	177	46
$(Me_4N)_2[Ru(SnCl_3)_2Cl_2]$	Orange	> 200 dec.	46
$(Me_4N)_2[Ru(SnCl_3)_2(CO)_2Cl_2]$	Yellow	—	50
$(Ph_4As)_2[Ru(SnBr_3)_2(CO)_2Br_2]$	Yellow	—	50
$[(Et_2S)_3Ru(SnCl_3)(CO)_2]Cl$	Green-yellow	145	51
$[(Et_2S)_3Ru(SnCl_3)(CO)_2]BPh_4$	Green-yellow	—	51
$py_2Ru(SnBr_3)_2(CO)_2$	Orange-yellow	> 250	51
$py_2Ru(SnCl_3)_2(CO)_2$	Yellow	184 dec.	51
$(Ph_3P)_4Ru_2Cl_3(SnCl_3)(CO)_2$	Pale yellow	—	53
$(Ph_3P)_3Ru_2Cl_3(SnCl_3)(CO)_2(Me_2CO)_2$§			
	Lemon-yellow	—	53
$Ru(SnPh_3)_2(CO)_4$	Pale yellow	180	56

* Isolated as the *trans* isomer. The complexes *trans*-$(Et_3P)_2Pt(SnCl_3)(p\text{-}FC_6H_4)$ and *trans*-$(Et_3P)_2Pt(SnCl_3)(m\text{-}FC_6H_4)$ are also mentioned in ref. 49.

† C_8H_{12} is 1,5-cyclo-octadiene.

‡ C_7H_8 is norbornadiene.

§ One acetone molecule coordinated, the other uncomplexed.

Organotin compounds $R'_xSnR''_{4-x}$ (R′, alkyl group; R″, vinyl or phenyl group) react with pentacarbonyl iron in refluxing ethylcyclohexane to give dialkyltin iron tetracarbonyl dimers $[R_2SnFe(CO)_4]_2$.[57] These same complexes can be obtained by treating dialkyldialkynyltin compounds with dodecacarbonyl triiron.[58] The fate of the vinyl, phenyl, or alkynyl groups in these reactions is unknown, but they are probably eliminated as hydrocarbons such as ethylene, benzene, acetylene, butadiene, etc. If this is so, the synthesis of the complexes $[R_2SnFe(CO)_4]_2$ by this route can be classified under method II (Table 1). Another elimination reaction which affords these same complexes is:[59]

$$2R_2SnH_2 + 2Fe(CO)_5 \rightarrow [R_2SnFe(CO)_4]_2 + 2CO + 2H_2$$

In refluxing pentane, octacarbonyl dicobalt and tetravinyltin afford $(CH_2:CH)_2Sn[Co(CO)_4]_2$. However, with tetrahydrofuran as solvent $CH_3C[Co(CO)_3]_3$ is produced, a vinyl group having been converted into an ethylidyne group.[38] In contrast, octacarbonyl dicobalt and tetravinylsilane react to give $CH_2:CHSi[Co(CO)_3]_3$, a complex assumed to have a structure based on a cluster of three Co atoms and one Si.[60] Two such clusters joined by a Si–Si bond may occur in $\{Si[Co(CO)_3]_3\}_2$, obtained by treating cobalt carbonyl with tetraphenylsilane.[61] By refluxing mixtures of organotin halides and $Fe(CO)_5$ several complexes with Sn–Fe bonds have been obtained, e.g. $[R_2SnFe(CO)_4]_2$, $[R_2SnCl]_2Fe(CO)_4$, $R_4Sn_3[Fe(CO)_4]_4$ and $Sn[Fe(CO)_4]_4$.[59] Under mild conditions, $Fe(CO)_5$ and $SnCl_4$ afford $(OC)_4Fe(SnCl_3)Cl$, the latter on heating giving $(OC)_4Fe(SnCl_3)_2$.[61a]

3. Structure and bonding

Covalent metal–metal bonding in the compounds can, in the majority of cases, be safely inferred from their methods of synthesis and their chemical compositions. The first direct physical proof of bonding between metal atoms of different elements in these complexes came from an X-ray crystallographic study of $Ph_3SnMn(CO)_4PPh_3$.[62] At the time of writing, structures have been established for the compounds listed in Table 3. Although only twelve have been elucidated so far,* each has revealed important structural features, and this will undoubtedly stimulate further studies.

The molecular configurations observed for various compounds in the crystalline state are not necessarily the same as those in solution.[65] The P·Mn·Sn bond angle is 176° in $Ph_3SnMn(CO)_4PPh_3$. This deviation from linearity, and other distortions observed in the structure, can probably be ascribed to the need to obtain optimal packing of the molecules in the crystal. In this context, it is also to be noted that the compound $Cl_2Ge[Fe(CO)_2(\pi\text{-}C_5H_5)]_2$ may change its molecular configuration on dissolution.[11, 69]

Crystallographic studies reveal that the Group IV atoms show distorted tetrahedral bonding while the iron and manganese atoms show basically octahedral bonding. In $Cl_2Ge[Fe(CO)_2(\pi\text{-}C_5H_5)]_2$

---

* *Note added in proof:* Several other molecular structures have recently been established by X-ray, e.g. $ClSn[Mn(CO)_5]_3$, $(OC)_4CoSnPh_2Mn(CO)_5$, $[\pi\text{-}C_5H_5Fe(CO)_2]_2Sn(ONO)_2$ and $[\pi\text{-}C_5H_5Fe(CO)_2]_2SnMe_2$. See *Chem. Comms.* (1967), pp. 702, 749, 750; (1968), p. 159.

TABLE 3. *X-ray crystallographic studies*

Compound	Metal–metal distance (Å)	Space group	Reference
$Ph_3SnMn(CO)_5$	$2\cdot674\pm0\cdot004$	$P\,2_1$	63
$Me_3SnMn(CO)_5$	$2\cdot674\pm0\cdot003$	$P\,2_1/n$	64
$Ph_3SnMn(CO)_4PPh_3$	$2\cdot627\pm0\cdot01$	$P\,2_1/n$	62, 65
$Ph_2Sn[Mn(CO)_5]_2$	$2\cdot70\pm0\cdot01$	$P\,2_1/c$	66
$Ph_3GeMn(CO)_5$	$2\cdot535\pm0\cdot02$	P_1	67
$Ph_3SnFe(CO)_2(\pi\text{-}C_5H_5)$	$2\cdot536\pm0\cdot003$	$P\,2_1/c$	68
$Cl_2Ge[Fe(CO)_2(\pi\text{-}C_5H_5)]_2$	$2\cdot36\pm0\cdot01$	$C\,2/c$	69
$Cl_2Sn[Fe(CO)_2(\pi\text{-}C_5H_5)]_2$	$2\cdot492\pm0\cdot008$	$C\,2/c$	70
$Sn[Fe(CO)_4]_4$*	Sn—Fe, $2\cdot54\pm0\cdot01$ Fe—Fe, $2\cdot87\pm0\cdot01$	$P\,2_1/n$	71
$Me_4Sn_3[Fe(CO)_4]_4$	Sn—Fe (interior), $2\cdot747\pm0\cdot008$ Sn—Fe (terminal), $2\cdot625\pm0\cdot008$	$P\,2_1/c$	44
$[Ph_3PMe]_3[Pt(SnCl_3)_5]$	$2\cdot54$	P_1	47
$(C_8H_{12})_2IrSnCl_3$†	$2\cdot642\pm0\cdot002$	—	72

* In this molecule there are two pairs of iron–iron distances (see text).
† C_8H_{12} = 1,5-cyclo-octadiene.

the Fe·Ge·Fe angle is expanded to 128° while the Cl·Ge·Cl angle is compressed to 96°. The crystal and molecular structure of $Cl_2Sn[Fe(CO)_2(\pi\text{-}C_5H_5)]_2$ is similar to that of the germanium analogue, the Fe·Sn·Fe and Cl·Sn·Cl bond angles being 128·6° and 94·1°, respectively.[70] In $Ph_3SnFe(CO)_2(\pi\text{-}C_5H_5)$ distortion of the Group IV metal valencies is less; the C·Sn·C angles are closed to 105° while the Fe·Sn·C angles are opened to 113°.[68]

The structures of $Sn[Fe(CO)_4]_4$[71] and the related complex $Me_4Sn_3[Fe(CO)_4]_4$[44] are shown in Fig. 1(*a*) and (*b*). The complex $Sn[Fe(CO)_4]_4$ is diamagnetic, as are the other metal–metal bonded complexes reviewed in this chapter. To account for the diamagnetism of $Sn[Fe(CO)_4]_4$, it is necessary to consider iron–iron bonding as well as tin–iron bonding. In $Sn[Fe(CO)_4]_4$ there are two iron–iron distances (4·65 and 2·87 Å). Whilst it is possible to envisage direct interaction between the pairs of iron

atoms separated by 2·87 Å this distance is longer than that
found in the majority of complexes that have been examined, and
in which normal covalent iron–iron bonding is likely.* Iron–iron

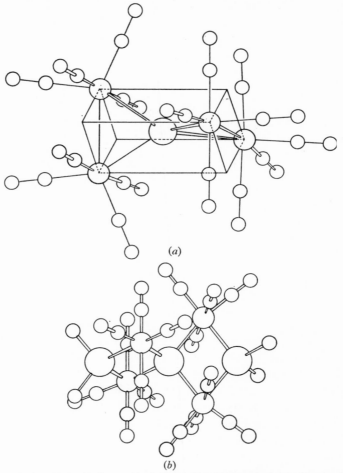

(a)

(b)

Fig. 1. (a) The SnFe$_4$(CO)$_{16}$ molecule. In order of increasing size, the atoms are
carbon and oxygen, iron and tin. (b) The Me$_4$Sn$_3$Fe$_4$(CO)$_{16}$ molecule. In order of
increasing size, the atoms are carbon and oxygen, iron and tin.

separations are 2·46 Å in Fe$_2$(CO)$_9$, 2·49 Å in [π-C$_5$H$_5$Fe(CO)$_2$]$_2$,
2·54 Å in [EtSFe(CO)$_3$]$_2$, 2·65 Å in Se$_2$Fe$_3$(CO)$_9$, and 2·88 Å in
[Fe$_2$(CO)$_8$]$^{2-}$. The latter species must have a direct metal–metal

* For leading references see reference no. 71.

bond since there are no bridging groups. In $Sn[Fe(CO)_4]_4$, however, interaction between the pairs of iron atoms separated by 2·87 Å is possible through two equivalent three-centre bonds involving a d orbital on the tin atom.[59, 71] Interestingly, establishment of the iron–iron distances of 2·87 Å leads to very considerable distortion of the regular tetrahedral valencies of tin. In this molecule one pair of angles has closed to 69°, and the other pair has opened out to 133°.

In the complex $Me_4Sn_3[Fe(CO)_4]_4$, the Sn_2Fe_2 quadrilaterals are planar, causing compression of the $Sn \cdot Fe \cdot Sn$ angles to 77·9°, and making the $Fe \cdot Sn \cdot Fe$ angles inside a quadrilateral 105·2° (at the terminal tin atoms) and 98·8° (at the central tin atom).

At present it is unwise to draw any firm conclusions about the nature of the metal–metal bonds from the bond lengths (Table 3), since it is difficult to estimate the relevant radii of the transition metals. It is generally accepted that transition metals form π as well as σ bonds with carbonyl, tertiary phosphine, cyclopentadienyl and other groups; these metals may also have π components in their bonds with metals such as tin. Indeed various i.r. spectroscopic studies seem to support this view (see below). Such metal–metal π bonding, if it occurs, could have an effect on bond lengths.

In $[Ph_3PMe]_3[Pt(SnCl_3)_5]$ there is a trigonal bipyramidal arrangement of tin atoms around the platinum.[47] There is little or no difference in the axial and radial platinum–tin distances. This complex is one of only eight penta-coordinate species for which precise solid-state structural data exists.[73] The ion $[Pt(SnCl_3)_5]^{3-}$ may not be capable of existence in solution.[46] Salts of two related complex anions, $[HPt(SnCl_3)_4]^{3-}$ and $[HPt(SnCl_3)_2(PEt_3)_2]^-$, have also been isolated,[47] and their structures in the solid state very probably involve five-coordinate platinum(II) species. They are also the first examples of anionic platinum hydrides.

For many of the complexes one can suggest likely molecular structures on the basis of their compositions and properties. A compound such as $Me_3SnCo(CO)_4$ would be expected to have a trigonal bipyramidal arrangement of ligands around the cobalt atom,* and a basically tetrahedral disposition of groups around the tin atom. An analysis of the infrared spectrum of $Me_3SnCo(CO)_4$

* This has been confirmed by X-ray for $Cl_3SiCo(CO)_4$. [73a]

in the CO stretching region is in accord with such a structure.[74] The i.r. spectrum of $(Ph_3Sn)_2Fe(CO)_4$, obtained from Ph_3SnCl and $Fe(CO)_4{}^{2-}$, shows four CO stretching bands, as expected if the iron atom is *cis*-substituted by the triphenyltin groups. In contrast, $(Ph_3Sn)_2Ru(CO)_4$, prepared by an analogous reaction using $Ru(CO)_4{}^{2-}$, shows one CO stretching band, as expected for a *trans* isomer.[56] I.r. spectral studies are playing an important role in this

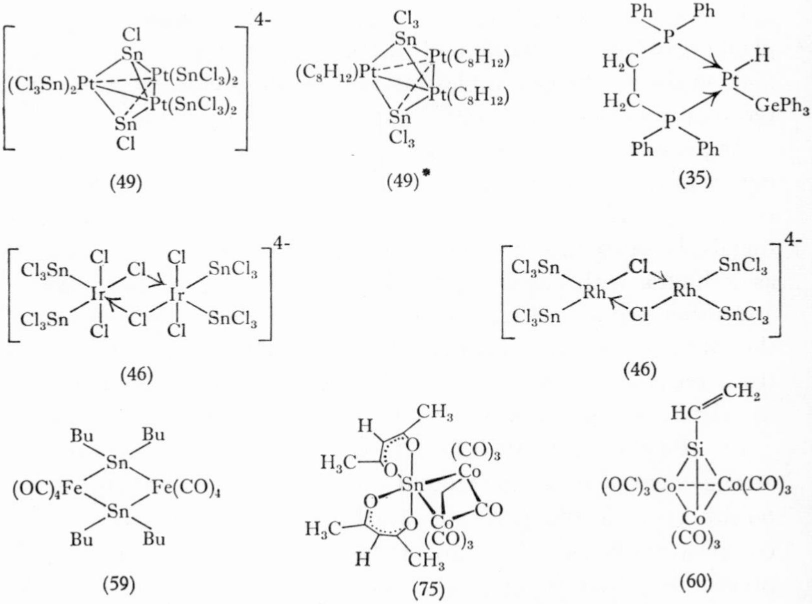

Fig. 2. Some probable molecular structures.

field both in assigning molecular structures and in providing information about the nature of the bonding.[15, 17, 74, 76-8]

Although crystallographic data are not yet available, probable molecular structures for a few novel species are given in Fig. 2. In $[Pt_3Sn_8Cl_{20}]^{4-}$ it is postulated that an SnCl group bridges three platinum atoms,[49] reminiscent of the bridging of three metal atoms by CO groups in $[\pi\text{-}C_5H_5Ni]_3(CO)_2$, $[\pi\text{-}C_6H_6Co]_3(CO)_2{}^+$, $Rh_6(CO)_{16}$ and $[\pi\text{-}C_5H_5FeCO]_4$. The suggested[45, 46] structure for $[Rh_2(SnCl_3)_4Cl_2]^{4-}$ is supported by the results of far infrared

* Structure confirmed recently by an X-ray study.[49a]

studies,[79] which reveal two bands ascribable to the Rh_2Cl_2 bridge and two bands assignable to the $SnCl_3$ groups.

The compound $CH_2:CHSi[Co(CO)_3]_3$ is notable because the tin analogues $RSn[Co(CO)_3]_3$ appear incapable of existence.[38] The structure proposed[60] for $CH_2:CHSi[Co(CO)_3]_3$, with cobalt–cobalt as well as silicon–cobalt bonds, is similar to that found with organocobalt complexes of formula $RC[Co(CO)_3]_3$.

Reaction between bis(acetylacetonato)dichlorotin(IV) and tetra-carbonylcobalt(−I) anion affords the red crystalline complex $(acac)_2SnCo_2(CO)_7$.[75] A cobalt–cobalt bond is invoked to preserve the effective atomic number formalism for the cobalt atoms. The structure proposed (Fig. 2) can be derived from that of octa-carbonyl dicobalt by replacing one bridging CO group with the $(acac)_2Sn$ group. This synthesis represents one of several notable contributions to this field by W. A. G. Graham and his co-workers.[12, 15, 29, 38, 40, 75, 76]

With few exceptions, stable complexes involving a Group IV metal bonded to a transition metal are formed only when the transition metal is simultaneously bonded to π-acceptor ligands. It is well known that this same condition operates when carbon atoms of alkyl groups are σ-bonded to the transition metals. However, with the Si, Ge, Sn and Pb complexes there is the possibility of π-acceptor as well as σ bonding with the transition elements since silicon and its congeners possess vacant d orbitals. Such π bonding will depend on the nature of the substituents on the Group IV element. Electronegative substituents contract the d_π orbitals making them more compatible with the filled d_π orbitals of the transition metal, enhancing back-bonding. However, the inductive effect of the substituents also has an influence on the electron distribution in the metal–metal bonds. These effects appear to be reflected in the infrared spectra of those complexes in which carbonyl groups are present.[74]

As would be expected, the thermal stability of the silicon, germanium, tin and lead complexes varies. Thus square planar Pt–Ge compounds are stable to 150 °C whereas the palladium complex $(Et_3P)_2Pd(GePh_3)_2$ decomposes in solution at 20 °C, and in the solid at 97 °C.[35] Interestingly, stabilities of alkylplatinum and alkylpalladium complexes vary in the same way, the platinum compounds being the more robust. There may well be a parallel between the relative stabilities of σ bonds between carbon and the various transition metals and the bonds formed by silicon and its congeners with these metals. Thus just as $MeMn(CO)_5$ is thermally more stable than $MeCo(CO)_4$, so $Ph_3SnMn(CO)_5$ (decomp. $\sim$ 200 °C) is more stable than $Ph_3SnCo(CO)_4$ (decomp. $\sim$ 110 °C).

The thermal stability of the transition metal derivatives of the Group IV B elements is greater than that of the simple organic derivatives in which groups such as alkyl are σ-bonded to the transition metals. The silicon–cobalt complex $H_3SiCo(CO)_4$ is thermally stable up to 85 °C,[10] whereas the simple cobalt complex $H_3CCo(CO)_4$ decomposes above -20 °C. Complexes such as $(Me_3Sn)_2Fe(CO)_4$ exist, whereas $(Me_3C)_2Fe(CO)_4$ does not. Moreover, it is noteworthy that in many instances attachment of electronegative groups to the Group IV atoms enhances the robustness of the metal–metal bonds. Thus $F_3SiCo(CO)_4$[80] is thermally more stable than $H_3SiCo(CO)_4$ just as $F_3CCo(CO)_4$ is more stable than $H_3CCo(CO)_4$. We have referred above to enhancement of stability of the metal–metal bonds by π-bonding and σ withdrawal effects, for which there is some spectroscopic evidence.[74, 80, 80a]

4. Chemical reactivity

Reactivity of the various complexes depends on the particular metals involved, the nature of the attached ligands, and on the reactants. It might be thought that the metal–metal bonds would always be the most reactive centres in these compounds, but this is not always so; for example,[9, 10]

$$Ph_3SnMn(CO)_5 + 3Cl_2 \rightarrow Cl_3SnMn(CO)_5 + 3PhCl$$

$$Ph_3SnFe(CO)_2(\pi\text{-}C_5H_5) + 3HCl$$
$$\rightarrow Cl_3SnFe(CO)_2(\pi\text{-}C_5H_5) + 3C_6H_6$$

$$H_3SiMn(CO)_5 + 3HCl \rightarrow Cl_3SiMn(CO)_5 + 3H_2$$

Organotin–rhenium carbonyl complexes undergo similar reactions with chlorine or bromine to afford compounds such as

$$Ph(Br_2)SnRe(CO)_5.[43]$$

It is interesting to contrast the stability of the metal–metal bonds in these reactions with the ease with which tin–tin bonds in hexa-organo-ditin compounds are cleaved by halogens. Moreover, the metal–metal bonds in $Et_2Pb[Mn(CO)_5]_2$ or $Ph_3SnFe(CO)_2(\pi-C_5H_5)$ are broken by chlorine,[9] showing that by no means all such bonds are inert to halogens.

Redistribution reactions, a characteristic feature of the chemistry of organotin halides, also occur with tin–manganese complexes:[9]

$$Ph_3SnMn(CO)_5 + Ph(Br_2)SnMn(CO)_5 \rightarrow 2Ph_2(Br)SnMn(CO)_5$$

The tin–manganese bond is preserved, as it is also in the nucleo-philic attack of tertiary phosphines on the carbonyl groups in these compounds. Thus treatment of $H_3SiCo(CO)_4$ with triphenyl-phosphine affords $H_3SiCo(CO)_3PPh_3$.[10] Similarly, the tin–iron complex $[Bu_2SnFe(CO)_4]_2$ with triethylphosphine yields both $[Bu_2SnFe(CO)_3PEt_3]_2$ and $(OC)_4Fe[SnBu_2]_2Fe(CO)_3PEt_3$;[77] and $X_3SnMn(CO)_5$ with $Ph_2PCH_2CH_2PPh_2$ reacts to give

$$X_3SnMn(CO)(diphos)_2, \quad where\ X = Cl,\ Br\ [81].$$

Preservation of the metal–metal bonds also occurs in numerous reactions involving Grignard reagents or complex hydrides, for example,

$$X_2Sn[Co(CO)_4]_2 \xrightarrow{RMgX} R_2Sn[Co(CO)_4]_2\ [23]$$

$$Cl_2Ge[Fe(CO)_2(\pi-C_5H_5)]_2 \xrightarrow{NaBH_4} H_2Ge[Fe(CO)_2(\pi-C_5H_5)]_2\ [11]$$

It will be evident from the above results that much functional group chemistry is possible for these compounds.

The carbon–metal bond in methylmanganese pentacarbonyl adds to tetrafluoroethylene to afford $MeCF_2CF_2Mn(CO)_5$. The analogous insertion of tetrafluoroethylene into germanium–manganese and tin–manganese bonds has also been observed:[82, 83]

$$Me_3MMn(CO)_5 + C_2F_4$$
$$\xrightarrow{h\nu} Me_3MCF_2CF_2Mn(CO)_5 \quad (M = Ge\ or\ Sn)$$

The tin complex is obtained in low yield, there being several other products. Trimethyltin fluoride is produced, probably via decom-position of the initially formed insertion product. Reaction between trifluoroethylene and $Me_3SnMn(CO)_5$ affords trimethyltin fluoride essentially quantitatively, as well as the *cis* and *trans* isomers of CFH:CFMn(CO)$_5$. Apparently migration of a fluorine atom on to tin prevents the isolation of Me$_3$SnC$_2$F$_3$HMn(CO)$_5$. The analogous complex Me$_3$GeC$_2$F$_3$HMn(CO)$_5$ also could not be prepared by insertion of trifluoroethylene into Me$_3$GeMn(CO)$_5$. However, the complex Me$_3$GeCF$_2$CF$_2$Mn(CO)$_5$ is formed in good yield from Me$_3$GeMn(CO)$_5$ and tetrafluoroethylene, only a trace of trimethyl-germanium fluoride being produced.

Recently it has been found that Me$_3$SnMn(CO)$_5$ reacts with perfluorobutadiene to give *cis*-CF$_2$:CFCF:CFMn(CO)$_5$ and Me$_3$SnF.[84] Reactions of this kind provide a route to novel fluoro-carbon–metal complexes.

Reference was made earlier to the thermal stability of many complexes described in this chapter. Many of the compounds are, however, highly reactive. For example the complexes represented by the formula (R$_3$P)$_2$Pt(GePh$_3$)$_2$, which are stable to air and water, readily react up to 150 °C with iodine, hydrogen chloride or carbon tetrachloride with cleavage of the germanium–platinum bonds.[34] One of the most remarkable reactions discovered by Glockling *et al.*[34, 35] is that which occurs with molecular hydrogen.

The hydrogenolysis takes place at atmospheric pressure or less and at room temperature. In spite of the metal–metal bond cleavage which occurs in the above reactions, anionic and neutral ligand exchange reactions are possible without rupture of the germanium–platinum bonds:

$$(Et_3P)_2Pt(GePh_3)_2 + 2KCN \rightarrow K_2[(Ph_3Ge)_2Pt(CN)_2] + 2Et_3P$$

$$(Et_3P)_2Pt(GePh_3)_2 + Ph_2PCH_2CH_2PPh_2$$
$$\rightarrow (diphos)Pt(GePh_3)_2 + 2Et_3P$$

A significant advance in homogeneous catalysis has been the recognition that various coordination compounds can catalyse certain reactions. The ability of transition metal ions and complexes to promote hydrogenation is well established. Some platinum–tin complexes mentioned earlier can function in this way.[32] Molecular hydrogen is cleaved by the catalyst with formation of a reactive transition metal–hydrogen bond, while an olefin or acetylene molecule also becomes coordinated to the metal. In this manner, a highly reactive intermediate is produced in which the transition metal is hexacoordinate rather than tetra- or penta-coordinate, the intermediate reverting to the original complex (the catalyst) after hydrogen transfer has saturated the double or triple bond. Although complexes in which $SnCl_3^-$ ligands are attached to platinum can be isolated in the crystalline state, in solution dissociation of some of the other groups (e.g. halide) on the metal occurs, leaving vacant sites for coordination of olefins and hydrogen. These solutions appear to involve interdependent equilibria which are affected by hydrogen ion concentration and temperature. In this connection the high *trans* activating ability of the $SnCl_3$ group is undoubtedly very important.[48] It is interesting that the origin of the *trans* effect for $SnCl_3$ appears to be by virtue of its π acceptor ability,[48] rather than by an inductive mechanism as it is with the group $SiPh_2Me$.[36]

Platinum–tin chloride complexes also catalyse double bond migration in olefins.[82] This suggests rapid exchange of coordinated and free olefin. Isomerization only occurs when hydrogen is present suggesting that a platinum hydride, $[PtCl_x(SnCl_3)_{3-x}H]^{2-}$, is involved. Addition of olefin affords an ion $[PtCl_x(SnCl_3)_{3-x}R]^{2-}$. Reversal of the process can bring about isomerization, while reaction with excess of hydrogen can give an alkane, in a hydrogenation reaction of the kind discussed above.

The hydrosilation of olefins catalysed by noble metal complexes undoubtedly involves mechanisms similar to those reviewed above.[26]

5. Conclusion

Compounds in which silicon or a Group IV metal are bonded to a transition element are being reported in ever-increasing numbers. This field could conceivably grow to the point where it could exceed that concerned with σ-bonded alkyl or aryl derivatives of the transition metals.

Complexes of the kind reviewed in this chapter may become industrially important. Reference was made previously to their use in homogeneous catalysis. Compounds possessing a polymeric metal skeleton will exhibit some of the properties of the free metal.

Unique solid state properties can be anticipated, arising from the presence of delocalized band-type orbitals.[86] Moreover, Group IV elements form stable bonds with the cobalt atoms of cobaloximes, the coordination analogues of vitamin B_{12}.[87] The study of model systems such as these may be important for biological chemistry.

REFERENCES

1   F. Hein and H. Pobloth, *Z. anorg. allgem. Chem.* 1941, **248**, 84.
2   W. Hieber and U. Teller, *Z. anorg. allgem. Chem.* 1942, **249**, 43.
3   F. Hein and E. Heuser, *Z. anorg. allgem. Chem.* 1947, **254**, 138.
4   G. E. Coates and F. Glockling, *Organometallic Chemistry* (ed. H. Zeiss), p. 426. Reinhold, A.C.S. Monograph, 1960.
5   J. Chatt and B. L. Shaw, *J. chem. Soc.* 1959, p. 705.
6   J. Chatt, *Rec. Chem. Prog.* 1960, **21**, 147.
7   T. S. Piper, D. Lemal and G. Wilkinson, *Naturwiss.* 1956, **43**, 129.
8   G. Wilkinson and T. S. Piper, *J. inorg. nucl. Chem.* 1956, **3**, 104.
9   R. D. Gorsich, *J. Am. chem. Soc.* 1962, **84**, 2486.
10   B. J. Aylett and J. M. Campbell, *Chem. Comms.* 1965, p. 217; 1967, p. 159; *Inorg. nucl. Chem. Letters*, 1967, **3**, 137.
11   N. Flitcroft, D. A. Harbourne, I. Paul, P. M. Tucker and F. G. A. Stone, *J. chem. Soc. (A)*, 1966, p. 1130.
12   H. R. H. Patil and W. A. G. Graham, *Inorg. Chem.* 1966, **5**, 1401.
13   W. Hieber and R. Breu, *Chem. Ber.* 1957, **90**, 1270.
14   F. Hein and W. Jehn, *Annalen*, 1965, **684**, 4.
15   W. Jetz, P. B. Simons, J. A. J. Thompson and W. A. G. Graham, *Inorg. Chem.* 1966, **5**, 2217.
16   J. P. Collman, F. D. Vastine and W. R. Roper, *J. Am. chem. Soc.* 1966, **88**, 5035.
17   O. Kahn and M. Bigorgne, *C. r. hebd. Séanc. Acad. Sci., Paris*, 1965, **261**, 2483.
18   A. J. Chalk and J. F. Harrod, *J. Am. chem. Soc.* 1965, **87**, 1133.
19   A. G. Massey, A. J. Park and F. G. A. Stone, *J. Am. chem. Soc.* 1963, **85**, 2021.
20   S. A. R. Knox and F. G. A. Stone, unpublished work.
21   A. N. Nesmeyanov, K. N. Anisimov, N. E. Kolobova and A. B. Antonova, *Izv. Akad. Nauk SSSR (Ser. Khim.)*, 1965, p. 1309.
22   F. Glockling and K. A. Hooton, *Chem. Comms.* 1966, p. 218.
23   F. Bonati, S. Cenini, D. Morelli and R. Ugo, *Inorg. nucl. chem. Letters*, 1966, **1**, 107; *J. chem. Soc. (A)*, 1966, p. 1052.
24   D. J. Cardin and M. F. Lappert, *Chem. Comms.* 1966, p. 506.
24a   J. K. Ruff, *Inorg. Chem.* 1967, **6**, 1502.
25   J. P. Collman and W. R. Roper, *J. Am. chem. Soc.* 1965, **87**, 4008; *Accounts Chem. Res.* 1968, **1**, 136.
26   A. J. Chalk and J. F. Harrod, *J. Am. chem. Soc.* 1965, **87**, 16.
27   R. C. Taylor, J. F. Young and G. Wilkinson, *Inorg. Chem.* 1966, **5**, 20.

28 A. J. Layton, R. S. Nyholm, G. A. Pneumaticakis and M. L. Tobe, *Chemy Ind.* 1967, p. 465.
29 D. J. Patmore and W. A. G. Graham, *Inorg. Chem.* 1966, **5**, 1405.
29a J. Hoyano, D. J. Patmore and W. A. G. Graham, *Inorg. Nucl. Chem. Letters*, 1968, **4**, 201.
30 F. Bonati and G. Wilkinson, *J. chem. Soc.* 1964, p. 179.
31 A. R. Manning, *Chem. Comms*, 1966, p. 906.
32 R. D. Cramer, E. L. Jenner, R. V. Lindsey and U. G. Stolberg, *J. Am. chem. Soc.* 1963, **85**, 1691.
33 J. C. Bailar and H. Itatani, *Inorg. Chem.* 1965, **4**, 1618.
33a J. K. Ruff, *Inorg. Chem.* 1967, **6**, 2080.
34 R. J. Cross and F. Glockling, *J. chem. Soc.* 1965, p. 5422.
35 E. H. Brooks and F. Glockling, *J. chem. Soc.* (*A*), 1966, p. 1241.
36 J. Chatt, C. Eaborn and S. Ibekwe, *Chem. Comms*, 1966, p. 700.
37 M. C. Baird, *J. inorg. nucl. Chem.* 1967, **29**, 367.
38 D. J. Patmore and W. A. G. Graham, *Inorg. nucl. chem. Letters*, 1966, **2**, 179; *Inorg. Chem.* 1966, **5**, 2222.
39 A. N. Nesmeyanov, K. N. Anisimov, N. E. Kolobova and V. V. Skripkin, *Izvest. Akad. Nauk SSSR* (*Ser. Khim.*), 1966, p. 1292.
40 H. R. H. Patil and W. A. G. Graham, *J. Am. chem. Soc.* 1965, **87**, 673.
41 A. N. Nesmeyanov, K. N. Anisimov, N. E. Kolobova and M. Ya Zakharova, *Izv. Akad. Nauk SSSR* (*Ser. Khim.*), 1965, p. 1122.
42 S. V. Dighe and M. Orchin, *J. Am. chem. Soc.* 1965, **87**, 1146.
43 A. N. Nesmeyanov, K. N. Anisimov, N. E. Kolobova and V. N. Khandozhko, *Dokl. Akad. Nauk SSSR*, 1964, **156**, 383.
44 C. J. Fritchie, R. M. Sweet and R. Schunn, *Inorg. Chem.* 1967, **6**, 749.
45 A. G. Davies, G. Wilkinson and J. F. Young, *J. Am. chem. Soc.* 1963, **85**, 1692.
46 J. F. Young, R. D. Gillard and G. Wilkinson, *J. chem. Soc.* 1964, p. 5176.
47 R. D. Cramer, R. V. Lindsey, C. T. Prewitt and U. G. Stolberg, *J. Am. chem. Soc.* 1965, **87**, 658.
48 R. V. Lindsey, G. W. Parshall and U. G. Stolberg, *J. Am. chem. Soc.* 1965, **87**, 658.
49 R. V. Lindsey, G. W. Parshall and U. G. Stolberg, *Inorg. Chem.* 1966, **5**, 109.
49a L. J. Guggenberger, *Chem. Comms.* 1968, p. 512.
50 J. V. Kingston and G. Wilkinson, *J. inorg. nucl. Chem.* 1966, **28**, 2709.
51 J. V. Kingston, J. W. S. Jamieson and G. Wilkinson, *J. inorg. nucl. Chem.* 1967, **29**, 133.
52 M. A. Khattak and R. J. Magee, *Chem. Comms.* 1965, p. 400.
53 J. A. Stephenson and G. Wilkinson, *J. inorg. nucl. Chem.* 1966, **28**, 945.
54 A. A. Vlček and F. Basolo, *Inorg. Chem.* 1966, **5**, 156.
55 R. C. Edmondson and M. J. Newlands, *Chemy Ind.* 1966, p. 1888.
56 J. D. Cotton, S. A. R. Knox and F. G. A. Stone, *J. chem. Soc.* (*A*), 1968 (in the Press).
57 R. B. King and F. G. A. Stone, *J. Am. chem. Soc.* 1960, **82**, 3833.
58 S. D. Ibekwe and M. J. Newlands, *J. chem. Soc.* (*A*), 1967, p. 1783.

59　J. D. Cotton, S. A. R. Knox, I. Paul and F. G. A. Stone, *J. Chem. Soc.* (*A*), 1967, p. 264.

60　S. F. A. Kettle and I. A. Khan, *Proc. chem. Soc.* 1962, p. 82.

61　S. F. A. Kettle and I. A. Khan, *J. organometal. Chem.* 1966, **5**, 588.

61*a* R. Kummer and W. A. G. Graham, *Inorg. Chem.* 1968, **7**, 1208.

62　R. F. Bryan, *Proc. chem. Soc.* 1964, p. 232.

63　H. P. Weber and R. F. Bryan, *Chem. Comms.* 1966, p. 443.

64　R. F. Bryan, *Chem. Comms.* 1967, p. 355.

65　R. F. Bryan, *J. chem. Soc.* (*A*), 1967, p. 172.

66　B. T. Kilbourn and H. M. Powell, *Chemy Ind.* 1964, p. 1578.

67　B. T. Kilbourn, T. L. Blundell and H. M. Powell, *Chem. Comms.* 1965, p. 444.

68　R. F. Bryan, *J. chem. Soc.* (*A*), 1967, p. 192.

69　M. A. Bush and P. Woodward, *J. chem. Soc.* 1967, p. 1883.

70　J. E. O'Connor and E. R. Corey, *Inorg. Chem.* 1967, **6**, 968.

71　P. F. Lindley and P. Woodward, *J. chem. Soc.* (*A*), 1967, p. 382.

72　P. Porta, H. M. Powell, R. J. Mawby and L. M. Venanzi, *J. chem. Soc.* (*A*), 1967, p. 455.

73　E. L. Muetterties and R. A. Schunn, *Q. Rev. chem. Soc.* 1966, **20**, 245.

73*a* W. T. Robinson and J. A. Ibers, *Inorg. Chem.* 1967, **6**, 1208.

74　J. Dalton, I. Paul, J. G. Smith and F. G. A. Stone, *J. chem. Soc.* (*A*), 1968, p. 1195, and following papers.

75　D. J. Patmore and W. A. G. Graham, *Inorg. Chem.* 1967, **6**, 1879.

76　D. J. Patmore and W. A. G. Graham, *Inorg. Chem.* 1967, **6**, 981.

77　O. Kahn and M. Bigorgne, *C. r. hebd. Séanc. Acad. Sci., Paris*, 1966, **262***c*, 906; 1966, **263***c*, 973.

78　R. Ugo, F. Cariati, F. Bonati, S. Cenini and D. Morelli, *Ricerca scient.* 1966, **36**, 253.

79　D. M. Adams and P. J. Chandler, *Chemy Ind.* 1965, p. 269.

80　A. P. Hagen and A. G. MacDiarmid, *Inorg. Chem.* 1967, **6**, 686.

80*a* W. A. G. Graham, *Inorg. Chem.* 1968, **7**, 315.

81　F. Bonati and R. Ugo, *Istituto Lombardo* (*Rend. Sc.*), A, 1964, **98**, 607.

82　H. C. Clark and J. H. Tsai, *Inorg. Chem.* 1966, **5**, 1407.

83　H. C. Clark, J. D. Cotton and J. H. Tsai, *Inorg. Chem.* 1966, **5**, 1582.

84　M. Green, N. Mayne and F. G. A. Stone, *J. chem. Soc.* (*A*), 1968, p. 902.

85　G. C. Bond and M. Hellier, *Chemy Ind.* 1965, p. 35.

86　C. G. Pitt, L. K. Monteith, L. F. Ballard, J. P. Collman, J. C. Morrow, W. R. Roper and D. Ulkii, *J. Am. chem. Soc.* 1966, **88**, 4286.

87　G. N. Schrauzer and G. Kratel, *Angew. Chem.* (*Int. Ed.*), 1965, **4**, 146.

13

THE CHEMISTRY OF COORDINATION COMPOUNDS OF SCHIFF BASES

B. O. WEST

1. Introduction

Transition metal complexes with Schiff bases as ligands have been amongst the most widely studied coordination compounds. In general the azomethine group $\diagup$C=N which is the functional group of a Schiff base is aided in forming a stable complex by either a second such group (Fig. 1), an acidic group like a phenolic OH (Fig. 2), or another donor group (Fig. 3). The formation of a chelate ring seems essential for the production of stable complexes with ligands containing the $\diagup$C=N group.

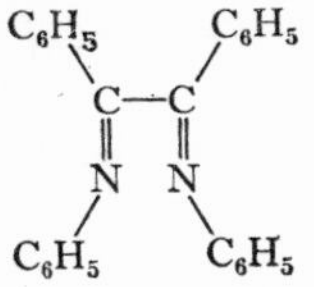

Fig. 1 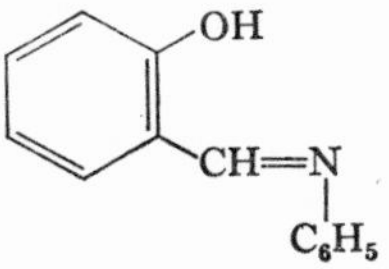   Fig. 2 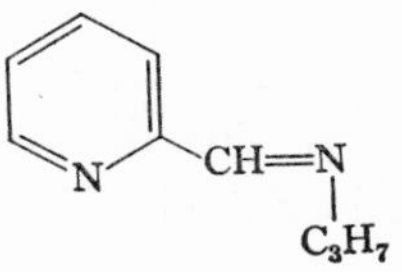   Fig. 3

Schiff[1] may be regarded as the first to have defined the composition of a metal complex with such a ligand by establishing the 1:2 metal–ligand ratio in copper complexes derived from N-aryl-salicylaldimines.

Subsequently studies have utilized Schiff bases derived from salicylaldehyde or related phenolic aldehydes to a very large extent owing to the ease with which they can be prepared. The pioneer work of Pfeiffer must be referred to in this regard. In a classic series of papers reviewed in 1940[2] he and his collaborators studied problems of synthesis, metal-exchange, ligand replacement, transamination, stereochemistry, and esterification, utilizing particularly salicylaldimine derivatives of copper(II), and these studies paved the way for much of the present-day interest in such metal complexes.

This review will be concerned primarily with problems of stereochemistry met in Schiff base complexes and with the behaviour of the complexes in solution. N-substituted-salicyl-aldimines and β-ketoamines (Fig. 4) will be the main classes of ligands referred to as these have been utilized to the greatest extent.

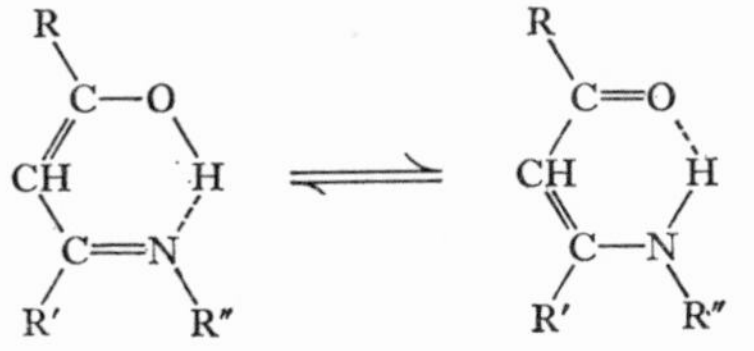

Fig. 4

A detailed account of the chemistry of Schiff base compounds containing tables of many of the complexes known to the end of 1964 has recently appeared.[3]

2. The synthesis of complexes

Metal complexes of these ligands have been synthesized by one of the following reactions.

(i) The direct reaction of a primary amine with a preformed salicylaldehyde-metal complex

$$Cu(OC_6H_4CHO)_2 + 2NH_2C_6H_5$$
$$\rightarrow Cu(OC_6H_4CH{=}NC_6H_5)_2 + 2H_2O$$

This method was developed by Pfeiffer[2] and is probably the most general method available. The two reactants are usually heated in a solvent such as an alcohol in which water is soluble and the product allowed to crystallize. Alternatively a solvent such as chloroform or benzene can be used and water distilled off during the course of reaction.

(ii) The reaction of a metal salt, usually an acetate, with a pre-formed Schiff base in aqueous ethanol or similar solvent. A base is often added to aid in the removal of an acidic proton from the ligand if required. The method is not convenient if the Schiff base is readily hydrolysed. A variation of this procedure is to prepare a strong alkoxide base, e.g. potassium t-butoxide in t-butanol, dis-solve the Schiff base in this solution and then add an appropriate anhydrous salt of the metal;[4] for example, $[Et_4N][NiBr_4]$. In each variation of the neutralization method the complex either precipitates or is obtained by evaporation of the solvent. The latter method is particularly useful if hydrolysis of the base or complex can occur. Other procedures involving the use of an anhydrous metal salt, the Schiff base and a proton acceptor in suitable solvents may be devised for particular circumstances.

(iii) The reverse reaction to (i)—namely reacting the primary amine complex of a metal with a ketone or aldehyde—has been termed 'the template synthesis'. The dimeric nickel salt tetrakis-(ethylenediamine)-μ-dichlorodinickel(II) chloride has been reacted with β-diketones or salicylaldehyde[5] to give nickel(II) complexes

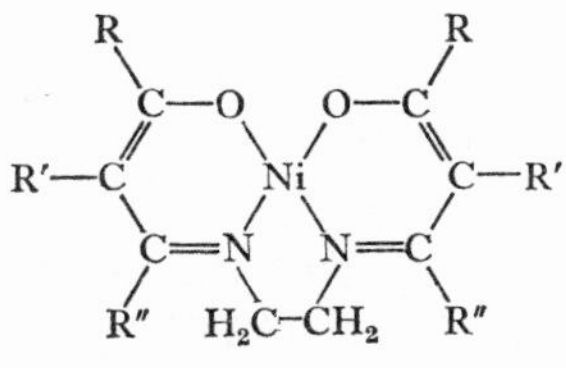

Fig. 5 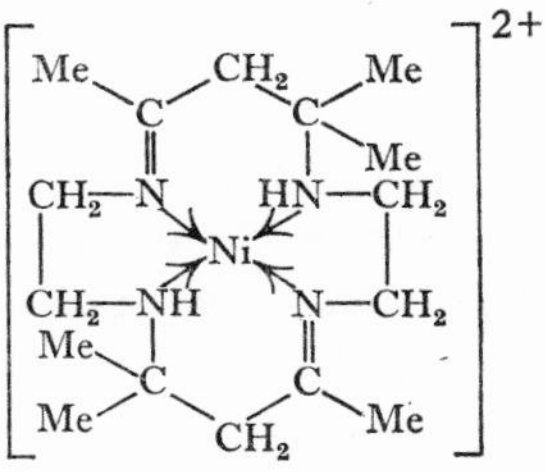   Fig. 6

of tetradentate ligands (Fig. 5). The reaction of metal diamine complexes with aliphatic aldehydes and ketones by Curtis[6] has led to the formation of macrocyclic metal complexes of considerable stability. Trisethylenediamine nickel perchlorate reacts with acetone at room temperature to yield a complex (Fig. 6). The reaction of $[Fe \, en_3]^{2+}$ with acetone has recently allowed the preparation of a labile Fe^{II} macrocycle from which other metal complexes can be prepared as well as the free ligand by displacement of Fe^{II}.[7]

3. The mechanism of the formation of Schiff base complexes

An attempt has been made by Nunez and Eichhorn[8] to study the formation of a metal Schiff base complex by measuring the kinetics of the reaction between salicylaldehyde, glycine, and nickel(II) or copper(II) ions. They concluded that the order of mixing the reactants materially effected the rate of formation of the final complex. If either nickel-glycine or nickel-salicylaldehyde complexes are formed before the third component is added (salicylaldehyde or glycine respectively) the reaction is much slower than when nickel reacts directly with the preformed Schiff base. The kinetic analysis suggests that the intermediate (Fig. 7) can form which further reacts to give the final complex (Fig. 8).

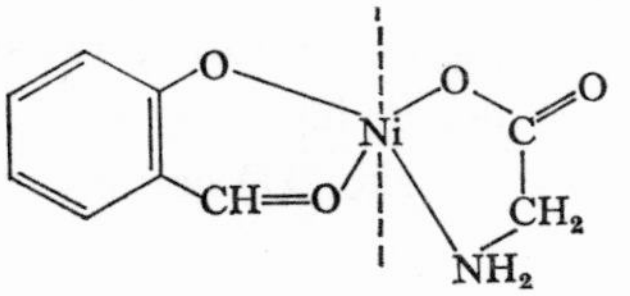

Fig. 7

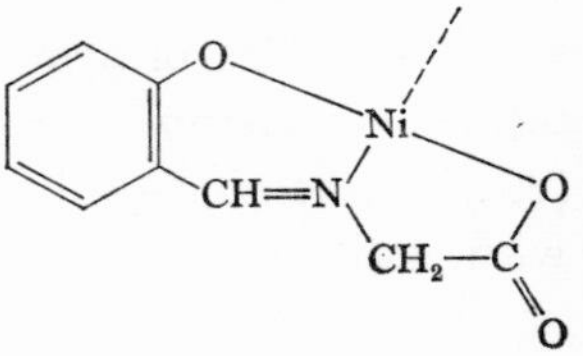

Fig. 8

A further study of the reaction of bis(salicylaldehydato)-copper(II) with cyclohexylamine has established the first-order rate dependence on both copper complex and amine.[9] The rate was much faster than between salicylaldehyde and the amine alone which suggests strongly that the reaction involved a direct attack by the amine on the carbon atom of a coordinated carbonyl group rather than requiring complete dissociation of a salicylaldehyde anion from the copper as a first step (Fig. 9). The polarizing power of the attached copper ion on the carbon atom would encourage nucleophilic attack at that point.

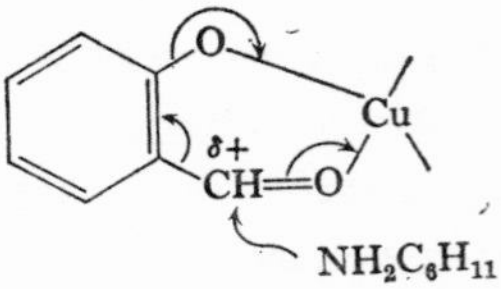

Fig. 9

No kinetic evidence (such as induction period) for the initial co-ordination of the amine with the copper salicylaldehyde complex was observed. The basicity of the amine will influence the rate of reaction. Thus aniline, a much weaker base than cyclohexylamine, gives a second-order rate constant of $1.43 \times 10^{-3} M^{-1} s^{-1}$ compared to cyclohexylamine $1.00 M^{-1} s^{-1}$, corresponding to a much slower rate. However, the importance of steric hindrance to the rate of reaction is shown by the value of $0.55 \times 10^{-3} \, M^{-1} \, s^{-1}$ given by tert-butylamine, which is as strong a base as cyclohexylamine.

The direct reaction of amines with β-diketone complexes is usually unsuccessful as a preparative method and in general no Schiff base complex is formed. Thus cyclohexylamine did not react with bis(acetylacetonato)copper(II) even at 116 °C.[9] The reaction between ethylenediamine and the copper complex in re-fluxing chloroform produced the bisethylenediaminecopper(II) cation and not the quadridentate Schiff base complex (Fig. 10). The salicylaldehyde–copper complex reacted readily even at room temperature to give the tetradentate complex (Fig. 11).

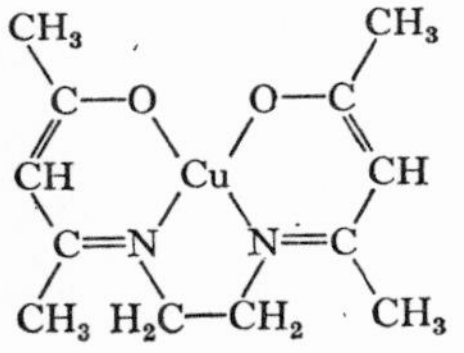

Fig. 10

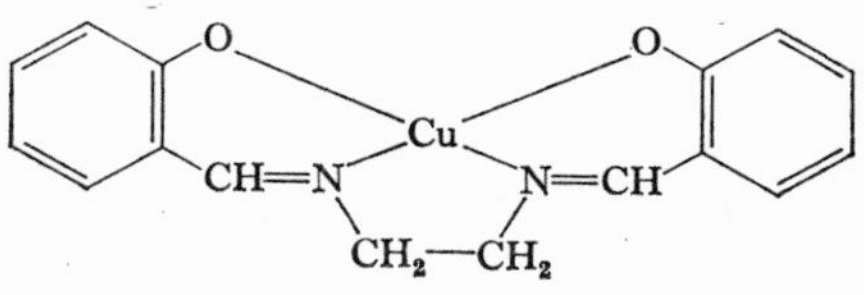

Fig. 11

Several factors no doubt contribute to the stabilization of the β-diketone complexes towards nucleophilic attack by amines. The most important is probably the delocalization of charge in the ligand with steric hindrance due to substituents on the carbon atoms next to the oxygens providing a secondary contribution.

4. Stereochemical properties of Schiff base complexes

The relative ease of preparing ligands based on the salicylaldimine or acetylacetone skeletons has meant that the influence of sub-stituents on the stereochemical properties of metal complexes containing these ligands has received considerable attention. X-ray diffraction studies have led to the discovery of a wide range of behaviour in the solid state including intermolecular interactions, distortion of metal stereochemistry by ligands or the requirements of a crystal lattice, while solution studies of magnetic susceptibility, spectra, dipole moments, and molecular weights have shown that certain groups of complexes can take part in equilibria of the form square-planar $\rightleftharpoons$ tetrahedral $\rightleftharpoons$ octahedral.

For convenience these results will be discussed in terms of the degree of chelation provided by the ligand.

5. Complexes with bidentate ligands

Structure determinations indicate that for the simplest Schiff base ligands with limited substitution the arrangement of bonds in the unit MO_2N_2 will conform to the basic arrangement found for the metal M in most of its compounds; that is, planar for Cu^{II}, Ni^{II}, Pd^{II}; tetrahedral for Co^{II} and Zn^{II}.[10] The planar complexes appear

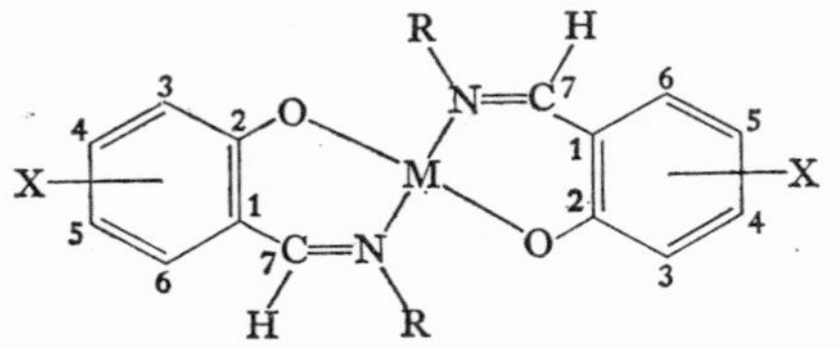

Fig. 12

without exception to have a *trans* configuration (Fig. 12). Distortion from planarity of the chelate rings and associated phenyl rings however seems relatively common. Thus the β form of bis(N-methylsalicylaldiminato)nickel(II) has its chelate rings distorted from planarity[11] by the grouping C_1—C_7—N—R being bent from the plane defined by the metal, oxygen, and the benzene ring. The bond C_1—C_7 makes an angle of 13° with the plane defined by C_1—C_2—C_6. Similar distortions have been observed with palladium complexes containing N-ethyl and N-n-butyl substituents.[12] When the substituent is larger as in the N-i-propyl nickel complex the distortion is greater and this complex[13] contains two planar chelate rings making an angle of 82° with each other. This structural change is paralleled by the magnetic properties of the compound, the virtually planar N-methyl derivative is diamagnetic[14] while the N-i-propyl complex is paramagnetic,[15] indicating that distortion from planarity has been sufficient to change the metal from a singlet to a triplet state. The stereochemistry of the N-i-propyl complex might be regarded as distorted tetrahedral, an angle of 90° between planes being expected for full tetrahedral symmetry.

The inspection of models of such complexes indicates that a major factor in deciding whether distortion of the chelate rings occurs, is the interaction between the R group attached to nitrogen and the oxygen and H_3 atom on the other ligand molecule. However, further subtleties of structure are apparent when it is observed that the N-i-propyl nickel complex having a methyl group at C_3 is diamagnetic and presumably has a planar arrangement of NiO_2N_2 whereas when an ethyl group is at C_3 the molecule is again paramagnetic.[16]

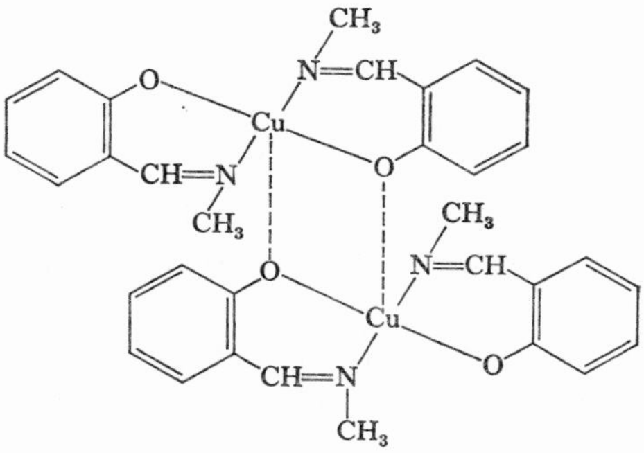

Fig. 13

Many more structural studies are required before these kinds of variations can be adequately explained.

Copper(II) Schiff base complexes also show an interestingly varied behaviour in the solid state. The N-methylsalicylaldimine complex of copper(II) exists in three modifications. The α-form[17] contains molecules having planar configurations of chelate rings about the metal but stacked one above the other so that the copper atoms, 3·33 Å apart, form chains running through the crystal. The β form[11] is isomorphous with the nickel(II) complex described above and has chelate rings distorted from planarity while the γ form[18] contains dimeric units of the complex with the copper from one molecule bonding to one of the oxygen atoms on the other ring (Fig. 13).

This variation in crystal structure for the same chemical compound could influence very notably other physical properties which depend on the type of ligand field surrounding the metal. Nickel for example would show changes in magnetic and spectral properties associated with singlet–triplet transformation depending on whether the metal was in a square-planar or octahedral environment. One form of the N-methylsalicylaldimine nickel complex is known which is paramagnetic and very sparingly soluble in organic solvents.[14, 19] It is believed to be polymerized by bonding between neighbouring chelate molecules so as to give an octahedral configuration to the nickel ions.[20]

The influence of the lattice requirements of a crystal as well as the small energy barriers separating the various possible crystal forms has been demonstrated by Chakravorty,[21] who crystallized the N-i-propyl 5-methylsalicylaldimine complex of Ni^{II} in a matrix of the corresponding Zn complex. The normally diamagnetic, presumably planar, nickel complex became paramagnetic in the presumed host lattice of tetrahedral zinc complex.

The relatively rare 5-coordinate stereochemistry has also been observed in metal complexes of N-methylsalicylaldimine.[22] The Zn^{II}, Co^{II}, and Mn^{II} complexes are all isomorphous and the structure of the Zn^{II} compound indicates that it contains dimeric units formed by the sharing of oxygen atoms between the two monomers (Fig. 14).

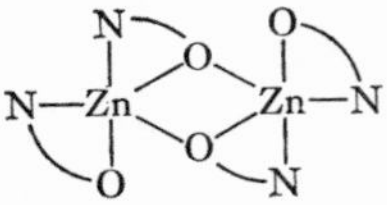

Fig. 14

In solution in non-coordinating solvents such as benzene a remarkable stereochemical change has been observed with nickel complexes in that square-planar $\rightleftharpoons$ tetrahedral equilibria or association phenomena can occur with salicylaldimine and β-ketoamine derivatives depending on the type of substituents present. N-R-salicylaldimine nickel(II) complexes are diamagnetic (square-planar) in the solid state when R = straight chain alkyl but solutions in benzene show varying degrees of paramagnetism, the values being less than would be associated with two unpaired electrons per nickel ion. This effect is due to the association of two or three molecules of complex in solution to give either 5- or 6-coordination around nickel ions,[23, 24] the Me derivative showing the greatest degree of association (see Fig. 13). The paramagnetism in solution is due to a mixture of diamagnetic square-planar derivatives and paramagnetic associated species. When R is an alkyl chain branched at the α carbon the nickel complexes are scarcely associated at 25° or higher in benzene but they show a much greater change from the diamagnetic solids to paramagnetic solutions than occurs with the straight chain complexes. Without association occurring this change can only be due to an equilibrium in solution between square-planar (singlet) and tetrahedral (triplet) species in solution.[25, 26, 16] The existence of tetrahedral species is confirmed by spectral measurements on solutions. The steric factor is thus of vital importance in promoting this conversion from square-planar to tetrahedral. The branched chain alkyl group being able to interact no doubt with the O atom and H_3 positions on the neighbouring chelate molecule. When R = tert-Bu the nickel complex becomes paramagnetic even in the solid state.[26] Little conversion to the tetrahedral form is observed when branching occurs at the β carbon.[23] Planar-tetrahedral equilibria also occur when R = m- or p-aryl in solutions above 100 °C,[20] but below this associated species can account for virtually all the observed paramagnetism in solution.[27] *Ortho*-substituted aryl groups can prevent either association or tetrahedral forms occurring in solution although the reason for any barrier to the latter configuration is not clear from models.

Similar planar-tetrahedral equilibria for nickel complexes with β-ketoamine ligands have been studied. Sec-alkyl substituents at R when R' = R" = CH_3 (Fig. 15) caused the complexes to be almost completely tetrahedral[28] whereas for R = aryl the equilibrium shifted towards the planar form.

The technique introduced by Eaton, Phillips and Caldwell[29] of utilizing the large isotropic proton hyperfine contact shifts found for NN'-disubstituted-aminotroponeimino Ni[II] complexes undergoing a planar-tetrahedral equilibration has been extended by Holm and his co-workers to examination of the nickel complexes of salicylaldimines and β-ketoamines.[3]

Complexes of cobalt(II) with salicylaldimines do not show

evidence for configurational equilibria in solution whatever
N-substituent is present. The complexes appear to be tetrahedral
in the solid state or in solution in benzene as evidenced by their
magnetic moments,[30, 31] lying in the range 4·3–4·5 B.M. expected
for such a configuration, their large dipole moments[26] and their
spectra.[31] The complex bis(N-n-butylsalicylaldiminato)cobalt(II)
is isomorphous with the corresponding Zn^{II} complex which was
found to be tetrahedral by X-ray structural analysis.

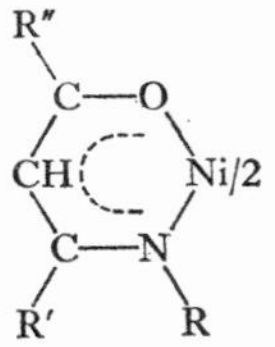

Fig. 15

An early report[32] that cobalt salicylaldimine complexes could not
be prepared when *ortho*-substituted aryl rings were attached to N
claimed that this was due to steric hindrance in a tetrahedral
arrangement of the ligands. It was subsequently shown[33] that
cobalt salicylaldimine complexes could be prepared containing
such *ortho*-substituted N-phenyl groups as methyl and the halogens F, Cl, Br.[34] The existence of an *ortho*-steric effect has subsequently been detected in the reaction of pyridine with various
cobalt complexes having aryl groups attached to nitrogen and
having *o*-, *m*-, or *p*-substituents.[34] The coordination of pyridine to
these complexes causes an increase in the magnetic moments of the
complexes in solution[30] towards the values found for octahedral
complexes and from spectral measurements equilibrium constants
for the attachment of two pyridine molecules may be determined:

$$CoL_2 + py \overset{K_1}{\rightleftharpoons} CoL_2py$$

$$CoL_2py + py \overset{K_1}{\rightleftharpoons} CoL_2py_2$$

The values given in Table 1 show that the attachment of a second
pyridine molecule to the Co(sal$N$-*o*-tolyl)$_2$ complex is associated
with a much smaller constant than the first whereas for both the
m- and *p*-complexes as well as the unsubstituted phenyl compound

$K_2 > K_1$. The '*ortho*' effect is associated with a large negative entropy concentrated in the second step of coordination. This indicates that considerable difficulty must be experienced in bringing up the second pyridine molecule to coordinate with the metal. Restricted rotation of pyridine about a Co—N bond due to interaction with the *o*-substituent can readily account for this. The *N*-cyclohexyl derivative has shown a similar spectrum in pyridine to that in benzene but with lowered band intensities. The data can only be made to agree with the formation of a monopyridinate even in pure pyridine.

TABLE I. *Equilibrium constants for the attachment of pyridine to N-arylsalicylaldimine cobalt(II) complexes in benzene at 15 °C*

Group	K_1	K_2	$-\Delta S_1^0$	$-\Delta S_2^0$
H	1·8	2·8	19	21
p-CH$_3$	0·87	2·1	24	13
m-CH$_3$	1·3	3·9	25	14
o-CH$_3$	2·4	0·68	16	47

K values are in litres/mole.

β-Ketoamine complexes of cobalt(II) show evidence for doublet–quartet state mixing in solution for some complexes[35] and this has been taken to indicate the presence of an equilibrium involving planar–tetrahedral interconversions.

Complexes of bidentate salicylaldimine ligands with trivalent transition metal ions are also known. The cobalt(III) derivatives show sufficient steric hindrance between groups attached to nitrogen that the *trans* type configuration (Fig. 16) is the one normally isolated.[36, 37] Holm[38] has confirmed this situation in several cases by the ingenious use of nuclear magnetic resonance measurements and further dipole moment data has shown the generality of the *trans* isomers.[39]

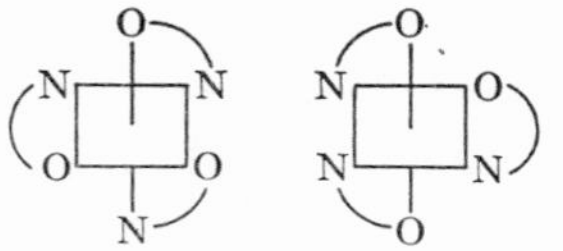

Fig. 16

Chromium(III) derivatives can be readily prepared by the reaction of an amine with the tris-salicylaldehydechromium(III) complex.[9] The complexes have spectra consisting of one well-defined band in the visible region and a poorly defined band in the near u.v. The band near 18000 cm^{-1} has been assigned to the transition $^4A_{2g} \rightarrow {}^4T_{2g}$ and taken to represent the octahedral splitting parameter $10D_q$. These values place the Schiff base ligands between oxalato and ethylenediamine in the spectrochemical series.[40] The second band is probably obscured by a ligand transition in view of its high ϵ values and is not considered to be a pure $^4A_{2g} \rightarrow {}^4T_{1g}$ transition.

TABLE 2. *Spectroscopic properties of N-substituted-salicylaldimine chromium(III) complexes in benzene*

R	Band position cm^{-1} (ϵ)			
n-propyl	25000	(9210)	18000	(120)
iso-butyl	24800	(9550)	17950	(124)
p-tolyl	23700	(5000)	17700	(90)
p-bromophenyl	23400	(10100)	17700	(180)

6. Complexes with tridentate ligands

The most noteworthy property of complexes derived from tridentate Schiff bases is that abnormally low magnetic moments (1·1–1·37 B.M. at 287 °K) are found for a number of copper complexes with such ligands[41, 42] (Fig. 17).

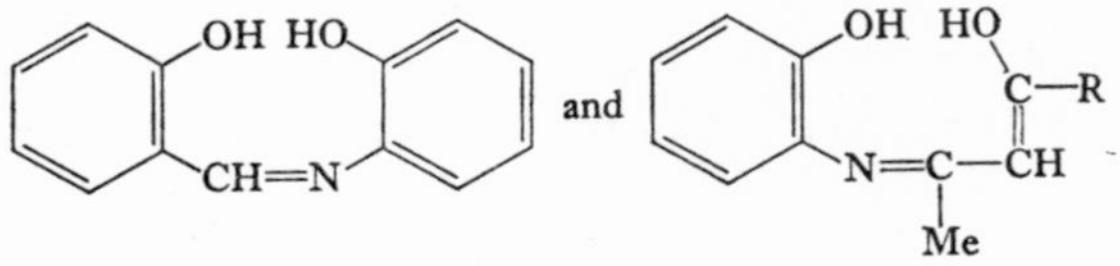

Fig. 17

The structure of (4-o-hydroxyphenylamino-3-penten-2-ono)-copper(II)[43] shows how a bridged dimeric structure can obviate the need for 3-coordinate copper in such compounds (Fig. 18). This type of structure is no doubt found in a number of cases where copper complexes show subnormal magnetic moments due to a spin-interaction mechanism.

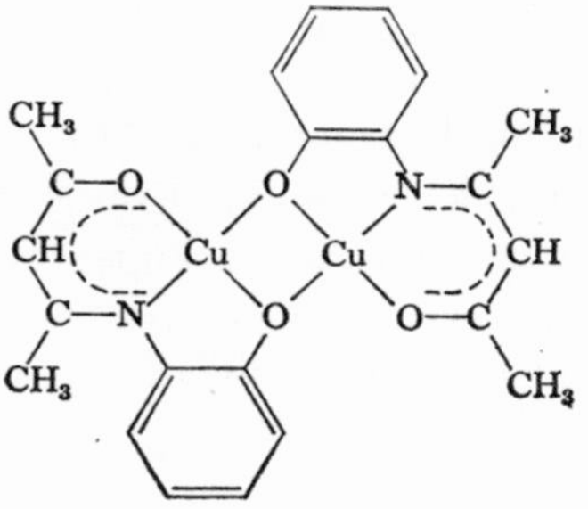

Fig. 18

Many of these complexes can react with pyridine or water and achieve a normal mononuclear, four-coordinate structure with the molecule of solvent occupying the fourth position.[44]

Other tridentate ligands have been synthesized by Sacconi and his co-workers based on the ethylenediaminesalicylaldimine skeleton (Fig. 19). These molecules can behave as bidentate or tridentate ligands and as the latter form octahedral complexes with divalent metals. The behaviour is defined by the substituents R′ and R″. Square-planar, diamagnetic nickel complexes are formed when the ligands coordinate as bidentate groups through the phenolic O and azomethinic N.[45] Square-pyramidal, 5-coordinate species were also detected when R′ = R″ = Et, and X = 3-Cl and 5-Cl.[46] The structure of the 5-Cl derivative[47] indicates that steric hindrance involving the ethyl groups of the final N atom prevents 6-coordination occurring. Equilibria between 4-, 5-, and 6-coordinate species has been detected in solution.[46] Similar types of cobalt complexes have been described.[48]

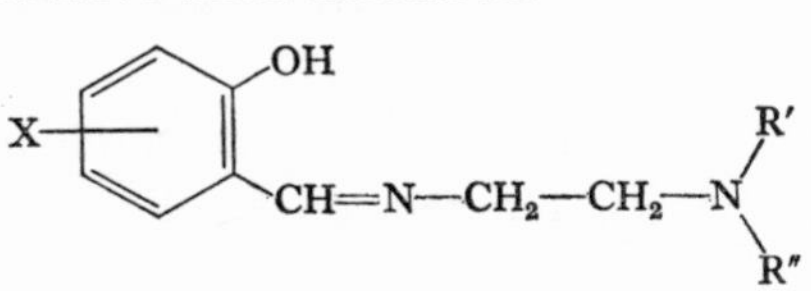

Fig. 19

7. Complexes with tetradentate ligands

The interaction of diamines with salicylaldehyde or acetylacetone or their substituted derivatives produces tetradentate ligands (Fig. 20). When the carbon chain linking the nitrogen atoms is short

(2–4 atoms long) the ligands appear constrained to coordinate only in a planar *cis* fashion and this indeed appears to be the gross stereochemistry of metal complexes with most such ligands. Second-order effects which cause slight variations from coplanarity are, however, well authenticated. Thus the *NN'*-ethylenebis-(salicylaldimine)copper(II) complex is dimeric in the solid state[49] with intermolecular Cu—O bonding at 2·41 Å causing the copper

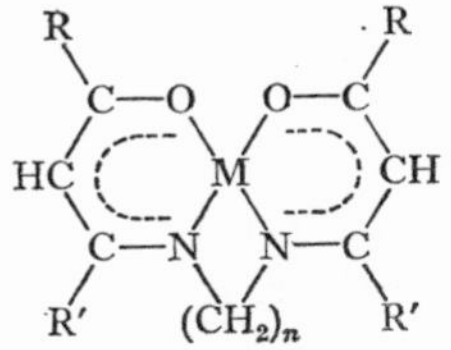

Fig. 20

atoms to exist in a pyramidal 5-coordinate environment. The ethylene bridge is twisted into the *gauche* position and the benzene rings and chelate rings slope slightly away from the copper atoms so that a separation of about 4·0 Å occurs at the extremities of the dimers which however approach to 3·18 Å at the molecular centre. This linkage via Cu—O intermolecular bonds does not, however, give rise to any significant magnetic exchange interaction between the copper ions[50] possibly due to the long Cu—O bonds or the Cu—O—Cu bond angle. The corresponding nickel complex is diamagnetic and therefore presumably planar although a full structural study has not yet been carried out.

Increasing the methylene chain length for such complexes may cause a change in stereochemistry from *cis*-planar to tetrahedral or less likely to *trans*-planar. Holm[51] examined nickel complexes up to tetramethylene but did not achieve a change in magnetic moment. Weigold[34] has prepared cobalt salicylaldimine complexes with methylene chains up to 10 units in length and has obtained magnetic moments in the range for tetrahedral cobalt for the complexes C_7–C_{10}. The spectra of the complexes are very similar to those found for *N*-alkylsalicylaldimine complexes which are tetrahedral. The *NN'*-ethylene cobalt(II) complex is spin-paired with a moment of 2·52 B.M. at room temperature.[52] The polymethylene cobalt complexes are poorly soluble in most solvents and their composition may be dimeric or more highly polymeric. The great length of the methylene chains may prevent both ends of the same ligand coordinating to a single metal. Pfeiffer has postulated a dimeric structure for the copper salicylaldimine complex with 2,2'-diMeO-benzidine where it appears certain that the two ends from the same molecule cannot meet[53] (Fig. 21).

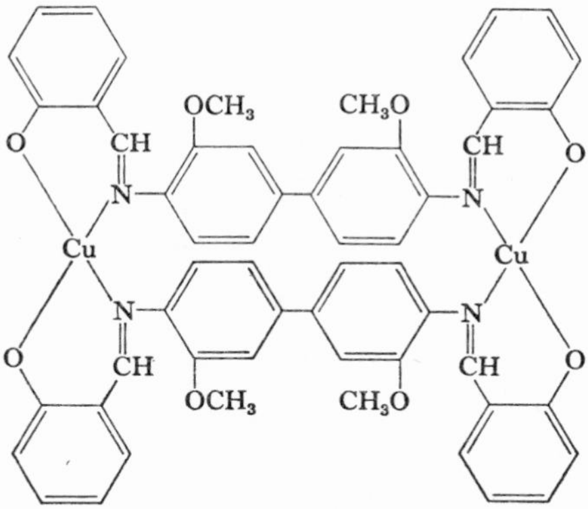

Fig. 21

An interesting example of competition between metal and ligand stereochemistries has been found in the structure of NN'-(2,2'-biphenyl)bis(salicylaldiminato)copper(II).[54] This complex was predicted[55] to be tetrahedral because of the considerable strain to be expected for a square-planar configuration with the biphenyl system remaining planar. In fact the complex is distorted only slightly as far as the CuO_2N_2 framework is concerned. The angle between the planes defined by CuO_1N_1 and CuO_2N_2 is 37° instead of the 90° expected for a tetrahedral array. This structure has been achieved by the twisting of the ligand at various points to accommodate the strain including twisting along the C—C bond joining the two phenyl rings.

Derivatives of Fe^{III} with tetradentate ligands have also shown unique behaviour among the metal complexes. The complex Fe salen Cl has been shown to possess a binuclear structure in the solid state[56] in which each Fe^{III} achieves 6-coordination by intermolecular Fe—O bonding (Fig. 22). The Fe—O 'in-plane' length is 1·98 Å, while the bridging length is greater (2·18 Å). This asymmetry is also present in the Cu^{II} complex.[49] The complex showed a slightly reduced high-spin moment (5·36 B.M. at 300 °K) suggested due to a super-exchange mechanism via the Fe—O—Fe grouping. A five-coordinate monomer was also produced by crystallizing the complex from CH_3NO_2 when the solvent was retained in the lattice without taking part in coordination but allowing monomeric, square-pyramidal Fe salen Cl molecules to exist. The moment of the monomeric form was increased to 5·9 B.M.

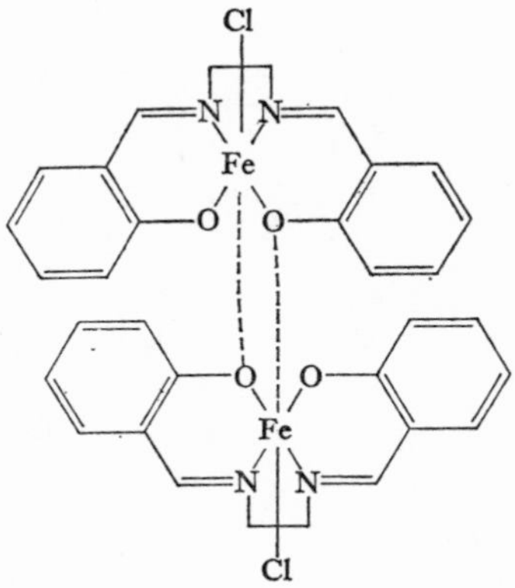

Fig. 22

The oxygen derivative [Fe salen]$_2$O[57] has an abnormally low magnetic moment 1·87 B.M. at 300 °K which is sensitive to temperature changes (0·68 B.M. at 80 °K). The magnetic properties of this complex were quantitatively explained in terms of spin interaction between the Fe atoms.

8. Complexes with pentadentate and hexadentate ligands

The chemistry of complexes containing hexadentate Schiff base ligands has been reviewed in detail by Goodwin[58] and these com-

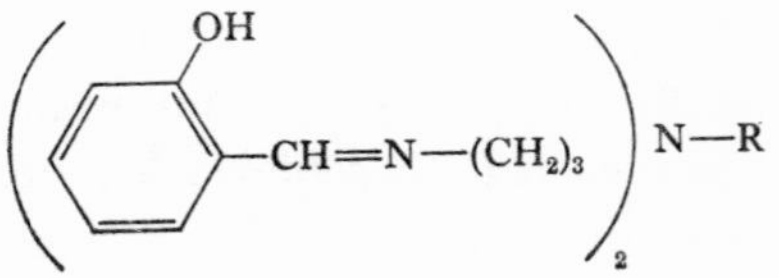

Fig. 23

plexes will not be discussed here. Pentadentate Schiff base ligands have only recently been reported.[59] Nickel complexes have been reported which are six-coordinate, the final position being filled by a pyridine molecule (Fig. 23).

9. The stability of Schiff base complexes

9.1. Dissociative stability

Stability constant measurements have only been made for a few ligands and metals. In particular may be mentioned the determinations by Lane and Kandathil[60] on N-arylsalicylaldimine complexes of Cu, Ni, and Cd, and Martin and Janusonis[61] on β-ketoamine derivatives. The general comment may be made that the N-phenyl ketoamine derivatives based on an acetylacetone skeleton are somewhat more stable than the simple acac complexes ($\log K_1 = 10\cdot8$, compared to $\log K_1 = 9\cdot55$) for the metals studied, namely Cu, Be, Ni, and Co. Also the beryllium complex has virtually the same stability as the copper complex ($\log K_1 = 10\cdot9$, $\log K_2 = 10\cdot5$ for Be, compared to $\log K_1 = 10\cdot8$, $\log K_2 = 10\cdot9$ for Cu). The ligands are neither extremely powerful nor very weak but are comparable in complexing ability to salicylaldehyde and acetylacetone respectively. The acid strengths of the ligands in fact suggested that they would have formed more stable complexes than in fact was found. The stability of the Be complexes with β-ketoamines contradicts an apparently traditional belief that Be complexed only weakly with a ligand containing O and N. Green has studied the extraction of Be by N-n-Bu-salicylaldimine[62] and the N-Et derivative[63] into toluene and finds favourable extraction ratios in each case, the n-Bu complex, perhaps because of favourable solvation, being the more readily extracted.

The use of Schiff base derivatives for the solvent extraction of metals can have considerable advantages for metal separation. Thus Table 3 sets out the pH values at which 50% of a standard solution of copper(II) and nickel(II) has been extracted by three common quadridentate Schiff base ligands. The figures for dithiazone are included for comparison. The wide separation in pH between salen and salophen extractions is reflected in the pH values at which extraction was first observed, namely 2·9 salen

TABLE 3. *pH values at which* 50% *extraction into chloroform has occurred from aqueous solutions of* Cu^{II} *and* Ni^{II}

Ligand	Cu^{II}	Ni^{II}
Salen	3·3	12·5
Salophen	2·4	11·9
Acacen	3·1	—
Dithiazone	1·0	3·0

Salen = *NN'*-ethylenebis(salicylaldimine).
Salophen = *NN'*-*o*-phenylenebis(salicylaldimine).
Acacen = *NN'*-ethylenebis(acetylacetoneimine).

Cu^{II}, 10·8 salen Ni^{II}. Other metals such as Fe^{II}, Co^{II}, and Mn^{II} are extracted at pH values intermediate between Cu and Ni, but owing to the ease with which these metal complexes can be oxidized to cations containing the trivalent form it becomes simple to keep them back in an analytical exercise as water soluble [M salen]$^+$, etc., species.[64] The failure of Ni^{II} to be extracted by acacen in chloroform even at high pH's is quite remarkable in view of the ease of direct preparation of the nickel complex.

9.2. Metal exchange

Work by Pfeiffer *et al.*[65] on the exchange of one metal ion for another bound in *NN'*-ethylene bis(salicylaldimine) complexes showed that an exchange series could be set up from experiments in boiling pyridine. The order of increasing ease of displacement being Cu > Ni > VO and Fe > Zn > Mg. Exchange studies using radio tracers established that metal ions exchanged with the same metal in tetradentate Schiff base ligands of the salen type in the order of increasing rate:[66] Ni < Cu < Co < Zn. The slow rate of exchange of square-planar nickel tetradentate complexes with copper ions was subsequently examined and a mechanism for exchange suggested[67] which involved partial dissociation of the tetradentate ligand and coordination of a free end to the exchanging metal ion. Resistance to twisting of the Schiff base ligands due to electronic or steric factors was suggested as the main factor controlling rates of exchange of copper with a series of tetradentate Schiff base nickel complexes. Complexes with bidentate ligands, exchange ligands or metal ions very rapidly in pyridine and other polar solvents. The slower rate of reaction of nickel complexes compared to copper is in line with crystal field theory which requires the activation energies for spin-paired d^8 systems to be larger than d^9.

9.3. Stability to hydrogenation of the C=N group

The $\diagdown$C$=$N linkage in N-substituted salicylaldimines is readily reduced by hydrogen in ethanol in the presence of Adams platinum catalyst to give the corresponding o-hydroxy benzylideneamine (Fig. 24). The metal complexes of these ligands with Cu^{II}, Co^{II}, and

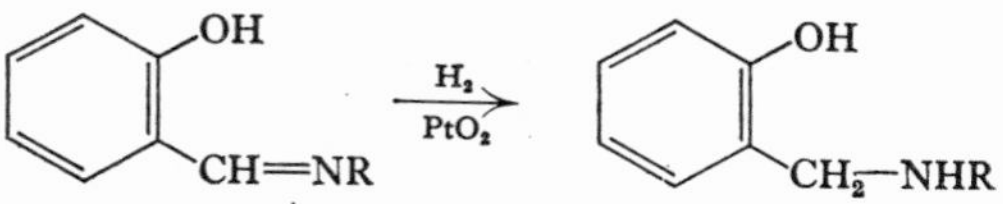

Fig. 24

Co^{III} are greatly resistant to such reduction however and require much larger amounts of catalyst to bring about reduction.[68] In the over-all process both the ligand and the metal ion are reduced to the benzylideneamine and free metal. Metal complexes of tetradentate Schiff base ligands could not be reduced even in the presence of a very large amount of catalyst. The existence of intermediate valency compounds was not detected except for Co^{II} species during the reduction of Co^{III} complexes. No evidence was found that selective reduction of the C=N groups in the complexes could be achieved forming a new type of complex in the process. Attempts to react o-hydroxy benzylideneamines with copper and cobalt ions directly only yielded complexes with copper.[9] These complexes are of the form CuLX, where L is a mole of ligand and

$$X = CH_3COO.$$

The magnetic moments of the complexes at room temperature are abnormally low for copper complexes[68] (1·27–1·35 B.M.) and although temperature versus magnetic susceptibility studies of the complexes have not yet been made the complexes are probably bridged compounds related to other copper complexes with subnormal moments.[69] The infrared spectra of the complexes show bands in the regions of 1540 cm^{-1} and 1340 cm^{-1} which give separations of the order of 205 cm^{-1}. These bands have been assigned to the asymmetric and symmetric C–O stretching modes associated with the acetate groups. The separation of 200 cm^{-1} is much larger than occurs in copper acetate itself where bridging acetate ions exist and suggest that the acetate ions are acting as unidentate groups in these complexes. A suggested structure is given (Fig. 25). The hydrogenated base derived from NN'-ethylenebis(salicylaldimine) reacts readily with Cu to give a complex which has a normal moment at room temperature. The nickel complex, however, is dimeric in benzene and paramagnetic.

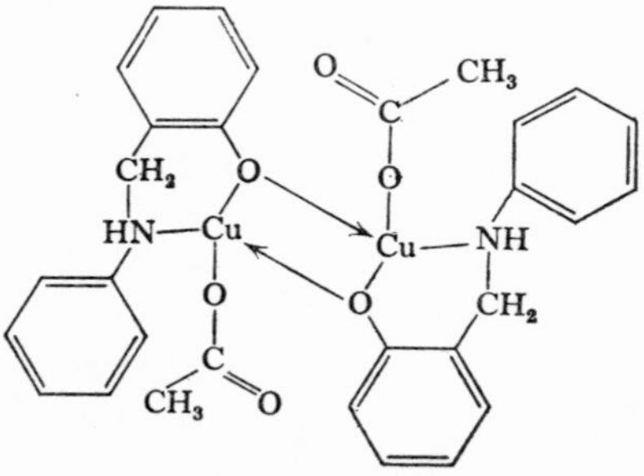

Fig. 25

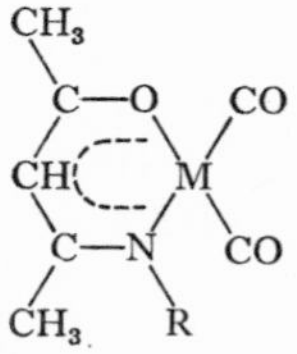

Fig. 26

10. Future developments

Schiff bases have been little used in reactions involving metals in low valence states. Rhodium and iridium complexes of the form shown in Fig. 26 have been prepared by the reaction of amine complexes of the carbonyl halides $M(CO)_2Cl$ amine with β-diketones.[70] It seems that an investigation of reactions involving metals in carbonyls or other low valence metal complexes and Schiff bases would prove of considerable interest. Work in this area is already being reported. Thus Schiff bases derived from aromatic aldehydes or ketones give cyclic products (Fig. 27) on high pressure carbonylation in the presence of cobalt carbonyl.[71] Pauson et al.[72] have isolated complexes from the reaction of $Fe_2(CO)_9$ and various Schiff bases. The derivative of N-p-tolybenzylaldimine has the structure[73] shown (Fig. 28).

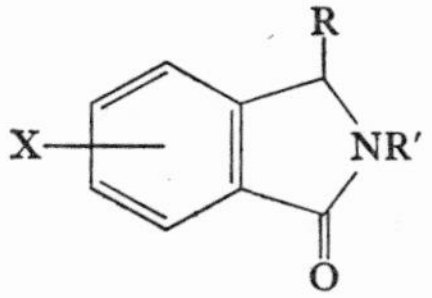

Fig. 27   Fig. 28

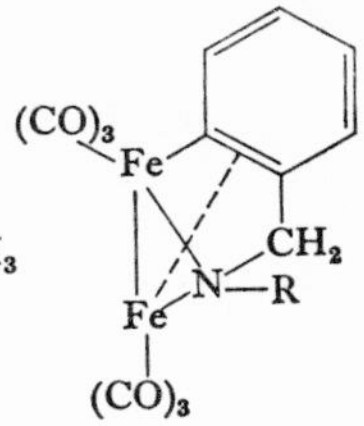

The possibilities still available in conventional complexes for the preparation of new types of compounds are well shown by recent reports on the syntheses of Co—C bonds from Schiff base complexes. Thus the Co^{III} derivative of NN'-ethylenebis(acetylacetone-imine), [Co acacen $(NH_3)_2$]Cl, reacts with Grignard reagents or lithium phenyl to give eventually the compounds [Co acacen LR], L being H_2O and R = Me, Et, Ph, containing a water stable Co—C bond.[74] Finally a low valence form of Co salen has been reported formed by the reduction of the divalent cobalt complex by sodium in T.H.F. to give Na[Co salen]. This derivative reacts with methyl iodide to form the complex Co salen Me . H_2O in which again a Co—CH_3 bond has been formed.[75]

REFERENCES

1   H. Schiff, *Annln Phys.* 1869, **150**, 193.
2   P. Pfeiffer, *Angew. Chem.* 1940, **53**, 93.
3   R. H. Holm, G. W. Everett, Jr., and A. Chakravorty, *Progress in Inorganic Chemistry*, vol. VII, pp. 83–214. New York: Interscience, 1966.
4   G. W. Everett and R. H. Holm, *J. Am. chem. Soc.* 1965, **87**, 2117; *Proc. chem. Soc.* 1964, p. 238.
5   E. J. Olszewski, L. J. Boucher, R. W. Oehmke, J. C. Bailar and D. F. Martin, *Inorg. Chem.* 1963, **2**, 661.
6   N. F. Curtis, *J. chem. Soc.* 1960, p. 4409. N. F. Curtis and D. A. House, *Chemy Ind.* 1961, **42**, 1708. N. F. Curtis, *J. chem. Soc.* 1965, p. 924.
7   N. Sadasivan and J. F. Endicott, *J. Am. chem. Soc.* 1966, **88**, 5468.
8   L. J. Nunez and G. L. Eichhorn, *J. Am. chem. Soc.* 1962, **84**, 901.
9   M. J. O'Connor and B. O. West, Ph.D. thesis, M. J. O'Connor, Monash University, 1966.
10 E. C. Lingafelter and R. L. Braun, *J. Am. chem. Soc.* 1966, **88**, 2951.
11 E. Frasson, C. Panattoni and L. Sacconi, *J. phys. Chem.* 1959, **63**, 1908.

12 E. Frasson, C. Panattoni and L. Sacconi, *Acta crystallogr.* 1964, **17**, 85, 477.

13 M. R. Fox, E. C. Lingafelter, P. L. Orioli and L. Sacconi, *Acta crystallogr.* 1964, **17**, 1159.

14 C. M. Harris, S. L. Lenger and R. L. Martin, *Aust. J. Chem.* 1958, **11**, 331.

15 R. H. Holm, A. Chakravorty and G. O. Dudek, *J. Am. chem. Soc.* 1964, **86**, 379.

16 R. H. Holm and K. Swaminathan, *Inorg. Chem.* 1963, **2**, 181.

17 E. C. Lingafelter, G. L. Simmons, B. Morosin, C. Scheringer and C. Freiburg, *Acta crystallogr.* 1961, **14**, 1222.

18 D. Hall, S. V. Sheat and T. N. Waters, *Chem. Commun.* 1966, p. 436.

19 L. Sacconi, P. Paoletti and R. Cini, *J. Am. chem. Soc.* 1958, **80**, 3583.

20 L. Sacconi and M. Ciampolini, *J. Am. chem. Soc.* 1963, **85**, 1750.

21 A. Chakravorty, *Inorg. Chem.* 1965, **4**, 127.

22 P. L. Orioli, M. Di Vaira and L. Sacconi, *Chem. Commun.* 1965, p. 103.

23 R. H. Holm, *J. Am. chem. Soc.* 1961, **83**, 4683.

24 H. C. Clark, K. Macvicar and R. J. O'Brien, *Can. J. Chem.* 1962, **40**, 822.

25 L. Sacconi, P. L. Orioli, P. Paoletti and M. Ciampolini, *Proc. chem. Soc.* 1962, p. 256.

26 L. Sacconi, P. Paoletti and M. Ciampolini, *J. Am. chem. Soc.* 1963, **85**, 411.

27 R. H. Holm and K. Swaminathan, *Inorg. Chem.* 1962, **1**, 599.

28 G. W. Everett, Jr. and R. H. Holm, *J. Am. chem. Soc.* 1965, **87**, 2117.

29 D. R. Eaton, W. D. Phillips and D. J. Caldwell, *J. Am. chem. Soc.* 1963, **85**, 397.

30 B. O. West, *J. chem. Soc.* 1962, p. 1374.

31 L. Sacconi, M. Ciampolini, F. Maggio and F. P. Cavasino, *J. Am. chem. Soc.* 1962, **84**, 3246.

32 B. O. West, *Nature, Lond.* 1964, **173**, 1187. B. O. West, W. G. P. Robertson and C. S. Hocking, *Nature, Lond.* 1955, **176**, 832.

33 H. Nishikawa and S. Yamada, *Bull. chem. Soc. Japan*, 1962, **35**, 1430.

34 B. O. West and H. Weigold, Ph.D. thesis, H. Weigold, University of Adelaide, 1965.

35 G. W. Everett and R. H. Holm, *J. Am. chem. Soc.* 1966, **88**, 2442.

36 B. O. West, *J. chem. Soc.* 1960, p. 4944.

37 F. P. Cavasino, M. Ciampolini and F. Maggio, *Attio Acad. Sci. Lett.*, Palermo, 1959–60, **20**, ser. IV, pt. I, p. 5; *Chem. Abstr.* 1963, **57**, 12086c.

38 A. Chakravorty and R. H. Holm, *Inorg. Chem.* 1964, **3**, 1521.

39 M. Ciampolini, F. Maggio and F. P. Cavasino, *Inorg. Chem.* 1964, **3**, 1188.

40 C. K. Jorgensen, *Absorption Spectra and Chemical Bonding in Complexes*, 1962, p. 110. London: Pergamon.

41 M. Kishita, Y. Muto and M. Kubo, *Aust. J. Chem.* 1957, **10**, 386.

42 M. Kishita, Y. Muto and M. Kubo, *Aust. J. Chem.* 1958, **11**, 309.

43 G. A. Barclay, C. M. Harris, B. F. Hoskins and E. Kokot, *Proc. chem. Soc.* 1961, p. 264.

44 M. Kubo, Y. Kuroda, M. Kishita and Y. Muto, *Aust. J. Chem.* 1963, **16**, 7.
45 L. Sacconi, P. Nannelli and U. Campigli, *Inorg. Chem.* 1965, **4**, 818.
46 L. Sacconi, P. Nannelli, N. Nardi and U. Campigli, *Inorg. Chem.* 1965, **4**, 943.
47 P. L. Orioli, M. Di Vaira and L. Sacconi, *J. Am. chem. Soc.* 1966, **88**, 4383.
48 L. Sacconi, M. Ciampolini and G. P. Speroni, *Inorg. Chem.* 1965, **4**, 1116.
49 D. Hall and T. N. Waters, *J. chem. Soc.* 1960, p. 2644.
50 J. Lewis and R. A. Walton, *J. chem. Soc.* 1966, p. 1559.
51 R. H. Holm, *J. Am. chem. Soc.* 1960, **82**, 5632.
52 M. Calvin and C. H. Barkelew, *J. Am. chem. Soc.* 1946, **68**, 2267.
53 P. Pfeiffer and H. Pfitzner, *J. prakt. Chem.* 1936, **145**, 243.
54 T. P. Cheeseman, D. Hall and T. N. Waters, *J. chem. Soc.* 1966, p. 1396.
55 F. Lions and K. V. Martin, *J. Am. chem. Soc.* 1957, **79**, 1273.
56 M. Gerloch, J. Lewis, F. E. Mabbs and A. Richards, *Nature, Lond.* 1966, **212**, 809.
57 J. Lewis, F. E. Mabbs and A. Richards, *Nature, Lond.* 1965, **207**, 855.
58 H. A. Goodwin, *Chelating Agents and Metal Chelates* (ed. F. P. Dwyer and D. P. Mellor), p. 143. New York: Academic Press, 1964.
59 L. Sacconi and I. Bertini, *J. Am. chem. Soc.* 1966, **88**, 5180.
60 T. J. Lane and A. J. Kandathil, *J. Am. chem. Soc.* 1961, **83**, 3782.
61 D. F. Martin, G. A. Janusonis and B. B. Martin, *J. Am. chem. Soc.* 1961, **83**, 73.
62 R. W. Green and P. W. Alexander, *Aust. J. Chem.* 1965, **18**, 1297.
63 R. W. Green and R. J. Sleet, *Aust. J. Chem.* 1966, **19**, 2101.
64 J. Beretka and B. O. West, unpublished observations.
65 H. Glaser, P. Pfeiffer and H. Thielert, *J. prakt. Chem.* 1939, **152**, 145.
66 F. Basolo and R. G. Pearson, *Mechanisms of Inorganic Reactions*, p. 200. New York: Wiley, 1958.
67 W. W. Fee and B. O. West, *Aust. J. Chem.* 1963, **16**, 788.
68 J. Beretka, B. O. West and M. J. O'Connor, *Aust. J. Chem.* 1964, **17**, 192.
69 M. Kato, H. B. Jonassen and J. C. Fanning, *Chem. Rev.* 1964, **64**, 99.
70 F. Bonati and R. Ugo, *Chem. Abstr.* 1965, **62**, 2488g; *Chimica Ind., Milano*, 1964, **46**, 1339.
71 S. Murahashi and S. Horrie, *J. Am. chem. Soc.* 1955, **77**, 6403; *Bull. chem. Soc. Japan*, 1960, **33**, 247.
72 P. L. Pauson, M. M. Bagga, F. J. Preston and R. I. Reed, *Chem. Commun.* 1965, p. 543.
73 P. E. Baikie and O. S. Mills, *Chem. Commun.* 1966, p. 707.
74 G. Costa, G. Maestroni, G. Taugher and L. Stefani, *J. organomet. Chem.* 1966, **6**, 181.
75 F. Calderazzo and C. Floriani, *Chem. Commun.* 1967, p. 139.

14

FLUOROSULPHATES

A. A. WOOLF

The binary systems of sulphur trioxide with oxides have been investigated far longer than the corresponding systems with fluorides. However, in recent years an upsurge of interest in the latter has demonstrated the wide range of systems available. Sulphur trioxide combines with fluorides ranging from the purely ionic alkali and alkaline earth fluorides, through the barely ionic interhalogen fluorides, to the covalent organic fluorides. The insertion of sulphur trioxide into fluorides markedly alters their chemical and physical behaviour. Thus hydrogen fluoride is converted from a liquid with a low electrical conductivity and boiling point, which rapidly dissolves oxide glasses, to a highly conducting and high boiling liquid in which the same glasses are stable. The reactivity of bromine trifluoride towards organic solvents is similarly moderated when combined with sulphur trioxide. Although a systematic study of this moderating influence has not yet been undertaken, enough data have accumulated for some generalizations to emerge. These are dealt with, together with a selective account of the different types of system, in this review.

1. Comparison of fluorosulphates with parent fluorides

The only properties which are known over the broad spectrum of fluorosulphates are transition points and vapour pressures. Even these are restricted because many fluorosulphates tend to solidify as glasses in which the liquid–solid transition cannot be observed. The fluorosulphates can be surveyed in broad outline by comparing their liquid ranges, or one end of the range, with those of the corresponding fluorides.[1] In Fig. 1 the elevation of boiling and freezing points of mono-fluorosulphates, XSO_3, over the parent fluoride X, is plotted against the mass ratio XSO_3/X. A smooth curve can be drawn through the boiling-point elevations, which can

A. A. WOOLF

be converted to an approximately linear relation by plotting against the inverse ratio X/XSO_3 (Fig. 2). The significance of the exceptions is discussed later. The freezing-point curve is less

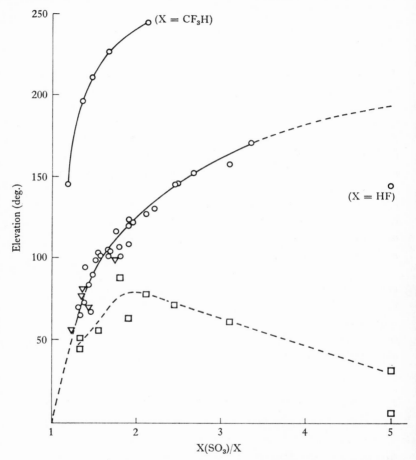

Fig. 1. Elevation of melting and boiling points on insertion of SO_3 into fluorides. ○, Elevation of b.p. (Table 1); ▽, elevation of b.p. (Table 2); □, elevation of m.p. (Table 1).

regular because of the sparsity of data, as well as the unreliability of some of them. (The melting points of some of the fluorosulphates which solidify as glasses have been identified with the movement of cracks on warming these glasses: a rate phenomenon rather than a true equilibrium between solid and liquid states.) In addition the solid–liquid transition is inherently less regular than the liquid–gas transition as shown by the less regular nature of fusion entropies compared with vaporization entropies. Nevertheless, the liquid range widens fairly regularly with the increasing proportion of sulphur trioxide in the fluorosulphate.

The exceptions to this regular behaviour can be correlated with changes in association and ionization when sulphur trioxide is inserted into fluorides. The most regular behaviour would be

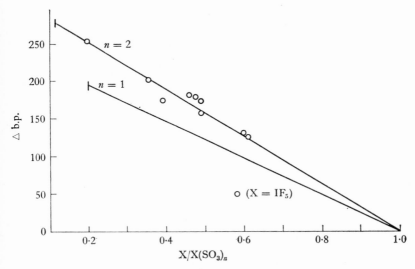

Fig. 2. Elevation of boiling point by insertion of two SO_3 into a fluoride compared with a single insertion.

expected when comparing fluoride and fluorosulphate with the same degree of ordering in a given state of matter. The only criterion of ordering available for these compounds is the molar entropy of vaporization (Trouton constant) which does not exceed about 22 cal./°C for non-associated liquids. Values above this indicate a greater ordering in the liquid, taking the gas phase as standard. The data (Table 1) show that the fluorosulphates are at most only slightly more associated than their parent fluorides taking into account uncertainties in the Trouton constants. (These arise because heats of evaporation are assumed temperature-

TABLE 1. *Mono-fluorosulphates compared with fluorides*

Reference to X(SO₃)	Fluoride X	X(SO₃)			X			1–3	2–4	X(SO₃)/X
		M.p. (1)	B.p. (2)	S_v	M.p. (3)	B.p. (4)	S_v			
2	$C_8F_{17}H$	.	259	.	.	114*	.	.	145	1·19
92	TeF_6	−89	56	.	−37·8	−38·9*	.	51	.	1·33
3	C_4F_{10}	−130	69·3	22·3	−84·5	4	.	45	65	1·34
2	C_4F_9H	.	211	.	.	15	.	.	196	1·36
4	$(CCl_2F)_2$	.	161·3	23·2	25	93	.	.	68	1·39
4	C_2F_5Br	.	73·4	23·2	.	−21	.	.	9·4	1·40
54	C_3F_8	.	46	.	.	−38	.	.	84	1·42
5	$(CF_3O)_2$	−99	31	.	.	−36	.	.	67	1·47
2	C_3F_7H	.	196	.	.	−14	.	.	210	1·47
6, 7	SF_5OF	.	54·1	21·9	−86	−35·1	.	.	89	1·49
6	C_2F_5Cl	.	55·4	23·8	.	−43	.	.	98	1·52
3	SF_6	−107	39·6	23·8	−51	−63·7*	.	56	103	1·55
8	C_2F_6	.	21·1	22·7	.	.	.	.	100	1·58
9	C_2F_6	.	22·1	23·2	−101	−79	.	.	101	1·58
10	C_2F_5H	.	56–57	.	.	.	.	.	105	1·67
2	C_2F_5H	.	178	.	.	−48·5	.	.	226	1·67

11	CF_3CH_2Cl		108		−105·5	6·9			101	1·67
12	CF_3COF	−134	46·3	25·1		−58			104	1·69
13	CF_3Cl		36·6			−81·2	20·5	88	118	1·75
14	SO_2F_2	−48	51	23·4	−136	−55	21·1		106	1·80
15	BrF		120·5	22·0	−33	20			101	1·81
16	C_6H_5F	−21	−180		−39	85		18	95	1·83
5	CF_3OF	−117	12·9	<	−215	−95	20·8		108	1·91
5	CF_4	−121	−4·2		−184	−128	21·5	63	124	1·91
11	CH_3CF_3		74·5	20·5	−11·3	−47·3	20·5		122	1·95
17	NF_3	−123·8	−2·5	21·8	−206·8	−129	19·2	78	127	2·13
18	CF_3H		162		−155	−82	20·9		244	2·14
19	CF_3H		34						116	2·14
20	COF_2		46·4	20·3†	−114	83·1	20·2		130	2·21
4	ClF	−84·3	45·1	24·0	−156	−100·1	30·4	72	145	2·46
21	F_2O		0		−223	−145	20·7		145	2·48
22	C_2H_5F		114		−143	−38			152	2·67
21	F_2	−158·5	−31·3	22·1	−219·6	−188	18·6	61	157	3·10
22	CH_3F		92		−142	−78	21·7		170	3·35
23	HF	−89	163·7	19·8	−83·4	19·5	6·1	6	144	5·00

* Sublimation point. S_v, Trouton's constant.
† Estimated from data.

independent, and from the very use of boiling points as corresponding temperatures.) It is therefore assumed that compounds the properties of which are closest to the curve (Fig. 1) are the nearest to unassociated fluorosulphates.

The exceptional behaviour of fluorosulphuric acid, which boils and freezes lower than predicted from the extrapolation of the curve, can be attributed to a significant decrease in ordering on converting hydrogen fluoride to the acid. Hydrogen fluoride is an exceptional liquid. It boils 109 °C and melts 29 °C higher than values obtained by extrapolating the other hydrogen halide values on a molecular weight basis. (The Trouton constant is exceptionally low because the liquid is being compared with an extensively associated gas phase.) The extent of association in fluorosulphuric acid is less certain. The presence of cyclic dimers is indicated by bands in the infrared spectra of liquid and gas [24] but an almost normal Trouton constant [22, 25] would also indicate little change in association between the two phases. The higher boiling point of fluorosulphuric compared with chlorosulphuric acid, in spite of the higher molecular weight of the latter, shows some association for the former. However, the smaller increment in boiling point between them (13 °C), compared with that between the hydrogen halides, is strongly indicative of the lower association in fluorosulphuric acid compared with hydrogen fluoride.

Deviations in the opposite sense are found with the sulphonic acids obtained when sulphur trioxide is inserted in the C—H link of perfluoroalkyl hydrides. Normal values are found for the isomers in which the trioxide is inserted in the C—F link. Again the exceptional behaviour, which is greatest with the lightest hydride, can be attributed to associated sulphates being compared with essentially unassociated fluorides. There is no direct evidence for association in trifluoromethyl sulphuric acid but this acid is closely related to fluorosulphuric acid (F replaced by CF_3). They boil within a degree of one another, are good electrical conductors, form stable hydrates and analogous products on electrolysis.[8] The positive deviation persists to the eighth member of the series. Not all isomers differ in boiling point. Thus insertion of sulphur trioxide in the C—C or the C—F link in hexafluoroethane produces liquids boiling at 21·1° and 22·1 °C respectively, because the association is little changed. (Trouton constant for $F_3CSO_3CF_3$ 22·7 cal./°C and $F_3CCF_2SO_3F$ 23·2 cal/°C.)

The difluorosulphates may be compared either with the mono-fluorosulphates on the same curve (Fig. 1) because there is little change in association on adding the second sulphur trioxide (Table 2), or with the parent fluoride (Fig. 2 and Table 3). The addition of the second trioxide molecule has the smaller effect. Again the outstanding exception, the $IF_3(SO_3F)_2$–IF_5 pair, can be explained by the decreased ordering when iodine pentafluoride is substituted. The latter is a highly associated liquid because the Trouton constant is high, the dipole moment decreases on vaporization, the liquid is appreciably conducting and the transition point to the gas phase is above that for the heavier heptafluoride. There is no information about ordering in the difluorosulphate. However, the negative deviation of 81 °C from the curve in Fig. 2 lies within the expected limits of boiling-point elevation caused by association in iodine pentafluoride (i.e. comparison of IF_5 with BrF_5 and IF_7 as part of the series CIF_5, BrF_5, IF_5; IF_3, IF_5, IF_7 suggests the limits 60–96 °C).

The elevation of melting points of fluorosulphates above the normal line (Fig. 1) enables the ionicity of fluorosulphates to be roughly assessed. The melting points of these fluorosulphates are compared in Table 4 with the values derived from Fig. 1 of essentially covalent fluorosulphates. Differences of less than about 25 °C are not significant in view of the uncertainty in the melting-point curve. Two fluorosulphates, $SbCl_4SO_3F$ and ClO_2SO_3F, have melting elevations higher than expected for covalent molecules although nowhere near as great as for nitronium and nitrosium fluorosulphates. There is abundant evidence for almost completely ionic structures in the latter, chloryl fluorosulphate has been formu-lated as an ionic compound expected to be much less stable than the nitronium salt.[33] The comparatively high melting point of adducts of chloryl fluoride with strong Lewis acids (SbF_5, AsF_5) also points to the existence of ClO_2^+ ions. The melting elevation of tetrachloro-antimony fluorosulphate also suggests it is a stibonium salt especially since it is being compared with $SbCl_4F$ (m.p. 82 °C) which is also presumably ionic. The corresponding fluorinated derivative SbF_4SO_3F is not significantly different from normal in

TABLE 2. *Difluorosulphates compared with monofluorosulphates*

Constants of compounds

| Reference | Fluoride | $X(SO_3)_2$ | | | $X(SO_3)$ | | | $1-3$ | $2-4$ | $\dfrac{X(SO_3)_2}{X(SO_3)}$ |
		M.p. (1)	B.p. (2)	S_v	M.p. (3)	B.p. (4)	S_v			
3	C_5F_{10}	·	148	24·3	·	92	23·8	·	56	1·24
3	C_4F_{10}	·	136	21·3	·	69	22·3	·	67	1·25
9	SF_6	−62·9	116·6	·	−107	39·6	23·8	44	77	1·35
3	C_2F_6	−28	102·7	24·1	·	22·1	23·2	·	69	1·44
26	SO_2F_2	·	120	·	−48	51	23·4	·	69	1·44
27	F_2	−158·5	−31·3	22·1	−55·4	67·1	22·4	103	98	1·74

TABLE 3. *Difluorosulphates compared with fluorides*

		Constants of compound								
		$X(SO_3)_2$			X					$\dfrac{X}{X(SO_3)_2}$
Reference	Fluoride	M.p. (1)	B.p. (2)	S_v	M.p. (3)	B.p. (4)	S_v	1–3	2–4	
3	C_5F_{10}	·	148	24·3	·	22·5	·	·	126	0·61
3	C_4F_{10}	·	136	21·3	−84·5	−4	·	·	132	0·60
15	IF_5	·	150	·	9·4	100·5	26·5	·	50	0·58
3	C_2F_5Cl	−66	132	22·0	·	−43	·	·	175	0·49
5	CF_3OCF_3	·	98	24·6	·	−59	·	·	157	0·49
9	SF_6	−62·9	116·6	·	−50·8	−63·7*	·	12	180	0·48
3	C_2F_6	−27·8	102·7	24·1	−101	−79	·	73	182	0·46
26	SO_2F_2	·	120	·	−136	−55·4	21·1	·	175	0·39
5	CF_4	·	76	·	−184	−128	21·5	·	204	0·35 (5)
27	F_2	−55·4	67·1	22·4	−219·6	−188	18·6	164	255	0·19

* B.p. figure for SbF_5.

TABLE 4. *Ionic fluorosulphates compared with fluorides*

| | | | M.p. of X(SO₃) | | | |
| | | | Actual* | Derived* | | |
Reference	Fluoride	M.p. of X	(1)	(2)	1–2	X(SO₃)/X
28, 29	SbCl₄F	83	210	123	87	1·28
30	SbF₅	141†	227†	206†	21	1·37
31	SeF₄	−9·5	70	45	25	1·51
32, 33	ClO₂F	−115	27	−43	70	1·92
34	NO₂F	−166	200	−91	291	2·25
15	NOF	−133	156	−66	222	2·63

* Figure 1.
† B.p. figures for SbF₅.

conformity with F^{19} n.m.r. evidence for a non-ionic structure with
fluorosulphate bridging.[30] (Behaviour is normal in spite of the
highly associated structure of SbF_4SO_3F because it is being compared with another highly associated liquid, SbF_5.) The reluctance
to form tetrafluoro-substituted cations compared with tetrachloro
ones seems general in Group V (e.g. $PCl_4^+PF_6^-$, $AsCl_4^+PF_6^-$ but
no $PF_4^+PF_6^-$ or $AsF_4^+AsF_6^-$). The fluoroselenium fluorosulphate

TABLE 5. *Estimated boiling points of unknown fluorosulphates*

Fluoride	B.p.	S_v	XSO₃/X	B.p. of X.SO₃*
PF₅	−84·6	21·8	1·64	10
SOF₄	−49	22·7	1·64	46
SF₄	−40	27·1	1·74	67
S₂F₂	−10·6	20·7	1·78	98
PF₃	−101	20·8	1·90	18
SOF₂	−43·8	22·7	1·93	75
N₂F₂(cis)	−106	22·0	2·21	29
NHF₂	−26·6	24·1	2·51	119
SNF	0·4	22·1	2·73	156
NH₂F	−77	†	3·00	84
			X(SO₃)₂/X	
WF₆	17	.	1·54	157

* Estimated in Fig. 1.
† Value expected between values of NHF₂ and NH₃: 24·1–28·1.

also appears non-ionic in conformity with the selenium n.m.r. chemical shift which is quite different from that of the ionic adduct $SeF_3^+BF_4^-$.[35]

Thus a rather naïve consideration of melting and boiling points provides a basis for differentiating ionic from covalent fluorosulphates, and for assessing change in ordering between fluoride and fluorosulphate. Conversely the boiling range, or at least the boiling point, of the as yet unprepared fluorosulphates can be predicted.

A few of the boiling points expected for covalent fluorosulphates are collected in Table 5. The direction of any deviation follows from the above discussion (e.g. SF_4SO_3F less associated than SF_4 and boiling lower than indicated unless ionization and salt formation occur: N_2F^+, NH_2^+, SF_3^+, SN^+ are all possibilities).

2. Preparation of fluorosulphates

2.1. Direct insertion of sulphur trioxide

Combination of a fluoride with sulphur trioxide or compounds of sulphur trioxide ($S_2O_7^{2-}$, HSO_3Cl, $C_5H_5N.SO_3$) was the basis of the original methods. The older work has been extended to some extent but the method is limited with ionic fluorides mainly to alkali and alkaline earth fluorides even under forcing conditions. Transition metal fluorides react incompletely. The thermal instability of polyvalent fluorosulphates (see p. 355) restricts the temperature range and it is not surprising that high melting fluorides of compact structures exposed to a non-polar liquid or gas in which the products are insoluble fail to react completely. Many covalent fluorides react completely at room temperatures. The reactions are summarized in Table 6. Alkali metal fluorides can extract sulphur trioxide from organic fluorosulphates.[36, 37]

2.2. Reactions in fluorosulphuric acid

Sulphur trioxide can be employed more effectively in combination with polar liquids. Fluorosulphuric acid is used in neutralization (see p. 344) or displacement reactions. The former are not of much preparative value because of the difficulty of separating products in this medium, but the latter have been used extensively with chlorides. Chlorides are difficult to displace and it is possible that

TABLE 6. *Reaction of fluorides with sulphur trioxide*

Fluoride	SO$_3$ added (moles)	Reference	Fluoride	SO$_3$ added (moles)	Reference
NH$_4$F, NaF	1·0	49			
AuF$_3$	2·7	47	AsF$_3$	1·5	41
AgF	0·9	40	SbF$_3$	3·0	38
CuF$_2$	0·4	39	SbF$_5$	1·0	30
BeF$_2$, ZnF$_2$, HgF$_2$	< 0·4	38, 40	BiF$_3$	0	38
			NOF	1·0	80
			NO$_2$F	1·0	
CaF$_2$	0·5, 2·0†		VF$_5$	(VOF$_3$)	42
BaF$_2$	2·0†	38, 40	SeF$_4$	1·0	31
SrF$_2$	0·7,* 2·0†		WF$_6$	4·5	43
TlF	1·0	40	ClO$_2$F	1·0	32, 33
PbF$_2$	1·5	40	BrF$_3$	1·5	4
(CH$_3$)$_3$SiF	1·0	46	BrF$_5$	1·0	44
NF$_3$	0	38	IF$_5$	1·2	45
PF$_3$	0	38	NiF$_2$	0·4	39

* At 50 °C.   † At 200 °C.

some of the mixed chloro-fluorosulphates claimed may be mixed products of partial displacement.[53] The displacement reaction has been extended to salts of the first transition metals.[48] Here separations are simplest because reactants and products are insoluble, or at most sparingly soluble in fluorosulphuric acid. The ease of displacement is in the order $CH_3COO^- > SO_4^{2-} > Cl^- > F^-$ and not in the order expected for a homogeneous solution reaction; that is, order of increasing strength of displaced acid

$$F^- \simeq CH_3COO^- > Cl^- > SO_4^{2-}$$

It has been suggested that displacement is easiest when the change in molecular volume on displacement is least. Acetates are very close to fluorosulphates in molecular volume, whereas fluorides are the farthest apart. Whatever the exact mechanism—a rate determining dissolution, a solid state diffusion—the choice of chlorides or fluorides for displacement reactants seems unpropitious. The reactions are summarized in Table 7.

Fluorosulphuric acid has also been used to prepare organic fluorosulphates by addition to fluorinated olefins[18] or by displacement of iodine from perfluoro-iodides.[54]

TABLE 7. *Displacement reactions in fluorosulphuric acid*

Reactant	Product	Reference
NaCl	NaSO$_3$F	50
NaF	NaSO$_3$F	51
BaCl$_2$	Ba(SO$_3$F)$_2$	52
MgCl$_2$	Mg(SO$_3$F)$_2$	53
Mg, Cd, Mn, Fe, Co, Ni, Zn, Cu(CH$_3$COO)$_2$	M(SO$_3$F)$_2$	48
FeSO$_4$	Fe(SO$_3$F)$_2$	48
TiCl$_4$	TiCl$_2$(SO$_3$F)$_2$	
ZrCl$_4$	ZrF$_3$(SO$_3$F)$_2$	
SnCl$_4$	SnCl$_3$(SO$_3$F)	53
	SnCl$_2$(SO$_3$F)$_2$	
SbCl$_5$	SbCl$_4$SO$_3$F	

2.3. Reactions in other solvents

Fluorosulphates are prepared in other solvents by reactions formulated as neutralizations. Thus, potassium fluorosulphate has been obtained by

$$K^+HF_2^- + H_2F^+SO_3F^- = KSO_3F + 2HF$$
$$K^+BrF_4^- + BrF_2^+SO_3F^- = KSO_3F + 2BrF_3$$
$$K^+CH_3COO^- + HSO_3F = KSO_3F + CH_3COOH$$

in hydrogen fluoride,[45] bromine trifluoride[55] and acetic acid[48] respectively. The formulation in hydrogen fluoride is probably an adequate representation, but that in bromine trifluoride is a gross oversimplification (see p. 353). Nevertheless, the 'neutralization' in this solvent is an efficient method for making fluorosulphates from compounds which dissolve as bases. The sulphur trioxide may also be added in the form of a salt provided the elements are present in appropriate amounts.

Neutralizations in acetic acid are more recent. Fluorosulphuric acid dissolves in acetic acid to form strongly acidic solutions which can be conductometrically or potentiometrically titrated against organic bases to give sharp inflections at equivalent amounts.[56] Some fluorosulphates can be precipitated from strong acetic acid solutions, a process which is not possible from fluorosulphuric acid itself in which similar neutralizations have been followed conductometrically.[47] Potassium fluorosulphate is made in this way although the salt is not thermally stable because of slight solvolysis. However, no precipitates are obtained with other soluble metal acetates; mixed acetate-fluorosulphate anions are postulated as stable species in solution.

Other solvents have been employed

$$N_2O_5 + HSO_3F \xrightarrow{CH_3NO_2} NO_2SO_3F + HNO_3 \text{ }[34]$$

$$NaF + SO_3 \xrightarrow{SO_2} NaSO_3F \text{ }[57]$$

Displacement reactions in bromine trifluoride and in acetonitrile have also proved feasible. In the former nitrosyl pyrosulphate can be displaced by metal ions when mixed in equivalent amounts:[55]

$$(NO)_2S_2O_7 \text{ Ag} \qquad\qquad\qquad NOF + BrF_3$$

$$\downarrow {\scriptstyle BrF_3} \quad \downarrow {\scriptstyle BrF_3} \qquad\qquad\qquad \uparrow$$

$$NOSO_3F + AgBrF_4 = AgSO_3F + NOBrF_4$$

Displacements in acetonitrile yield solvated fluorosulphates; for example,[40]

$$4AgSO_3F + SiCl_4 \xrightarrow{CH_3CN} Si(SO_3F)_4 . 2CH_3CN + 4AgCl$$

Similarly in toluene:[58]

$$AgSO_3F + Cu \xrightarrow{C_6H_5CH_3} Cu(SO_3F)\text{solvent} + Ag$$

2.4. Radical reactions

The reactions in solvents proceed by ionic mechanisms. It is also possible to prepare fluorosulphates by reactions with radicals thermally generated from peroxy-disulphuryl difluoride. This compound is formed by the action of fluorine with sulphur trioxide on a catalytic surface of silver difluoride.[27] The metal fluorosulphate is considered to be a reaction intermediate which is then oxidized by fluorine:

$$SO_3 + F^- \rightarrow SO_3F^-; \quad SO_3F^- + \tfrac{1}{2}F_2 \rightarrow SO_3F\cdot + F^-;$$
$$2SO_3F\cdot \rightarrow S_2O_6F_2$$

If an excess of fluorine is used the radicals combine to form FSO_3F.

Dissociation of the peroxy fluoride is already detectable at 100 °C by the yellow colour of the radical. The absorption is linearly related to the square root of the peroxy disulphuryl difluoride concentration at any temperature, in accordance with a process $R_2 \rightleftharpoons 2R\cdot$. The heat of dissociation 22–23 kcal/mole can be calculated from the temperature dependence of the equilibrium measured absorptiometrically or manometrically.[59] Chemical evidence can only be reconciled with a symmetrical dissociation. The heat of dissociation is close to that for N_2F_4, which also undergoes symmetrical dissociation, but is much less than that of the most easily dissociated halogen. (Heat of dissociation of F_2 is 38 kcal/mole.) Alternatively the compound can be made from other metal fluorosulphates and fluorine, or by anodic oxidation of fluorosulphuric acid.[39] This is a convenient procedure when elemental fluorine is not available:

at cathode $\qquad$ $H^+ + e \rightarrow H\cdot \rightarrow \frac{1}{2}H_2$

at anode $\qquad$ $SO_3F^- - e \rightarrow SO_3F\cdot \rightarrow \frac{1}{2}S_2O_6F_2$

The oxidizing ability of the radical can be used *in situ* by anodically oxidizing substances in solution, a process akin to Simon's electrochemical fluorination in anhydrous hydrogen fluoride. A list of preparations is tabulated in Table 8.

2.5. Other methods

Aromatic fluorosulphates can be prepared by the 'Balz–Schiemann' decomposition of diazonium fluorosulphates. The mechanism of the decomposition is unknown because it is not clear whether the original workers had orientated their substituted derivatives.[60]

Ethyl fluorosulphate has been used as a mild fluorosulphating agent:[53]

$$TaCl_5 + 2C_2H_5SO_3F = TaCl_3(SO_3F)_2 + 2C_2H_5Cl$$

The anodic oxidation of metals in a methyl cyanide solution of silver fluorosulphate has enabled the following solvates to be isolated:

$$CuSO_3F.CH_3CN, \quad Al(SO_3F)_{3}3CH_3.CN \quad \text{and}$$

$$Sb(SO_3F)_3.2CH_3CN$$

These cannot be desolvated without decomposition.[40]

TABLE 8. *Reactions of $S_2O_6F_2$*

Reactant	Product	Reference
Hg, HgO	$Hg(SO_3F)_2$	15
$BH_2[N(CH_3)_3]_2BH_4$	$B(SO_3F)_2[N(CH_3)_3]_2BH_4$	101
$SnCl_4$	$SnCl(SO_3F)_3$	79
N_2F_4	NF_2SO_3F	17
NO	$NOSO_3F$	15
N_2O_4	$NOSO_3F, NO_2SO_3F$	15
$POBr_3$	$POF_2(SO_3F)$ $POF(SO_3F)_2$ $PO(SO_3F)_3$	88
$VOCl_3$	$VO(SO_3F)_3$	102
$Nb(Ta)_2Cl_{10}$	$Nb(Ta)O(SO_3F)_3$	102
SF_4	$SF_4(SO_3F)_2$	9
SF_5OSF_5	SF_5OSO_3F	6
CrO_2Cl_2	$CrO_2(SO_3F)_2$	9
$Mo(CO)_6$	$MoO_2(SO_3F)_2$	9
Re	$ReO_3(SO_3F)$	102
	$ReO_2(SO_3F)_2$	102
CF_3X	CF_3SO_3F halogen SO_3F	17
$CF_2{=}CF_2$	$(CF_2SO_3F)_2$	17
$CF_2{=}CFCl$	$CF_2SO_3FCFClSO_3F$	17
$CH_2{=}CHF$	$CH_2SO_3FCHFSO_3F$	17
$CF_3CF{=}CFCF_3$	$(CF_3CFSO_3F)_2$	17
[F-substituted cyclopentene]	[cyclopentane with SO_3F and SO_3F and F]	17
CF_3COBr	CF_3COSO_3F, CF_3SO_3F	12
$CF_3CFBrCOCF_3$	$CF_3CFSO_3FCOCF_3$	103
$(R_fCO)_2O$	R_fSO_3F, R_fCOSO_3F	13

See Table 10 for halogen fluorosulphates.

Amine complexes of transition metal fluorosulphates have been isolated from the metal hydroxide in aqueous solution of amines to which soluble alkali fluorosulphates have been added.[61]

3. The $HF-SO_3$ systems

Although the 1:1 compound was discovered in 1892 by Thorpe and Kirman its investigation as a solvent is of comparatively recent origin. The initial approach was to study its applicability as a general solvent for preparing simple or complex fluorides,[47] the later approach was to examine quantitatively the analogies with the well-known H_2O–SO_3 system.[62–65] This dichotomy of approach emphasizes the uniqueness of HSO_3F *vis-à-vis* H_2SO_4 and HF (Table 9).

TABLE 9. *Comparison in hydrofluoric, fluorosulphuric and sulphuric acids*

Property	HF	HSO_3F	H_2SO_4
Melting point (°C)	−89·37	−88·98	10·37
Boiling point (°C)	19·51	163	290–317
Trouton constant	6·1	19·8	—
Cryoscopic constant	1·43	3·93	6·12
Heat of formation (kcal at 25 °C)	65·5	191	194
Viscosity (cP)	0·959 (19·5°)	1·56, 1·72 (25°)	25·54 (25°)
Density (g/ml at 25 °C)	0·9546	1·7264 1·7292	1·8269
Dielectric constant	84 (0°)	120–150 (25°)	100 (25°)
Electrical conductivity ($\Omega^{-1}.cm^{-1}$)	$\sim 10^{-5}$ (0 °C)	$1·09 \times 10^{-4}$ (25 °C)	$1·04 \times 10^{-2}$ (25 °C)
Log (auto-protolysis constant)	9·7	7	3·6
Hammett acidity function	10·2	12·8	11
Mobility of proton/anion of auto-protolysis	1·3	1·37	1·45

The 1:1 system is appreciably ionized as shown by the electrical conductivity of the most carefully purified specimen. The identification of the ions follows most directly from the electrolysis in which the solvated proton and fluorosulphate ions are discharged as hydrogen and $S_2O_6F_2$ (p. 341). Other evidence derives from acid-base behaviour and analogies with prototropic and fluorotropic systems. Thus alkali and alkaline earth fluorosulphates, which enhance the solvent conductivity by increasing the SO_3F^- concentration, are bases. Their conductivities vary approximately

linearly with concentration and decrease slightly in the order $NH_4^+ > Rb^+ \sim K^+ > Na^+ > Li^+$ the order expected for increased solvation. However, the similarity in conductivity of univalent fluorosulphates show that most of the current is carried by a mobile anion. Measurements of the small cation transfer numbers confirm this.

Another extensive class of bases consist of organic species which can be protonated in fluorosulphuric acid. The degree of protonation is assessed by a comparison of their conductivities in solution with potassium fluorosulphate solutions which are taken to be fully ionized. Nitrobenzene, m-nitrotoluene, acetic and benzoic acids are as strongly basic as the potassium salt.

The stronger acidity of HSO_3F compared with H_2SO_4 enhances the dissociation of any given base in the former; it also prevents proton donation from inorganic acids. Thus, HF, H_2SO_4 and $HClO_4$ are themselves weakly protonated in HSO_3F. The inability of even the most extreme proton acid, $HClO_4$, to manifest its acidity in HSO_3F led to the alternative approach of generating acidic behaviour by fluoride ion acceptance rather than proton donation. Antimony pentafluoride enhances the conductivity of many fluoride solvents in this way; its behaviour in HSO_3F is closely analogous. Solutions of the pentafluoride in HSO_3F can be titrated conductometrically with the above alkali or nitro bases. Other fluorides are classified as acids or bases by titrating them against KSO_3F or SbF_5 as the standard base and acid respectively. An initial decrease in conductivity indicates neutralization. Thus AsF_3, SbF_3, BrF_3, IF_5 are basic whereas AuF_3, TaF_5, PtF_4 are acidic. The acidity of antimony pentafluoride solutions is demonstrated by the dissolution of metals which are unattacked by the fluorosulphuric acid alone.[100]

Only antimony pentafluoride solutions have been examined in detail.[63] The pentafluoride is only a weak acid, the minimum conductivity occurs at $KSO_3F/SbF_5 \simeq 0.4$ although there is a sharp increase near equivalence.

$$SbF_5 + 2HSO_3F \rightleftharpoons H_2SO_3F^+ + SbF_5(SO_3F)^-$$

The acidity increases with addition of sulphur trioxide which slowly inserts in the Sb—F bonds. The fluorosulphates formed,

$SbF_{5-n}SO_3F_{1+n}$ where $n = 1, 2, 3$, ionize similarly to antimony pentafluoride, but with increasing displacement to the ionic side. This increased acidity is shown by the minimum conductivity at $1:1$ for the titration of $H[SbF_2(SO_3F)_4]$ with potassium fluorosulphate. The mobility of $H_2SO_3F^+$ in this acid solution, like SO_3F^- in base solutions, is abnormally high.

The solution of SbF_5 in HSO_3F is more complicated than indicated by the above equilibrium because the polymeric tendency of the pentafluoride persists in solution (see Fig. 3). This is

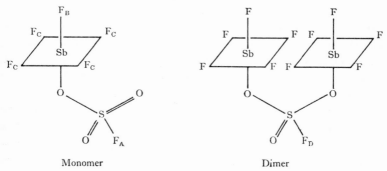

Monomer   Dimer

Fig. 3. Structures in SbF_5/HSO_3F solutions.

deduced from the ^{19}F n.m.r. spectra resolved at low temperatures when fluorine exchange is frozen. (Hydrogen exchange occurs at all temperatures.) In the simple acid three types of fluorine are present: $S-F_A$, $Sb-F_B$, $Sb-F_C$. The first produces a peak to the low field of the solvent peak, the latter constitute an AX_4 system. An additional satellite peak close to $S-F_A$, and an additional quintet, is ascribed to $S-F_D$ and the AX_4 units in a dimer structure present to about a fifth the extent of the monomer.

More complicated spectra are interpreted for the higher fluorosulphated acids and reasonable agreement with conductivity data can be obtained assuming equilibrium amounts of bridged structures.

Cryoscopic measurements, similar to those carried out in sulphuric acid, confirm the ionizations in fluorosulphuric acid. The cryoscopic constant is established with such non-electrolytes as SO_3, $S_2O_5F_2$ and $S_2O_6F_2$. The freezing point of the solvent, prepared by distillation in glass, is variable according to the excess of sulphur trioxide present. The extent of dissociation can be roughly estimated from available data[67, 24] to be of the order observed, although uncertainty arises because of the values to be given to polymerized species. The deficiency of hydrogen fluoride is presumably removed by attack on the borosilicate glass before recondensation. The presence of a strong absorption due to SiF_4 in the Raman spectra of HSO_3F confirms this attack. The absence of any other type of dissociation is shown by Ruff's experiment in which the acid was recovered unchanged after passage through a platinum tube at 800 °C. The 1 : 1 compound has the maximum freezing point at $-88.98(0)$ °C. The presence of excess of sulphur trioxide in distilled HSO_3F can also be estimated calorimetrically because of the large heat (13.6 kcal/mol.) for the combination of hydrogen fluoride and the trioxide. Solutions of metal fluorosulphates and benzoic acid behave cryoscopically as fully ionized binary electrolytes, whereas aromatic nitro compounds may be only partly ionized in agreement with the results of conductivity measurements. The ionization of hydrogen fluoride in fluorosulphuric acid is too small to affect cryoscopic measurements although it is detectable from conductivities.

Redox reactions can be carried out in fluorosulphuric acid. There is a similarity with ionic redox potentials in aqueous solutions.[100] Thus manganic and cobaltic salts are reduced to divalent fluorosulphates whereas chromous and cuprous salts are oxidized. Certain metals dissolve to form green solutions (Nb, Ta, U, Pb) or others give white precipitates and green supernatants (Na, K, Ca, In, Tl, Sn). The colour and e.s.r. spectra of these solutions are identical with solutions of sulphur in fluorosulphuric acid or oleum. Other metals dissolve (Ag, As, Sb) or produce insoluble fluorosulphates and colourless supernatants (Cu, Bi). The sulphur in HSO_3F is only reduced as far as sulphur dioxide in these instances. A similarity with metal potentials in aqueous solution seems likely because metals more noble than hydrogen on the aqueous scale produce colourless solutions in fluorosulphuric acid whereas the less noble metals are those which further reduce the acid.

Spectroscopic evidence exists for other distinctive compositions in the HF—SO_3 system although other evidence is not conclusive. Thus the conductivity of HSO_3F decreases with addition of the trioxide through the 1:2 composition without any inflection; cryoscopic measurements show no change in the number of particles in solution and the heat of solution is small. However, the isolation of a salt $KF,2SO_3$[68] from KSO_3F and the trioxide implies the existence of the corresponding 1:2 acid in solution. The viscosity-composition isotherms around this ratio indicate some discontinuity[69] but the new lines which appear in the Raman spectra when the trioxide is added to HSO_3F supply the only reliable evidence:[70]

New lines	Assignment	$S_2O_5F_2$
300		290
311		301
325	Disulphuryl structure	323
458		455
721		733
1074	SO_3	
1210	SO_2 symmetric stretch	
1250		
1394	SO_3	
1412	SO_2 asymmetric stretch	
1489		

There is also evidence for a 1:3 acid from the spectra at higher sulphur trioxide concentrations. The ^{19}F n.m.r. spectra consist of single resonances probably because of rapid exchange via ionic equilibria (compare the BrF_3–SO_3 system, p. 353). The chemical shift decreases linearly through the 1:2 and 1:3 compositions. Addition of $S_2O_5F_2$ produces a separate resonance which demonstrates that the new Raman lines of sulphur trioxide in HSO_3F are not those of $S_2O_5F_2$ formed *in situ*. However, some polysulphuryl fluorides can be produced from SO_3–HSO_3F mixtures if they are heated at 100 °C.

The weakening of the fluoro-substituted polysulphuric acids compared with the increasing strength of the unsubstituted acids is a distinctive difference between the HF–SO_3 and H_2O–SO_3 systems.

4. Carbonium fluorosulphates

The generation of carbonium ions has been mentioned in connection with conductivity and cryoscopic measurements on nitro aromatics in fluorosulphuric acid. Considerable interest has arisen in the stabilization of other carbonium ions in fluorosulphuric acid media, especially when they had been postulated as short-lived intermediates in reaction mechanisms. This development stems in direct line from studies in other non-aqueous solvents, especially those directed towards explaining 'Friedel–Crafts' type catalysis. The special advantages of fluorosulphuric acid solutions for stabilizing ions lie in their extreme acidity and in their wide liquid range which enables solutions to be examined at low temperatures. This is important both for preventing chemical rearrangement, and, even more, for resolving ^{1}H n.m.r. spectra which are liable to collapse by hydrogen exchange at higher temperatures.

A simple example is provided by amides in acid solutions, where it is not always easy to distinguish whether protonation occurs at oxygen or nitrogen sites. The ^{1}H n.m.r. spectrum of acetamide in HSO_3F at 25 °C consists of a sharp peak of methyl hydrogens and a peak broadened by quadrupole relaxation of the hydrogens on nitrogen. At −80 °C the latter splits into a doublet and a new peak appears on the high field side of the solvent peak which becomes fully resolved at −92 °C. The relative peak areas of 1, 2, 3 show that protonation occurs on oxygen; that is, C=O$^+$—H (Fig. 4). Similar sulphur protonation occurs in the corresponding thioamides.[71, 72]

Protonation of methyl benzenes has been well authenticated in acid media such as BF_3/HF or $BF_3.H_2O/CF_3COOH$. Their p.m.r. spectra in HSO_3F confirm the nature of the carbonium ions. The aromatic hydrocarbon is converted into a conjugate acid containing aliphatic protons. At a low enough temperature the spectra can be resolved. Thus with pentamethyl benzene four peaks of relative areas 2, 3, 6, 6 correspond to 'aliphatic' CH_2, p-CH_3, o-CH_3 and m-CH_3 respectively.[73]

The spectra of the more weakly basic hydrocarbons can be resolved by increasing the acidity of the solvent with antimony pentafluoride. Thus protonation of *m*-xylene occurs ortho to one methyl group and para to the other. The strong base hexamethyl benzene is so firmly protonated by SbF_5/HSO_3F that no appreciable exchange broadening occurs at 25 °C.

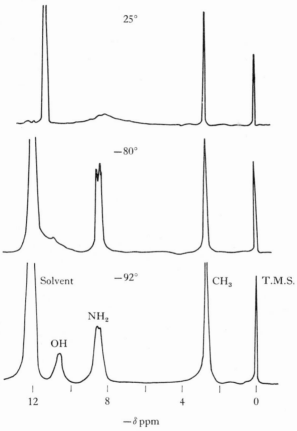

Fig. 4. ¹H n.m.r. spectra of acetamide in HSO_3F.

Stable carbonium ions can also be generated from alcohols in SbF_5/HSO_3F.[74] Sulphur dioxide is often used as a diluent. At −60 °C p.m.r. spectra served to identify the tertiary ions $(CH_3)_3C^+$, $C_6H_5CCH_3C_2H_5^+$, $(C_6H_5)_2CC_2H_5^+$, $(C_6H_5)_2C^+$ and

Cyclopropyl carbonium ions are similarly generated from alcohols and identified at $-60°$ to -78 °C.[75] They decompose at -25 °C. The structure of the dimethyl cyclopropylium ion shows the non-equivalence of the methyl groups. They are perpendicular to the plane of the ring but one methyl is *cis* to the ring. This arrangement allows maximum overlap of the vacant p-orbital on carbon with the ring orbitals. There is charge delocalization into α and β hydrogens. Reasonable spectra can be observed with some of the alcohols in HSO_3F alone at -50 °C.

Bridged phenonium ions have been proposed and opposed as intermediates in solvolysis of symmetrically substituted β-phenyl ethyl derivatives. Either of the alcohols I, III give the same p.m.r. spectra at -60 °C with four equivalent methyls when dissolved in SbF_5/HSO_3F.[76] This is not consistent with the position of methyl groups in the equilibrated ion (IV) which should appear at a position in the spectrum midway between that of methyls in t-butyl and t-pentyl cations. The existence of the bridged structure II in such an extremely acid medium does not necessarily show such a species under the more basic conditions employed in kinetic studies.

5. Halogen fluorosulphates

Much of the chemistry of the interhalogen compounds can be simulated with the fluorosulphate group behaving as a pseudo-halogen. Thus the halogens can be mixed with peroxydisulphuryl difluoride, the preferred source of fluorosulphate radicals, or with other fluorosulphates to produce inter-halogen fluorosulphates (Table 10).

TABLE 10. *Pseudo-interhalogens*

Reactants	Product	Reference
Excess $F_2 + SO_3$	FSO_3F	21
Excess $Cl_2 + S_2O_6F_2$	$ClSO_3F$	4
Excess $Br_2 + S_2O_6F_2$	$BrSO_3F$	15
$Br_2 +$ excess $S_2O_6F_2$	$Br(SO_3F)_3$	15
$I_2 + S_2O_6F_2$	$I(SO_3F)$	77
$I_2 +$ excess $S_2O_6F_2$	$I(SO_3F)_3$	15
$ICl + S_2O_6F_2$	$I(SO_3F)_3$	78
$I_2 + FSO_3F$	$IF_3(SO_3F)_2$	15
Excess $Cl_2 + ISO_3F$	$ICl_2(SO_3F)$	77
Excess $I_2 + ISO_3F$	I_3SO_3F	77

Whereas BrF_3 and IF_3 disproportionate to a halogen and a higher halogen fluoride, the corresponding fluorosulphates are quite stable at room temperature. The monohalogen fluorosulphates resemble halogens and halogen fluorides in adding across unsaturated bonds (Table 11). The analogy with halogen fluorides can be extended farther to the reactions of alkali halides with $S_2O_6F_2$ which produce tetrapseudohalogen halites $KBr(SO_3F)_4$ and $KI(SO_3F)_4$ analogous to $KBrF_4$ and KIF_4.[79]

Iodine trifluorosulphates have been examined in fluorosulphuric acid.[80] Mixtures of iodine and $S_2O_6F_2$ in this acid can be shown by cryoscopic and conductivity measurements to form $I(SO_3F)_3$ which behaves as a weakly dissociated base

$$\tfrac{1}{2}(I_2 + 3S_2O_6F_2) \rightleftharpoons I(SO_3F)_3 \overset{K_b}{\rightleftharpoons} I(SO_3F)_2{}^+ + SO_3F^-$$

Addition of the strong base, potassium fluorosulphate, suppresses this ionization until $I(SO_3F)_3$ begins to act as an acid:

$$I(SO_3F)_3 + SO_3F^- \overset{K_a}{\rightleftharpoons} I(SO_3F)_4{}^-$$

TABLE 11. *Addition of halogen fluorosulphates across double bonds*

Halogen fluorosulphate	Reactant	Product	Reference
FSO_3F	$CF_2{=}CF_2$	$CF_3CF_2SO_3F$	4
	$CF_3CF{=}CFCF_3$	$CF_3CF_2CF(SO_3F)CF_3$	12
			4
	$CCl_2{=}CCl_2$	$FCCl_2CCl_2SO_3F$	4
$ClSO_3F$	$CF_2{=}CF_2$	$CF_2Cl{-}CF_2SO_3F$	4
$BrSO_3F$	$CF_2{=}CF_2$	$CF_2BrCF_2SO_3F$	4
	$CF_3CF{=}CFCF_3$	$CF_3CFBrCF(SO_3F)CF_3$	12
	$CFCl{=}CF_2$	$CFClBrCF_2SO_3F$	103
		$CFCl(SO_3F)CF_2Br$	

that is, equivalent to a solution of $KI(SO_3F)_4$ in HSO_3F. Thus iodine tri-fluorosulphate is amphoteric $(K_b \sim 10^{-5}$ mole/kg; $K_a \sim 10$ moles/kg).

The other iodine fluorosulphates are sources of iodonium ions in solution. The mono-fluorosulphate forms a blue solution in fluorosulphuric acid. The absorption spectrum resembles those obtained with solutions of iodine in iodine monochloride, oleum or iodine pentafluoride, which have been attributed to the I^+ cation. Solid iodine fluorosulphate is diamagnetic and unlikely to have an ionic lattice. The formation of the I^+ cation in solution has been disputed, the evidence being more compatible with an I_2^+ species.[81] This is shown from measurements on I_2–$S_2O_6F_2$ mixtures in HSO_3F. It can be seen from Table 12 that there is satisfactory correlation with the proposed ionization for the I_5^+ and I_3^+ cations. The latter can also be produced directly from I_3SO_3F in HSO_3F or H_2SO_4.[77] However, the observed values for the 1:1 mix are widely divergent from the expected $\nu = 4$ and $\gamma = 2$ for an I^+ ion. The results are explained by the ionization

$$5I_2 + 5S_2O_6F_2 \rightarrow 4I_2^+ + 4SO_3F^- + 2I(SO_3F)_3$$

at low concentrations, together with disproportionation

$$8I_2^+ \rightarrow 5I_3^+ + I^{3+}$$

shown by the appearance of the I_3^+ spectrum.

TABLE 12. *Properties of I_2–$S_2O_6F_2$ mixtures in HSO_3F*

$I_2/S_2O_6F_2$	Ionization	Observed ν	Observed γ	Spectrum max. $(m\mu)$
5	$5I_2 + S_2O_6F_2 \rightarrow 2I_5^+ + 2SO_3F^-$	0·82	Variable	450, 345, 270, 245
3	$3I_2 + S_2O_6F_2 \rightarrow 2I_3^+ + 2SO_3F^-$	1·26	0·66	470, 305, 210
1	$I_2 + S_2O_6F_2 \rightarrow 2I^+ + 2SO_3F^-$	1·21	1	640, 490, 410

ν = number of particles formed/iodine mole from cryoscopy.
γ = number of SO_3F^- formed/iodine mole from conductometry.

An iodyl fluorosulphate has been prepared by the reaction
$$I_2O_5 + S_2O_6F_2 = 2IO_2SO_3F + \tfrac{1}{2}O_2.^{[77]}$$
The iodyl compound, insoluble in fluorosulphuric acid, is probably polymeric with bridging fluorosulphate groups.

The Br_2–$S_2O_6F_2$ system in HSO_3F may be a source of bromonium ions. Three other systems, which have been given a cursory examination, apparently give the same unique composition. Bromine and excess fluorine fluorosulphate form an adduct

$$Br_2 \cdot 3SO_3F_2$$

identical with a equimolecular mixture of BrF_3 and $Br(SO_3F)_3$.[4] Similarly the products isolated from mixtures of sulphur trioxide and bromine trifluoride approximate the composition

$$SO_3 \cdot o\cdot67BrF_3 \;^{[55]}$$

All these are equivalent to $2BrF_3 \cdot 3SO_3$, the analogue of the sulphur trioxide adduct with arsenic trifluoride $2AsF_3 \cdot 3SO_3$.[41] The ^{19}F n.m.r. spectrum shows this is a unique composition because the chemical shift of the fluorine attached to bromine is a maximum at this point and decreases linearly to either side. Only two resonances are observed in the bromine system, compared with the three in the arsenic one, presumably because of easier exchange via ionic intermediates (cf. the greater electrical conductivity of BrF_3 than AsF_3). A bridged structure, not necessarily the one illustrated, probably exists in equilibrium with fluorosulphate ions. Certainly, solutions of sulphur trioxide in bromine trifluoride are a source of these ions and the trioxide can be converted quantitatively into ionic fluorosulphate (see p. 339). It is unnecessary to regard mixtures of BrF_3 and $Br(SO_3F)_3$ as non-stoicheiometric on account of the linear variation in chemical shift (cf. SO_3–HSO_3F system), unless a non-equilibrium mixture varying with the mode of formation is produced. Further work is required to elucidate the equilibrium mixture of ions and polymers in these systems.

6. Hydrolytic stability of fluorosulphates

Fluorosulphuric acid was initially believed to hydrolyse completely in water, but it was found that with a limited amount of water an equilibrium was set up $HSO_3F + H_2O \rightleftharpoons H_2SO_4 + HF$. A small increase in water concentration removed more fluorosulphuric acid than expected by mass action.[82] However, this earlier work can be criticized because of the method of 'freezing' the equilibrium by addition of strong alkali and also the analytical methods. The equilibrium has now been followed physically by cryoscopic and conductivity techniques.[83] It is proposed that the water is protonated and that the hydroxonium ion exists in equilibrium

$$H_3O^+ + SO_3F^- \overset{K}{\rightleftharpoons} HF + H_2SO_4.$$

The two particles formed by each molecule of water cannot serve to distinguish ionic from non-ionized forms, but the conductivity measurements show that less than one fluorosulphate ion remains. Assuming that K^+ and H_3O^+ ions have about the same small mobility, the amount of fluorosulphate in equilibrium is found by comparing the conductivity of water solutions and equiconducting potassium fluorosulphate solutions ($K = 0·12$ at 25 °C; 0·042 at 0 °C).

A large excess of water does not ensure complete hydrolysis. The hydrolysis is kinetically controlled in that it depends on the rate of mixing. The hydrolysis subsequent to the initial rapid reaction is slow. A faster rate of fission of the S—F bond in the covalent structure of the anhydrous acid than in the fluorosulphate ion is postulated.[84] If the fluorosulphate ion is preformed as in an alkali salt, it can be combined with the hydroxonium ion by cation exchange without hydrolysis. The aqueous acid so formed is of the same order of stability as aqueous tetrafluoroboric acid and more stable than the Group V fluoro-complex acids (e.g. $H_3O^+PF_6^-$). The kinetics of the slow hydrolysis of the fluorosulphate ion are similar to that of the fluoroborate ion.[85] Bivalent metal fluorosulphates are more extensively hydrolysed but the measurements are not precise enough to correlate any covalency in, for example, the transition metal salts with amount hydrolysed.[48]

The covalent perfluoro fluorosulphates are reasonably stable to hydrolysis as are aryl fluorosulphates. Slow hydrolysis occurs with hot alkali solutions. The lower alkyl fluorosulphates are hydrolysed fairly rapidly by water. Pyrosulphuryl fluoride behaves as the anhydride of fluorosulphuric acid and the slow hydrolysis of this fluoride affords another preparation of the aqueous acid.[86]

7. Thermal stability of fluorosulphates

The mono and divalent metal fluorosulphates, although stable up to comparatively high temperatures, are less stable than the corresponding sulphates. More possibilities are available for decomposition:

$$M(SO_3F)_2 \begin{cases} MF_2 + 2SO_3 & (1) \\ \rightarrow MO + S_2O_5F_2 & (2) \\ MSO_4 + SO_2F_2 & (3) \end{cases}$$

(The fluorosulphates of K^+, Cu^+, Mg^{2+}, Ca^{2+}, Ni^{2+} decompose to products of route (1); Ag^+, Sr^{2+}, Ba^{2+} of (3); Na^+ of (1) and (3).)

The relationship between cation size and decomposition temperature with sulphates has encouraged speculation on structural correlations with decomposition products of fluorosulphates. Preliminary values for the free energies of fluorosulphates and sulphur oxyfluorides, obtained by measuring their heats of formation and estimating entropies, show that (3) is the equilibrium situation. Other products must arise from kinetic control. However, until the reaction mechanism is known and the rate determining processes have been elucidated it is unprofitable to attempt to correlate fluorosulphate structure and decomposition products.

The tri- and tetra-fluorosulphates are less stable. Thus in displacement reactions of tri- and tetra-acetates, or chlorides, in fluorosulphuric acid, the corresponding fluorosulphates are not obtained pure but can be admixed with oxy-sulphates, sulphates or oxy-fluorosulphates.[87] The composition of solids isolated from titanic chloride corresponds to mixtures of $TiO(SO_3F)_2$, $TiOSO_4$ and $Ti(SO_4)_2$ which could arise by the sequence

$$TiCl_4 \xrightarrow{HSO_3F} \text{complex acid} \to Ti(SO_3F)_4 \to TiO(SO_3F)_2 + S_2O_5F_2$$
$$\downarrow \qquad\qquad\qquad \downarrow$$
$$Ti(SO_4)_2 + 2SO_2F_2;\ TiOSO_4 + SO_2F_2$$

Titanic nitrate undergoes an analogous stepwise decomposition. Tin polyfluorosulphates are even less stable. Other examples of unstable polyfluorosulphates are the tetra fluorosulphato-antimony acid in fluorosulphuric acid:[63]

$$SbF_3(SO_3F)_4^- \to SbF_4(SO_3F)_3^- + SO_3$$

and the disproportionation of the tri-fluorosulphate:[77]

$$2I(SO_3F)_3 = I(SO_3F) + IF_3(SO_3F)_2 + 3SO_3$$

A pentafluorosulphate cannot be made from $IF_3(SO_3F)_2$ and excess sulphur trioxide.

It is reasonable to assume that substances which dissolve in fluorosulphuric acid are solvolysed to simple or complex fluorosulphates. If some of these decompose to volatile products then the thermal instability of these polyfluorosulphates affords a route to simple fluorides, often in good yield. Thus phosphoric oxide is converted to phosphorus oxy-trifluoride. The equation proposed was:[41]

$$P_2O_5 + 3HSO_3F = POF_3 + HPO_3 + 2SO_3 + H_2SO_4$$

Recently the corresponding fluorosulphate has been isolated from phosphorus oxybromide:[88]

$$POBr_3 + 6S_2O_6F_2 = PO(SO_3F)_3 + 3Br(SO_3F)_3$$

and its thermal instability demonstrated; that is,

$$PO(SO_3F)_3 \to POF_x(SO_3F)_{3-x} + xSO_3 \quad (x = 0,\ 1,\ 2)$$

The above reaction can be reformulated

$$P_2O_5 + 6HSO_3F = 2PO(SO_3F)_3 + 3H_2O$$

followed by thermal decomposition to lower fluorosulphates. The fate of the water in such a complex mixture is not evident. The reaction between sulphur trioxide and xenon difluoride at 120 °C[99]

$$XeF_2 + 2SO_3 = Xe + S_2O_5F_2 + \tfrac{1}{2}O_2$$

may proceed via an unstable fluorosulphate (route 2).

A few oxyfluorides can be conveniently prepared by warming solutions of oxy-acid salts in fluorosulphuric acid. These can be regarded as thermal decompositions of acidium fluorosulphates. The very stable perchloryl fluoride is prepared by heating potassium perchlorate solutions at 50 °C:[89]

$$KClO_4 + 2HSO_3F = H_2ClO_4^+ + 2SO_3F^- + K^+$$

Perchloric acid is a much weaker acid than fluorosulphuric and hence the perchlorate is solvolysed, on heating

$$H_2ClO_4^+SO_3F^- = ClO_3F + H_2SO_4$$

Similarly the production of selenium oxyfluoride from barium selenate[91] can be formulated:

$$BaSeO_4 + 3HSO_3F = Ba^{2+} + H_3SeO_4^+ + 3SO_3F^-$$

$$H_3SeO_4^+SO_3F^- = HSeO_3F + H_2SO_4$$

$$HSeO_3F + HSO_3F = H_2SeO_3F^+SO_3F^- = SeO_2F_2 + H_2SO_4$$

Fluoroselenic acid is known[91] and could be dissolved in fluorosulphuric to examine the latter stages of the proposed mechanism.

The reactions of barium tellurates with fluorosulphuric acid are more complicated.[92] The products are TeF_6, TeF_5OH, TeF_5SO_3F, $(TeF_5O)_2SO_2$, TeF_5OSO_3H and SO_3. A summary of reactions leading to volatile fluorides is given in Table 13.

8. Metal fluorosulphates

Much of the earlier work on fluorosulphates stressed their similarities with perchlorates, perrhenates and fluoroborates. (For summary see reference no. 97.) Thus solubilities of the less soluble salts are in the same order. Potassium salts are sparingly soluble in water, silver salts are very soluble in water as well as in organic solvents. Many salts are isomorphous and molecular volumes of

TABLE 13. *Preparation of volatile fluorides with HSO_3F*

Source material	Fluoride	Reference
H_3BO_3	BF_3 ⎫	
$SiO_2.xH_2O$	SiF_4 ⎬	95
As_2O_3	AsF_3 ⎭	
As_2O_5	AsF_3 and $S_2O_5F_2$	41
D_2O	DF	96
$BaSeO_4$	SeO_2F_2	90
$BaTeO_4$	TeF_5OH	
BaH_4TeO_6	TeF_5SO_3F	92
K_2CrO_4	CrO_2F_2	94
$KClO_4$	ClO_3F	89
$KMnO_4$	MnO_3F	93

ions in aqueous solutions,[98] and their mobilities[84] are also similar. However, these similarities should not be over-emphasized, especially with regard to chemical behaviour.

Analogies can also be drawn between solubilities of fluorosulphates in fluorosulphuric acid and of sulphates in sulphuric acid. Thus salts of the bivalent transition metals are almost insoluble, alkali and alkali earth salts most soluble, and silver and lead salts are also soluble.

The magnetic and spectral behaviour of anhydrous transition metal fluorosulphates is fairly normal.[87] These salts have normal moments tending towards the maximum values encountered with fluorides. The electronic spectra show that the fluorosulphate ion fits spectrochemically between fluoride and sulphate. However, the infrared spectra would indicate that the fluorosulphate group is

TABLE 14. *S—F stretching frequencies (cm^{-1}) in fluorosulphates*

Na^+	786, 740	Fe^{2+}	840	H	850, 837
K^+	732	Co^{2+}	865	CF_3	843
Rb^+	729	Ni^{2+}	865	C_2F_5	843
NH_4^+	737	Cu^{2+}	850	C_5F_9	849
Cs^+	728	Zn^{2+}	860	NF_2	840
Ag^+	767	Cd^{2+}	825	F	852
$(C_6H_5)_3C^+$	710	$Cu(NH_3)_6^{2+}$	735	SO_3F	848
Mn^{2+}	835	$Co(NH_3)_6^{2+}$	715	SF_5O	848

behaving as a bridging unit. There is considerable increase in the double bonding between S and F in these salts compared with the alkali fluorosulphates. Insertion of ligands reduces this frequency to normal ionic values (Table 14).

9. Conclusions

It can be seen that our knowledge of fluoride–sulphur trioxide systems is fragmentary and incomplete. While it is impossible to predict future advances, the following topics seem worthy of fuller investigation:

(i) The stabilization of fluorides by sulphur trioxide and property correlations between fluoride and fluorosulphate.

(ii) Transition metal polysulphates, especially in different valency states.

(iii) The behaviour of the fluorosulphate radical as a halogen, and substitution into organometallic and carbonyl compounds.

(iv) The use of miscible systems (e.g. AsF_3–SO_3 or BrF_3–SO_3) as controllable reaction media by varying the sulphur trioxide content.

(v) The further characterization of solutions and reactions in the HF–SO_3 system, especially of organic bases and inorganic fluoroacids.

(vi) Kinetic studies on the thermal and hydrolytic stability of fluorosulphates.

REFERENCES

1 A. A. Woolf, *J. chem. Soc. (A)*, 1967, p. 401.
2 T. Gramstad and R. N. Haszeldine, *J. chem. Soc.* 1957, p. 2640.
3 C. J. Ratcliffe and J. M. Shreeve, *Inorg. Chem.* 1963, **2**, 631.
4 W. P. Gilbreath and G. H. Cady, *Inorg. Chem.* 1963, **2**, 496.
5 W. van Meter and G. H. Cady, *J. Am. chem. Soc.* 1961, **82**, 8005.
6 C. J. Merrill and G. H. Cady, *J. Am. chem. Soc.* 1963, **85**, 909.
7 H. J. Eméleus and K. J. Packer, *J. chem. Soc.* 1962, p. 771.
8 R. E. Noftle and G. H. Cady, *Inorg. Chem.* 1965, **4**, 1010.
9 J. M. Shreeve and G. H. Cady, *J. Am. chem. Soc.* 1961, **83**, 4521.
10 H. H. Gibbs, W. L. Edens and R. N. Griffin, *J. org. Chem.* 1961, **26**, 4140.
11 J. D. Calfee and P. A. Florio, *Chem. Abstr.* 1954, **48**, 1413f.
12 J. J. Delfino and J. M. Shreeve, *Inorg. Chem.* 1966, **5**, 308.
13 D. D. DesMarteau and G. H. Cady, *Inorg. Chem.* 1966, **5**, 169.
14 E. Hayek and W. Koller, *Mh. Chem.* 1951, **82**, 942.

15 J. E. Roberts and G. H. Cady, *J. Am. chem. Soc.* 1960, **82**, 352.
16 R. Cramer and D. D. Coffmann, *J. org. Chem.* 1961, **26**, 4164.
17 M. Lustig and G. H. Cady, *Inorg. Chem.* 1963, **2**, 388.
18 R. N. Haszeldine and J. M. Kidd, *J. chem. Soc.* 1954, p. 4228.
19 G. A. Sokolskiï and M. A. Dimitriev, *Zhur. obsch. Khim.* 1961, **31**, 706.
20 W. B. Fox and G. Franz, *Inorg. Chem.* 1966, **5**, 946.
21 F. B. Dudley, G. H. Cady and D. J. Eggers, *J. Am. chem. Soc.* 1956, **78**, 290.
22 J. Meyer and G. Schramm, *Z. anorg. Chem.* 1932, **206**, 24.
23 J. Barr, R. J. Gillespie and R. C. Thompson, *Inorg. Chem.* 1964, **3**, 1149.
24 R. Savoie and P. A. Giguère, *Can. J. Chem.* 1964, **42**, 277.
25 A. S. Lenskii, A. D. Shaposhnikova and E. S. Sokolova, *Zhur. neorg. Khim.* 1964, **9**, 1147.
26 R. J. Gillespie, J. V. Oubridge and E. A. Robinson, *Proc. chem. Soc.* 1961, p. 428.
27 F. B. Dudley and G. H. Cady, *J. Am. chem. Soc.* 1957, **79**, 513.
28 E. Hayek, J. Puschmann and A. Czaloun, *Mh. Chem.* 1954, **85**, 359.
29 L. Kolditz, *Z. anorg. Chem.* 1957, **289**, 128.
30 R. J. Gillespie and R. A. Rothenbury, *Can. J. Chem.* 1964, **42**, 416.
31 R. D. Peacock, *J. chem. Soc.* 1953, p. 3617.
32 M. Schmeisser and W. Fink, *Angew. Chem.* 1957, **69**, 780.
33 A. A. Woolf, *J. chem. Soc.* 1954, p. 4113.
34 D. R. Goddard, E. D. Hughes and C. K. Ingold, *J. chem. Soc.* 1950, p. 2570.
35 T. Birchall, R. J. Gillespie and S. L. Vekris, *Can. J. Chem.* 1965, **43**, 1672.
36 A. Zappel and H. Jonas, *Chem. Abstr.* 1963, **59**, P 5021 c.
37 M. Lustig and J. K. Ruff, *Inorg. Chem.* 1964, **3**, 287.
38 E. L. Muetterties and D. D. Coffmann, *J. Am. chem. Soc.* 1958, **80**, 5914.
39 F. B. Dudley, *J. chem. Soc.* 1963, p. 3407.
40 E. Hayek, A. Czaloun and B. Krismer, *Mh. Chem.* 1956, **87**, 741.
41 A. Engelbrecht, A. Aignesberger and E. Hayek, *Mh. Chem.* 1955, **86**, 469.
42 H. C. Clark and H. J. Emeléus, *J. chem. Soc.* 1958, p. 190.
43 H. C. Clark and H. J. Emeléus, *J. chem. Soc.* 1957, p. 4778.
44 M. Schmeisser and E. Pammer, *Angew. Chem.* 1955, **6**, 7156; 1957, **69**, 781.
45 A. A. Woolf, *J. chem. Soc.* 1950, p. 3678.
46 K. Schmidbaur, *Ber. dtsch. chem. Ges.* 1965, **98**, 83.
47 A. A. Woolf, *J. chem. Soc.* 1954, p. 433.
48 A. A. Woolf, *J. chem. Soc.* (*A*), 1967, p. 355.
49 W. Traube, *Ber. dtsch. chem. Ges.* 1913, **46**, 2513.
50 O. Ruff, *Ber. dtsch. chem. Ges.* 1914, **47**, 656.
51 W. Traube, J. Hoerenz and F. Wunderlich. *Ber. dtsch. chem. Ges.* 1919, **12**, 1272.
52 M. Trautz and K. Ehrmann, *J. prakt. Chem.* 1935, **142**, 79.
53 E. Hayek, J. Puschmann and A. Czaloun, *Mh. Chem.* 1954, **85**, 360.

54 M. Hauptstein and M. Braid, *J. Am. chem. Soc.* 1961, **83**, 2500.
55 A. A. Woolf, *J. chem. Soc.* 1950, p. 1053.
56 R. C. Paul, S. K. Vasisht, K. C. Malhotra and S. S. Pahil, *Analyt. Chem.* 1962, **34**, 820.
57 H. Jonas, *Chem. Abstr.* 1960, **54**, P 3891f.
58 D. W. A. Sharp and A. G. Sharpe, *J. chem. Soc.* 1956, p. 1858.
59 F. B. Dudley and G. H. Cady, *J. Am. chem. Soc.* 1963, **85**, 3375.
60 W. Lange and E. Muller, *Ber. dtsch. chem. Ges.* 1930, **63**, 2653.
61 W. Lange, *Ber. dtsch. chem. Ges.* 1927, **60**, 962. E. Wilke-Doerfurt *et al. Z. anorg. Chem.* 1929, **185**, 417; **184**, 121, 145.
62 J. Barr, R. J. Gillespie and R. C. Thompson, *Inorg. Chem.* 1964, **3**, 1149.
63 R. C. Thompson, J. Barr, R. J. Gillespie, J. B. Milne and R. A. Rothenbury, *Inorg. Chem.* 1965, **4**, 1641.
64 R. J. Gillespie, J. B. Milne and R. C. Thompson, *Inorg. Chem.* 1966, **5**, 468.
65 R. J. Gillespie, J. B. Milne and J. B. Senior, *Inorg. Chem.* 1966, **5**, 1233.
66 R. J. Gillespie and J. B. Milne, *Inorg. Chem.* 1966, **5**, 1236.
67 A. A. Woolf, *J. inorg. nucl. Chem.* 1950, **14**, 21; A. A. Woolf and G. W. Richards, *J. chem. Soc. (A)*, 1967, p. 1118.
68 H. A. Lehmann and L. Kolditz, *Z. anorg. Chem.* 1953, **272**, 69.
69 A. S. Lenskii, A. D. Shaposhnikova and E. A. Sokolova, *Zhur. neorg. Khim.* 1963, **8**, 2716.
70 R. J. Gillespie and E. A. Robinson, *Can. J. Chem.* 1962, **40**, 675.
71 R. J. Gillespie and T. Birchall, *Can. J. Chem.* 1963, **41**, 148.
72 T. Birchall and R. J. Gillespie, *Can. J. Chem.* 1963, **41**, 2642.
73 T. Birchall and R. J. Gillespie, *Can. J. Chem.* 1964, **42**, 502.
74 G. A. Olah, M. B. Comisarow, C. A. Cupas and C. U. Pittman (Jr.), *J. Am. chem. Soc.* 1965, **87**, 2997.
75 N. C. Deno, J. S. Liu, J. A. Turner, D. N. Lincoln and R. E. Fruit (Jr.), *J. Am. chem. Soc.* 1965, **87**, 3001.
76 G. A. Olah and C. U. Pittman (Jr.), *J. Am. chem. Soc.* 1965, **87**, 3507.
77 F. Aubke and G. H. Cady, *Inorg. Chem.* 1965, **4**, 269.
78 J. M. Shreeve and G. H. Cady, *J. Am. chem. Soc.* 1961, **83**, 452.
79 M. Lustig and G. H. Cady, *Inorg. Chem.* 1962, **1**, 714.
80 E. E. Aynsley, G. Hetherington and P. L. Robinson, *J. chem. Soc.* 1954, p. 1119.
81 R. J. Gillespie and J. B. Milne, *Inorg. Chem.* 1966, **5**, 1577.
82 W. Traube and E. Reubke, *Ber. dtsch. chem. Ges.* 1921, **54**, 1618; 1923, **56**, 1656.
83 R. J. Gillespie, J. B. Milne and J. B. Senior, *Inorg. Chem.* 1966, **7**, 1233.
84 A. A. Woolf, *J. chem. Soc.* 1954, p. 2840.
85 T. G. Ryss and T. A. Gribonova, *Zhur. fiz. Khim.* 1955, **29**, 1822.
86 E. Hayek and A. Czaloun, *Mh. Chem.* 1956, **87**, 790.
87 Unpublished observations.
88 D. D. DesMarteau and G. H. Cady, *Inorg. Chem.* 1966, **5**, 1829.
89 G. Barth-Wehrenalp, *J. inorg. nucl. Chem.* 1956, **2**, 266.

90 A. Engelbrecht and B. Stall, *Z. anorg. Chem.* 1957, **292**, 20.

91 H. Bartels and E. Class, *Helv. chim. Acta*, 1962, **45**, 179.

92 A. Engelbrecht and F. Sladky, *Mh. Chem.* 1965, **96**, 159.

93 A. Engelbrecht and A. V. Grosse, *J. Am. chem. Soc.* 1954, **76**, 2042.

94 H. Atzwanger, quoted by A. Engelbrecht, *Angew. Chem. (Int. ed.)*, 1965, **4**, 643.

95 L. J. Belf, *Chemy Ind.* 1955, p. 1296.

96 G. A. Olah and S. J. Kahn, *Chem. Abstr.* 1961, **55**, P 16924h.

97 W. Lange, ch. 3 in *Fluorine Chemistry*, vol. 1, ed. J. H. Simons. Academic Press, 1950.

98 J. R. Maurey and J. Wolff, *J. inorg. nucl. Chem.* 1963, **25**, 312.

99 G. L. Gard, F. B. Dudley and G. H. Cady, in *Noble-gas Compounds* (ed. H. H. Hyman), p. 109. Chicago University Press, 1963.

100 J. N. Brazier and A. A. Woolf, *J. chem. Soc. (A)*, 1967, p. 99.

101 N. E. Miller and E. L. Muetterties, *J. Am. chem. Soc.* 1964, **86**, 1033.

102 G. C. Kleinkopf and J. M. Shreeve, *Inorg. Chem.* 1964, **3**, 607.

103 B. L. Earl, B. K. Hill and J. M. Shreeve, *Inorg. Chem.* 1966, **5**, 2184.

AUTHOR INDEX

Figures in bold type indicate pages on which references are listed

Abedini, M., 151, 156, 157, 162, **171**, **172**
Abragam, A., 221, **231**
Adams, D. M., 295, **302**
Adams, R. W., 222, **231**
Addison, C. C., 9, **13**, 234, **260**
Aignesberger, A., 338, 353, 356, 358, **360**
Akitt, J. W., 38, **63**
Albrecht, A. H., 92, **112**
Alexander, P. W., 319, **325**
Amma, E. L., 21, **36**
Amster, R. L., 40, 41, **63**
Anderson, F. A., 90, **111**
Anderson, F. E., 21, **35**
Anderson, H. H., 93, **112**, 158, 160, 169, **172**, **174**
Anderson, J. S., xxi, **xxviii**, 150, **170**, 263, 264, 266, 267, 270, 271, **282**
Anderson, P. W., 176, 194, 195, 196, 199, **229**
Andreades, S., 106, **113**
Andrews, T. D., 151, 152, **171**
Ang, H. G., xxvii, **xxx**
Angelov, S., 67, **85**
Anisimov, K. N., 284, 285, 286, 296, **300**, **301**
Antipin, P. F., 153, **171**
Antonova, A. B., 284, **300**
Appel, R., 96, **112**
Ariya, S. M., 270, **282**
Attaway, J. A., 92, 94, **112**
Atwell, W. H., 155, 165, **172**, **173**
Atzwanger, H., 358, **362**
Aubke, F., 351, 352, 353, 356, **361**
Aubrey, N. E., 168, **174**
Aylett, B. J., xxiv, **xxix**, 149, **170**, 284, 296, 297, **300**
Aynsley, E. E., 338, 351, **361**
Ayscough, P. B., xxiv, **xxix**

Baay, Y. L., 153, 169, **171**, **174**
Baaz, M., 66, 75, 76, 78, 82, **84**, **85**, 86
Babko, A. K., 80, **86**
Bagga, M. M., 322, **325**
Baikie, P. E., 322, **325**
Bailar, J. C., 283, 288, **301**, 305, **323**
Baird, M. C., 285, 288, **301**

Bak, B., 90, **111**
Bald, J. F., jun., 153, **171**
Ballard, L. F., 300, **302**
Ballhausen, C. J., 184, **229**
Banford, T. A., 169, **174**
Banks, A. A., xxiii, **xxix**, 115, **135**
Banks, R. E., 106, **113**, 137, **147**
Bannister, E., 212, 213, **230**
Banus, J., xxiv, **xxix**
Barclay, G. A., 28, **36**, 215, **230**, 314, **324**
Barkelew, C. H., 316, **325**
Barr, D. A., 92, 98, 99, **112**
Barr, J., 331, 343, 344, 356, **360**, **361**
Barraclough, C. G., 176, 196, 222, **228**, **231**
Barrat, S., 81, **86**
Barrinok, M. S., 80, **86**
Bartell, L. S., 29, **36**
Bartels, H., 357, **362**
Barth-Wehrenalp, G., 358, **361**
Bashford, L. A., xxi, **xxviii**
Basolo, F., 31, **36**, 287, **302**, 320, **325**
Basushkim, A. A., 238, **260**
Bayles, J. W., 75, **85**
Beattie, I. R., 6, 8, **12**
Belf, L. J., 358, **362**
Bell, T. N., 122, 125, 126, **136**
Bennett, F. W., xxiv, **xxix**, 96, **112**, 115, **135**, 143, 145, **147**
Beretka, J., 320, 321, **325**
Bergman, J. G., jun., 34, **36**
Bernal, J. D., 273, **282**
Bernstein, H. J., 23, **36**
Berry, R. S., 20, **35**
Bertini, I., 319, **325**
Bertrand, J. A., 190, **229**
Besson, A., 152, **171**
Bethe, H. A., 192, **229**
Bevan, W. I., 125, 126, 129, 130, 131, 132, **136**
Bigelow, L. A., 92, 94, **112**
Bigorgne, M., 284, 294, 297, **300**, **302**
Birchall, J. M., 93, **112**, 132, 133, 134, **136**
Birchall, T., 238, **261**, 337, 348, **360**, **361**
Birckenbach, L., 88, **111**

Blake, A. B., 190, **229**
Bleaney, B., 177, **229**
Blundell, T. L., 291, **302**
Bobtelsky, M., 80, 81, **86**
Bockris, J. M., 247, **261**
Bohunovsky, O., 78, 79, 81, 82, 83, **86**
Boisbaudran, F. lecoq de, 37, 40, **63**
Bolles, T. F., 67, **85**
Bonati, F., 58, **64**, 284, 285, 294, 297, **301**, **302**, 322, **325**
Bond, A. C., 38, **63**, 156, **172**
Bond, G. C., **302**
Boorman, P. M., 214, **230**
Borelius, G., 271, **282**
Borer, K., 152, **171**
Bormann, D., 90, **111**
Bossert, E. C., 143, 145, 146, **147**
Bott, R. W., 149, **170**
Boucher, L. J., 305, **323**
Bowers, K. D., 177, **229**
Boyer, S., 37, **63**
Bragg, W. L., 263, **282**
Braid, M., 330, 338, **361**
Brandt, A., 151, **170**
Brandt, G. R. A., 115, **135**
Braun, R. L., 308, **323**
Brazier, J. N., 344, 346, **362**
Breisacher, P., 60, **64**
Bresler, S., 272, **282**
Breu, R., 284, **300**
Briegleb, G., 65, **84**
Brinckman, F. E., 166, **173**
Brintzinger, H., 103, **113**
Briscoe, H. V. A., xxi, **xxviii**
Britton, D., 7, **12**
Brode, W. R., 80, **86**
Brooker, R. E., 234, **260**
Brooks, E. H., 285, 295, 298, **301**
Brosset, C., 176, 214, **228**
Brown, H. C., 61, **64**
Brunneck, E., **85**
Bryan, R. F., 290, 291, **302**
Buffagny, S., 77, **85**
Bullen, G. J., 187, 188, **229**
Bunch, G. M., 143, 145, **147**
Bunger, F. L., 215, 216, **230**
Burg, A. B., 44, 48, 50, 52, **63**, **64**, 74, **85**
Burton, R. A., 134, **136**
Bush, M. A., 290, 291, **302**
Buswell, A. M., 234, **260**
Butler, G. B., 116, 119, **135**

Cady, G. H., xxvi, **xxx**, 330, 331, 332, 334, 335, 336, 338, 340, 341, 342,

351, 352, 353, 356, 357, **359**, **360**, **361**, **362**
Calfee, J. D., 331, **359**
Calderazzo, F., 323, **325**
Caldwell, D. J., 311, **324**
Calloman, H. J., 90, **111**
Calvin, M., 316, **325**
Campbell, G. W., 74, **85**
Campbell, J. M., 284, 296, 297, **300**
Campigli, U., 315, **325**
Carberth, E., 165, **173**
Cardin, D. J., 284, 288, **301**
Cariati, F., 294, **302**
Caron, A., 21, **36**
Carter, R. P., jun., 20, **35**
Cass, R. C., 140, **147**
Cavasino, F. P., 312, 313, **324**
Cavell, K. G., xxv, **xxx**
Cenini, S., 284, 285, 294, 297, **301**, **302**
Chadha, S. L., 238, 240, 241, **260**
Chakravorty, A., 304, 308, 310, 311, 313, **323**, **324**
Chalk, A. J., 118, **135**, 284, 299, **300**, **301**
Chambers, R. D., 143, **147**
Chandler, P. J., 295, **302**
Chandrasekar, S., 276, **282**
Chao, T. H., 168, **174**
Chao, T. S., 116, **135**
Charbnau, H. P., 281, **282**
Chatt, J., 283, 285, 299, **300**, **301**
Chaus, I. S., 38, **63**
Cheeseman, T. P., 317, **325**
Cheesman, G. H., xxi, **xxviii**
Cherneyshev, E. A., 116, 119, **135**
Chetham-Strode, A., jun., 21, **36**
Child, H. R., 194, **229**
Ciampolini, M., 26, **36**, 310, 311, 312, 313, 315, **324**, **325**
Cini, R., 310, **324**
Clark, H. C., xxv, **xxx**, 4, 6, 8, 9, 10, **12**, **13**, 297, 299, **302**, 311, **324**, 338, **360**
Class, E., 357, **362**
Clearfield, A., 239, **261**
Clifford, A. F., xxiv, **xxix**
Coates, G. E., 25, **36**, 92, **112**, 140, 143, **147**, 283, **300**
Cocking, P. A., 6, **12**
Coetzee, J. F., 251, **261**
Coffmann, D. D., 94, 95, **112**, 331, 338, **360**
Collmann, J. P., 23, **36**, 284, 288, 289, 300, **300**, **301**
Colten, E., 234, **260**

Comisarow, M. B., 349, **361**
Connett, J. E., 143, **147**
Connor, J. A., 164, **173**
Conway, B. E., 247, **261**
Cook, C. D., 19, **35**
Cook, D. I., 120, **136**
Cooper, G. D., 154, 155, **171**
Coppley, M. J., 234, **260**
Corey, E. R., 291, **302**
Corrigan, J. F., 37, **63**
Coslett, V. E., 90, **111**
Costa, G., 323, **325**
Cotton, F. A., 19, 34, **35**, **36**, 78, **85**, 91, **111**, 190, 213, 227, **229**, **230**, **231**
Cotton, J. D., 289, 290, 293, 294, 297, **302**
Cotton, J. L., 75, **85**
Cottrell, J. L., 164, **173**
Cox, A. P., 165, **173**
Coyle, T. D., 166, **173**
Craig, A. D., 151, 156, 157, 158, **170**, **172**
Craig, D. P., 33, **36**, 161, 163, **173**
Cramer, R. D., 285, 287, 288, 291, 293, 299, **301**, 331, **360**
Cross, R. J., 285, 298, **301**
Crowfoot, D. M., 4, **12**
Cullen, W. R., xxvi, **xxx**
Cupas, C. A., 349, **361**
Curtis, N. F., 305, **323**
Czaloun, A., 336, 338, 339, 340, 341, 355, **360**, **361**

Daggett, H. M., 256, **261**
Dale, J. W., xxiv, **xxix**, 92, 96, **112**
Dalton, J., **302**
Damerell, A. G. H., xxi, **xxviii**
Danders, C., 150, **170**
Danielian, A., 184, **229**
Davidson, I. M. T., 164, **173**
Davies, A. G., 143, **147**, 287, 294, **301**
Dawson, L. R., 253, 256, **261**
Deacon, G. B., 5, **12**, 143, **147**
Delfino, J. J., 331, 342, 352, **359**
Deno, N. C., 350, **361**
DesMarteau, D. D., 331, 342, 356, **359**, **361**
Dev, R., 238, 239, **260**
deWet, J. F., **230**
Di Vaira, M., 310, 315, **324**, **325**
Didtschenko, R., 153, **171**
Diehl, P., 238, **260**
Dighe, S. V., 286, **301**
Dimitriev, M. A., 331, **360**
Dirac, P. A. M., 176, 179, **228**

Dittmann, O., 38, **63**
Doak, G. O., 10, **13**
Dobbie, R. C., xxvii, **xxx**
Donohue, J., 21, **36**
Douglas, C. M., 97, **112**
Downey, W. E., xvii, **xxviii**
Downs, A. J., xxv, **xxx**, 25, **36**, 88, 104, 105, **111**, **113**, 140, **147**
Drago, R. S., 4, 6, 9, **12**, **13**, 66, 67, **84**, **85**, 142, **147**, 238, **260**
Drake, J. E., 151, 157, 163, **171**, **172**
Drako, D. F., 80, **86**
Dresdener, R. D., 92, **112**
Dubicki, L., 211, 212, 213, **229**, **230**
Dudek, G. O., 308, **324**
Dudley, F. B., 331, 334, 335, 338, 340, 341, 351, 357, **360**, **361**, **362**
Dunitz, J. D., 25, **36**, 230
Dunlap, R. D., 116, **135**
Dunmire, R., 116, 119, **135**
Dunn, M. J., xxvi, **xxx**
Dunn, T. M., 77, **85**
Durrell, W. S., 92, **112**
Dyckes, G. W., 116, 119, **135**
Dyer, G., 31, **36**

Eaborn, C., 149, 161, 169, **170**, 285, 299, **301**
Earl, B. L., 342, 352, **362**
Earnshaw, A., 217, 218, 221, **230**
Eaton, D. R., 311, **324**
Ebsworth, E. A. V., xxiv, xxv, xxvi, **xxix**, **xxx**, 21, **36**, 149, 162, 163, 166, 167, 169, **170**, **173**, **174**
Eckstrom, H. C., 253, **261**
Edens, W. L., 330, **359**
Edmondson, R. C., 287, **302**
Eggers, D. J., 331, 351, **360**
Ehlers, K., 155, **172**
Ehlert, T. C., 153, **171**
Ehrmann, K., 339, **360**
Eichhorn, G. L., 305, **323**
El-Shamy, K. H., xxiv, **xxix**
Elder, R. C., 190, 227, **229**
Elias, L., 256, **261**
Elliott, N., 217, **230**
Eméléus, H. J., xvii–xxx, 74, **85**, 92, 96, 97, 98, 101, 104, **112**, **113**, 115, **135**, 137, 140, 143, 145, **147**, 150, 151, 152, 156, 168, **170**, **171**, **172**, 330, 338, **359**, **360**
Endicott, J. F., 305, **323**
Engelbrecht, A., 330, 338, 353, 356, 358, **360**, **362**
Englin, M. A., 90, **111**

Evans, A. G., 75, **85**
Evans, J. C., 23, **36**
Evans, P. R., xxvi, **xxx**
Everett, G. W., jun., 304, 305, 311, 313, **323, 324**

Fanning, J. C., 176, 210, 212, 213, 214, 218, **228**, 321, **325**
Farrar, T. C., 166, **173**
Farrow, S. G., 134, **136**
Faulks, J. N. G., 58, **64**
Fawcett, F. S., 90, **111**
Feder, R., 281, **282**
Fee, W. W., 320, **325**
Fehér, F., 152, 156, **171, 172**
Fenkart, K., 78, 79, 81, 82, **86**
Fessenden, J. S., 149, **170**
Fessenden, R., 149, **170**
Fields, R., 120, 134, **136**
Fieser, L. F., 168, **174**
Fieser, M., 168, **174**
Figgis, B. N., 176, 189, 210, 218, 219, 221, **228, 229**
Fild, M., 97, **112**
Findeiss, W., 60, **64**
Fine, D. A., 80, **86**
Finholt, A. E., 38, **63**, 155, **172**
Fink, W., 336, 338, **360**
Finney, G., 164, **173**
Fischer, H., 151, **170**
Fishwick, G., 125, 127, **136**
Flitcroft, N., 284, 285, 290, 297, **300**
Floriani, C., 323, **325**
Florio, P. A., 331, **359**
Foex, G., 218, 221, **230, 231**
Forbes, G. S., 93, **112**, 158, 160, **172**
Forrester, J. D., 35, **36**
Fournier, L., 152, **171**
Fowler, R. H., 264, **282**
Fox, A. P., 154, **171**
Fox, M. R., 308, **324**
Fox, W. B., 331, **360**
Fraenkel, G., 238, **261**
Franconi, C., 238, **261**
Frank, F. C., 272, **282**
Frank, S., 93, **112**
Frankiss, S. G., 167, **174**
Franz, G., 331, **360**
Fraser, G. W., 41, 42, **63**
Frasson, E., 308, 309, **323, 324**
Freedman, L. D., 10, **13**
Freeman, L. P., 152, 154, **171**
Freiburg, C., 309, **324**
Friedberg, S. A., 218, **230**
Fritchie, C. J., 286, 291, **301**

Fritz, G., 151, **170**
Fruit, R. E., 350, **361**
Fujimoto, H., 118, **136**

Gamble, E. L., 151, 169, **170, 174**
Gard, G. L., 357, **362**
Gardner, E. R., xxii, **xxviii**
Garner, C. S., 212, **230**
Garton, W. R. S., 37, **63**
Gast, E., 60, **64**
Gaylord, N. G., 39, **63**
Gaziera, G. B., 92, 97, **112**
Gerding, H., 168, **174**
Gerloch, M., 317, **325**
Gershzon, T. P., 109, **113**
Geyer, A. M., 116, 119, **135**
Gibbon, G. A., 151, **171**
Gibbs, H. H., 330, **359**
Giguère, P. A., 332, 346, **360**
Gijsman, H. M., 218, 221, **230, 231**
Gilbert, A. R., 154, 155, **171**
Gilbreath, W. P., 330, 331, 338, 351, 352, 353, **359**
Gill, N. S., 77, **85**
Gillard, R. D., 287, 288, 289, 293, 294, **301**
Gillespie, R. J., 29, **36**, 238, **261**, 331, 334, 335, 336, 337, 338, 343, 344, 347, 348, 352, 354, 356, **360, 361**
Gilman, H., 165, **173**
Gilmore, G. N., 134, **136**
Gilson, T., 8, **12**
Ginsberg, A. P., 214, 215, 216, 224, 226, 227, **230, 231**
Ginter, M. L., 37, **63**
Glaser, H., 320, **325**
Glemser, O., 90, 97, **111, 112**
Glen, G. L., 19, 21, 26, 33, **35**
Glockling, F., 283, 284, 285, 295, 298, **300, 301**
Goddard, D. R., 336, 340, **360**
Goel, R. G., 6, 8, 9, 10, **12, 13**
Gokhale, S. D., 151, 152, 156, 163, 166, 168, **171**
Goldwhite, H., 143, 145, **147**
Goodenough, J. B., 193, 194, 201, 202, **229**
Goodwin, H. A., 318, **325**
Gopal, R., 234, **260**
Gorsich, R. D., 283, 290, 297, **300**
Graff, H., 217, **230**
Graham, W. A. G., 9, **13**, 48, **64**, 284, 285, 287, 290, 294, 295, 296, **300, 301, 302**
Grakauskas, V., 90, **111**

Gramstad, T., 330, **359**
Gray, B. F., 29, **36**
Green, J. H. S., 5, **12**, 143, **147**
Green, L. G., 164, **173**
Green, N., 298, **302**
Green, R. W., 319, **325**
Greenwood, N. N., xxiv, **xxix**, 37, 38, 39, 40, 41, 42, 43, 44, 45, 46, 47, 48, 49, 50, 51, 52, 53, 54, 55, 56, 57, 58, 59, 60, 61, **62**, **63**, **64**, 214, **230**
Greer, W. N., 2, **12**
Grewe, F., 111, **113**
Gribonova, T. A., 355, **361**
Griffin, R. N., 330, **359**
Griffiths, J. E., 11, **13**, 20, **35**
Grobe, J., xxvi, **xxx**
Grosse, A. V., 358, **362**
Groth, H. H., 92, 94, **112**
Guggenberger, L. J., 294n., **301**
Guggenheim, E. A., 264, **282**
Gunn, S. R., 164, **173**
Gunning, H. E., 151, **171**
Gustin, S. T., 216, **230**
Gutmann, V., xxiii, **xxix**, 66, 67, 68, 69, 71, 74, 75, **76**, 77, 78, 79, 81, 82, 83, **84**, **85**, **86**

Haas, A., xxv, xxvi, **xxx**, 92, 97, 98, 99, 100, 101, 103, 104, 106, 108, 111, **112**, **113**
Haber, C. P., 97, **112**
Hagen, A. P., 296, **302**
Hall, D., 309, 316, 317, 318, **324**, **325**
Hall, J. R., 21, **36**
Haller, J. F., 90, **111**
Hamilton, W. C., 7, **12**, 23, **36**
Hampel, G., 75, 78, 82, 83, **85**, **86**
Harbourne, D. A., 284, 285, 290, 297, **300**
Harckness, A. C., 256, **261**
Harris, C. M., 211, 212, 213, 215, **229**, **230**, 308, 310, 314, **324**
Harris, G. S., xxv, **xxx**
Harris, J., 92, **112**
Harris, J. F., 109, **113**
Harrod, J. F., 118, **135**, 284, 299, **300**, **301**
Haseda, T., 217, **230**
Hass, H. B., 168, **174**
Haszeldine, R. N., xxiii, xxiv, **xxix**, 92, 93, 96, 97, 99, 106, **112**, **113**, 115, 116, 117, 119, 120, 122, 123, 125, 126, 127, 129, 130, 131, 132, 133, 134, **135**, 137, 143, 145, **147**, 164, **173**, 330, 331, 338, **359**

Hatfield, W. E., 215, 216, **230**
Hauptstein, M., 330, 338, **361**
Hawthorne, M. F., 44, 49, **64**
Hayek, E., 331, 336, 338, 339, 340, 341, 353, 355, 356, 358, **359**, **360**, **361**
Hayter, R. G., 140, **147**
Heal, H. G., xxii, **xxviii**
Hein, F., 2, **12**, 283, 284, **300**
Heine, V., 202, **229**
Heinrich, F., 154, 168, **171**
Heisenberg, W., 176, 177, 179, **228**
Heitler, W., 193, **229**
Hellier, M., **302**
Henle, W., 39, **63**
Herber, R. H., 7, **12**
Herrmann, W., 151, **170**
Hertwig, K. A., 152, **171**
Herzberg, G., 156, **170**
Hess, G. G., 164, **173**
Hetherington, G., 338, 351, **361**
Heuser, E., 283, **300**
Hieber, W., 283, 284, **300**
Hill, B. K., 342, 352, **362**
Hirsch, A., 156, **172**
Hirschmann, E., 151, **170**
Hoard, J. L., 19, 21, 26, 29, 33, **35**, **36**
Hocking, C. S., 312, **324**
Hoerenz, J., 339, **360**
Hoffmann, E. G., 58, **64**
Hoffmann, H., 7, **12**
Hollandsworth, R. P., 157, **172**
Hollenberg, I., 97, **112**
Holm, R. H., 227, **231**, 304, 305, 308, 309, 311, 313, 315, **323**, **324**, **325**
Holmes, R. R., 20, **35**
Hooton, K. A., 284, **300**
Horowitz, J., 221, **231**
Horrie, S., 322, **325**
Hoskins, B. F., 213, 215, **230**, 314, **324**
House, D. A., 305, **323**
Hoyano, J., 287, **301**
Hrotstowski, H. J., 38, **63**
Hübner, L., 83, **86**
Hughes, E. D., 336, 340, **360**
Hurst, G. L., xxvi, **xxx**
Husain, M. M., 234, **260**
Husk, G. R., 165, **173**
Husted, D. R., 92, **112**
Hütte, H. H., 38, **63**
Huttner, K., 88 n.

Ibekwe, S., 285, 289, 299, **301**, **302**
Ibers, J. A., 21, 22, 23, **36**, 293, **302**
Ijdo, D. J. W., 176, 214, **228**

Iles, B. R., 120, **136**
Ingle, W. M., 157, **172**
Ingold, C. K., 336, 340, **360**
Innes, K. K., 37, **63**
Irish, D. E., 11, **13**
Isenberg, S., 152, **171**
Itatani, H., 285, 288, **301**

Jaffé, H. H., 161, 163, **173**
James, F. W., xx, **xxviii**
Jamieson, J. W. S., 287, 289, **302**
Jander, G., 234, **260**
Janssen, M. J., 8, **13**
Janusonis, G. A., 319, **325**
Janz, G. J., 78, **85**
Jehn, W., 284, **300**
Jellinek, H. H. G., xxii, **xxviii**
Jenner, E. L., 285, 287, 288, 294, 299, **301**
Jetz, W., 284, 295, **300**
Job, P., 80, **86**
Joesten, M. D., 238, **260**
Johannesen, R. B., 166, **173**
Johannsen, T., 37, **63**
Johnson, B. F. G., 78, **85**
Johnson, W. C., 152, **171**
Jolley, L. J., xxi, **xxviii**
Jolly, W. L., 151, 152, 156, 163, 166, 168, **171**
Jonas, H., 337, 340, **360**, **361**
Jonassen, H. B., 176, 210, 212, 213, 215, 218, **228**, 321, **325**
Jones, A., 120, **136**
Jones, J. R., 75, **85**
Jorgensen, C. K., 314, **324**
Judd, G. F., 116, **135**

Kaczmarczyk, A., 155, **172**, **174**
Kahn, O., 284, 294, 297, **300**, **302**
Kahn, S. J., 358, **362**
Kahovec, L., 234, **260**
Kambe, K., 181, 182, **229**
Kanamori, J., 194, **229**
Kanda, E., 217, **230**
Kandathil, A. J., 319, **325**
Karantassis, T., 221, **231**
Kasai, N., 7, **12**
Kato, M., 176, 210, 212, 213, 215, 218, **228**, 321, **325**
Kautsky, H., 156, **172**
Keller, H., 156, **172**
Keller, O. L., jun., 21, **36**
Kellermann, K., 88, **111**
Kennedy, R. C., 152, 154, **171**
Kent, R. A., 153, **171**

Kepert, D. L., 34, **36**
Kerk, G. J. M. van der, 8, **13**
Kerrigan, V., xxiii, **xxix**, 115, **135**, 137, **147**
Kettle, S. F. A., xxv, **xxx**, 28, 29, **36**, 290, 295, **302**
Khan, I. A., 290, 295, **302**
Khandozhko, V. N., 286, 296, **302**
Khattak, M. A., 287, 288, **301**
Kida, S., 142, **147**
Kidd, J. M., 97, **112**, 331, 338, **360**
Kilbourn, B. T., 291, **302**
Kimura, K., 53, **64**
King, A., xxi, **xxviii**
King, R. B., 289, **302**
King, W. R., 212, **230**
Kingston, J. V., 287, 289, **301**, **302**
Kishita, M., 215, **230**, 314, 315, **324**, **325**
Kivenskaya, L. I., 103, **113**
Kiyoshi, Y., 7, **12**
Klauke, E., 111, **113**
Klein, C. H., 153, **171**
Klein, H. F., 60, 61, **64**
Kleinkopf, G. C., 342, **362**
Klemm, W., 176, 214, **228**
Kling, K. E., 103, **113**
Klug, W., 106, **113**
Knollmüller, K., 234, **260**
Knox, S. A. R., 284, 289, 290, 293, 294, **300**, **302**
Kobayashi, H., 217, **230**
Kober, E., 109, **113**
Koddebusch, H., 103, **113**
Koehler, W. C., 194, **229**
Kokot, E., 176, 211, 212, 213, 215, **228**, **229**, 314, **324**
Kolditz, L., 336, 347, **360**, **361**
Koller, W., 331, **359**
Kolobova, N. E., 284, 285, 286, 296, **300**, **301**
König, E., 218, **230**
Koubek, E., 215, 216, **230**
Kramers, H. A., 176, 194, **228**
Kratel, G., 300, **302**
Kraus, C. A., 2, **12**
Kriegsman, H., 7, **12**
Krismer, B., 340, 341, **360**
Kubo, M., 53, **64**, 215, **230**, 314, 315, **324**, **325**
Kuchen, W., xxiv, **xxix**
Kuhlbörsch, G., 156, **172**
Kühle, E., 111, **113**
Kuhn, St J., 95, **112**
Kumada, M., 149, 165, 169, **170**, **173**

Kumler, W. D., 234, **260**
Kummer, R., 290, **302**
Kunze, O., 76, **85**
Kuroda, Y., 315, **325**

Lacher, J. R., 263, **282**
Lagowski, J. J., xxv, **xxx**, 88, **111**, 138, 140, 141, 143, 144, 146, **147**
Lampe, F. W., 164, **173**
Lane, T. J., 319, **325**
Lange, W., 341, 342, **361**
LaPlaca, S. J., 23, **36**
Lappert, M. F., 284, 288, **301**
LaRoche, L., 238, **260**
Lauder, A., 140, 143, 145, 146, **147**
Laussegger, H., 71, **85**
Lawton, D., 213, **230**
Layton, A. J., 284, 288, **301**
Lee, B., 26, 29, **35**, **36**
Leedham, K., 116, 117, 119, **135**
Lefever, R. A., 158, **172**
Lehl, H., 37, **63**
Lehmann, H. A., 347, **361**
Lehné, M., 80, **86**
Leigh, G. J., 164, **173**
Leitmann, O., 78, 82, 83, **86**
Lemal, D., 283, **300**
Lenger, S. L., 308, 310, **324**
Lenskii, A. S., 332, 347, **360**, **361**
Lewis, J., 189, 210, 213, 217, 218, 219, **229**, **230**, 316, 317, 318, **325**
Libus, W., 78, 81, **86**
Lidiard, A. B., 184, **229**
Liehr, A. D., 20, **35**
Lincoln, D. N., 350, **361**
Lind, M. D., 26, 29, **36**
Lindley, P. F., 291, 292n., 293, **302**
Lindquist, I., 65, **84**
Lindsey, R. V., 285, 287, 288, 291, 293, 294, 299, **301**
Lingafelter, E. C., 308, 309, **323**, **324**
Linston, H., 247, **261**
Lions, F., 317, **325**
Lipscomb, R. D., 90, **111**
Liu, J. S., 350, **361**
Livingston, J. G., 143, **147**
Loeb, A. L., 199, 202, **229**
London, F., 193, **229**
Long, G. G., 10, **13**
Long, L. H., xxii, **xxviii**
Longuet-Higgins, H. C., 20, **35**
Lord, R. C., 24, **36**
Luhleich, H., 156, **172**
Luijten, J. G. A., 8, **13**

Lustig, M., 331, 337, 342, 351, **360**, **361**
Lythgoe, S., 119, **136**

Mabbs, F. E., 210, 213, **229**, 317, 318, **325**
McBee, E. T., 116, 119, **135**, 168, **174**
Maccoll, A., 161, 163, **173**
MacDiarmid, A. G., xxiv, **xxix**, 149, 151, 153, 155, 156, 157, 158, 159, 160, 162, 163, 166, 167, 168, 169, **170**, **171**, **172**, **173**, **174**, 296, **302**
MacDuffie, D. E., xxv, **xxx**
McGinnety, J. A., 61, **64**
Machin, D. J., 194, **229**
Mackay, K. M., xxv, **xxx**, 166, **173**
McKenzie, D. E., 74, **85**
Mackey, D., 213, **230**
Macvicar, K., 311, **324**
Macwalter, R. J., 81, **86**
Maddock, A. G., xxii, xxiii, xxiv, **xxviii**, **xxix**, 152, 156, 168, **171**, **172**
Maestroni, G., 323, **325**
Magee, R. J., 287, 288, **302**
Maggio, F., 312, 313, **324**
Magnusson, E. A., 33, **36**
Mains, G. J., 151, 154, **170**, **171**
Mairinger, F., 66, **85**
Malde, M. de, 58, **64**
Malhotra, K. C., 238, **260**, 339, **361**
Malkiewich, E. J., 239, **261**
Manning, A. R., 285, 286, **301**
Marcinkovsky, A. E., 78, **85**
Marconi, W., 58, **64**
Margrave, J. L., 153, 154, 168, **171**
Marklow, R. J., 115, 119, **135**
Martin, B. B., 319, **325**
Martin, D. F., 21, 35, 305, 319, **323**, **325**
Martin, K. V., 317, **325**
Martin, R. L., xxiv, **xxix**, 176, 194, 210, 211, 212, 213, 214, 222, 224, 226, 227, **228**, **229**, **230**, **231**, 308, 310, **324**
Marvel, C. S., 234, **260**
Masaguer, J. R., 75, 78, 82, 83, **85**, **86**
Mason, R., 187, 188, 213, **229**, **230**
Massey, A. G., 284, **300**
Mathieson, A. McL., 216, 217, **230**
Maung, M. T., 140, 141, **147**
Maurey, J. R., 358, **362**
Mawby, R. J., 291, **302**
Mayer, U., 71, **85**
Mayne, N., 298, **302**

Mazzei, A., 58, **64**
Meckbach, H., 150, **170**
Meek, D. W., 142, **147**, 238, **260**
Meininger, H., 2, **12**
Mellow, D. P., 216, 217, **230**
Merrill, C. J., 330, 342, **359**
Meter, W. van, 330, 331, 335, **359**
Meyer, J., 331, 332, **360**
Meyers, M. D., 93, **112**
Michael, K. W., 118, **136**
Michaud, H., 54, **64**
Middleton, J., 130, 131, **136**
Miles, G. L., xxiii, **xxix**
Miller, J. M., xxv, **xxx**
Miller, N., xxii, **xxviii**
Miller, N. E., 342, **362**
Mills, O. S., 322, **325**
Milne, J. B., 343, 344, 352, 354, 356, **361**
Mironov, V. F., 116, 119, **135**
Mitchell, J. W., 270, **282**
Mityureva, T. T., 38, **63**
Mödritzer, K., 50, **64**
Monteith, K. L., 300, **302**
Morelli, D., 284, 285, 294, 297, **301**, **302**
Mori, M., 217, **230**
Moriya, T., 212, **229**
Morosin, B., 309, **324**
Morozova, M. P., 270, **282**
Morrison, J. A., 154, **171**
Morrow, J. C., 300, **302**
Morris, J. H., 45, 58, **64**
Moss, J. H., xxiv, **xxix**
Motornyi, S. P., 103, 109, **113**
Mouneyrat, A., 168, **174**
Muetterties, E. L., 15, 18, 19, 20, 21, 22, 26, 33, 34, **35**, 76, **85**, 94, **112**, 293, **302**, 338, 342, **360**, **362**
Muller, E., 341, **361**
Mulliken, R. S., 165, **173**
Muney, W. S., 251, **261**
Murahashi, S., 322, **325**
Musgrave, W. K. R., 143, **147**
Muto, Y., 215, **230**, 314, 315, **324**, **325**

Nabi, S. N., xxv, **xxx**, 104, 108, **113**
Nannelli, P., 315, **325**
Nardi, N., 26, **36**, 315, **325**
Nay, M. A., 151, **170**
Nebergall, N. H., 156, **172**
Neimyshewa, A. A., 103, **113**
Nesbet, R. K., 199, 202, **229**
Nesmeyanov, A. N., 284, 285, 286, 296, **300**, **301**

Neuhaus, H., 37, **63**
Newlands, M. J., 116, 119, 120, 122, 125, 126, **135**, 287, 289, **302**
Ng, C. F., 176, 196, **228**
Nichols, L. D., 164, **173**
Nicolini, M., 80, **86**
Niki, K., 151, 154, **170**, **171**
Nishikawa, H., 312, **324**
Noftle, R. E., 330, 332, **359**
Nordman, C. E., 40, **63**
Nordwig, A., 39, **63**
Nöth, H., 39, **63**
Nunez, L. J., 305, **323**
Nuss, J. W., 155, **172**
Nyholm, R. S., 5, **12**, 19, 27, 28, 29, 31, **35**, **36**, 77, **85**, 161, 163, **173**, 194, **229**, 284, 288, **301**

O'Brien, R. J., 6, 8, 9, **12**, **13**, 311, **324**
O'Brien, T. D., 15, **35**
O'Connor, J. E., 291, **302**
O'Connor, M. J., 306, 307, 314, 321, **323**
Oehmke, R. W., 305, **323**
Ogg, R. A., 238, **260**
Oh, D. Y., 102, 108, **113**
Okawara, R., 7, 10, **12**, **13**
Olah, G. A., 95, **112**, 349, 350, 358, **361**, **362**
Oldham, C., 213, **230**
Olin, J. F., 103, **113**
Olszewski, E. J., 305, **323**
Onak, T. P., xxvii, **xxx**
Onyszchuk, M., xxiv, **xxix**, 161, 160, 163, **173**
Opitz, H. E., 156, **172**
Orchin, M., 286, **301**
Orgel, L. E., 25, **36**, 161, 163, **173**, 217, 221, **230**, **231**
Orioli, P. L., 308, 310, 311, 315, **324**, **325**
Orloff, D., 165, **173**
Orloff, H., 165, **173**
Oubridge, J. V., 334, 335, **360**

Packer, K. J., xxv, xxvi, **xxx**, 330, **359**
Pamil, S. S., 339, **361**
Pammer, E., 338, **360**
Panattoni, C., 308, 309, **323**, **324**
Pankratov, A. V., 90, **111**
Paoletti, P., 310, 311, **324**
Parish, R. V., 15, 18, 28, 31, **35**
Parkinson, A. R., 93, **112**
Parkinson, C., 125, 127, **136**

Parry, R. W., 37, 38, 40, 41, 42, 53, 54, **62**, **63**, 94, **112**
Parshall, G. W., 287, 288, 294, 299, **301**
Passino, H. J., 116, 140, **135**
Patil, H. R. H., 284, 285, 295, **300**, **301**
Patmore, D. J., 285, 287, 290, 294, 295, **301**, **302**
Patton, R. H., 96, **112**
Paul, I., 284, 285, 289, 290, 293, 294, 295, 296, 297, **300**, **302**
Paul, R. C., xxiv, **xxix**, 75, **85**, 115, **135**, 238, 239, 240, 241, **260**, 339, **361**
Pauling, L., 26, **36**
Pauling, P., 187, 188, **229**
Pauson, P. L., 322, **325**
Payne, D. S., xxii, **xxviii**
Peach, M. E., 99, 100, 104, **112**, **113**
Peachey, S. J., 2, **12**
Peacock, R. D., 338, **360**
Peake, J. S., 156, **172**
Pearson, R. G., 31, **36**, 320, **325**
Pearson, T. G., xxi, **xxviii**
Pecile, C., 80, **86**
Pejic, R., 155, **172**
Petrov, A. D., 116, 119, **135**
Petrov, K. A., 103, **113**
Pfannstiel, K., 103, **113**
Pfeiffer, P., 303, 304, 317, 320, **323**, **325**
Pfieffer, P., 238, **260**
Pfitzner, H., 317, **325**
Pflugmacher, A., 151, 155, 168, **170**, **172**
Phillips, C. S. G., 33, **36**, 151, 152, **170**, **171**
Phillips, W. D., 311, **324**
Pickard, A. L., 4, 5, 9, **12**, **13**
Pierce, O. R., 116, **135**
Pietsch, E., 37, **63**
Pietsch, G., 152, **171**
Piper, T. S., 283, **300**
Pischtschan, S., 7, **12**
Pitt, C. G., 300, **302**
Pittman, C. U., jun., 349, 350, **361**
Pitzer, K. S., 165, **173**
Plumb, J. B., 116, 119, 120, 122, 125, 126, **135**
Plummer, W. J., 93, **112**
Pneumaticakis, G. A., 284, 288, **301**
Pobloth, H., 283, **300**
Polynova, T. N., 10, **13**
Ponomarenko, V. A., 116, 119, **135**
Pope, W. J., 2, **12**

Popov, Y. G., 270, **282**
Porai-Koshits, M. A., 10, **13**
Porta, P., 291, **302**
Powell, H. B., 138, 140, 141, **147**
Powell, H. M., 4, **12**, 291, **302**
Powell, P., 152, **171**
Pratt, G. W., 199, 202, **229**
Pratt, J. M., 77, **85**
Preston, F. J., 322, **325**
Prewitt, C. T., 287, 288, 291, 293, **301**
Pritchard, H. O., 29, **36**
Proskow, St., 95, **112**
Pruitt, M. E., 253, **261**
Puerckhauer, G. W. R., 116, 119, **135**
Pugh, H., xxv, **xxx**
Purcell, K. F., 66, **84**
Purcell, R. H., xxi, **xxviii**
Puschmann, J., 336, 338, 339, 341, **360**
Pussol, M., 168, **174**

Quagliano, J. V., 142, **147**

Racky, G., 168, **174**
Raksha, M. A., 109, **113**
Ramaiah, K., 21, **35**
Randolph, C. L., 44, **63**
Rao, G. S., xxv, **xxx**
Ratcliffe, C. J., 330, 334, 335, **359**
Reed, R. L., 322, **325**
Rees, A. L. G., 272, **282**
Reichle, W. T., 6, 7, **12**
Reid, C., xxii, **xxviii**, 156, **172**
Renning, J., 168, **174**
Reubke, E., 354, **361**
Richards, A., 210, 213, **229**, 317, 318, **325**
Richards, G. W., 346, **361**
Richards, T. W., 37, **63**
Richter, T., 156, **172**
Riecke, C. A., 165, **173**
Rijnders, 168, **174**
Riley, H. L., xxi, **xxviii**
Ring, M. A., 152, 154, 157, 169, **171**, **172**, **174**
Ritter, D. M., 169, **174**
Rittersbacher, H., 96, **112**
Roberts, C. W., 116, 119, **135**
Roberts, E. R., xxi, **xxviii**
Roberts, J. E., 331, 335, 336, 342, 351, **360**
Roberts, P. D., 25, **36**
Robertson, G. B., 218, 219, **230**
Robertson, W. G. P., 312, **324**
Robin, M. B., 214, **230**
Robinson, D. W., 24, **36**

Robinson, E. A., 334, 335, 347, **360**, **361**
Robinson, P., 166, **173**
Robinson, P. J., 119, 120, 123, 125, 127, **136**, 164, **173**
Robinson, P. L., 338, 351, **361**
Robinson, S. R., xxii, **xxviii**
Robinson, W. R., 213, **230**
Robinson, W. T., 293, **302**
Rochow, E. G., 151, 153, **170**, **171**
Rocktäschel, C., 152, **171**
Rodebush, W. H., 234, **260**
Roesky, H. W., 90, **111**
Rohrman, I., 155, **172**
Roman, W., 37, **63**
Roper, W. R., 23, **26**, 284, 288, 289, 300, **300**, **301**
Ross, E. J. F., 38, 39, 43, 44, 45, 47, 48, 49, 50, 51, 52, 53, **63**, **64**
Roth, W. L., 265, 270, **282**
Rothenbury, R. A., 336, 338, 343, 344, 356, **360**, **361**
Rousseau, Y., 151, **171**
Roy, M. F., 234, **260**
Royen, P., 152, **171**
Rubin, L. C., 116, **135**
Rudolph, R. W., 94, **112**
Ruff, J. K., 44, 49, 61, 62, **64**, 284, **301**, 337, **360**
Ruff, O., 339, **360**
Rundle, R. E., 21, 24, 29, **36**
Ryan, J. W., 118, **135**
Ryss, T. G., 355, **361**

Saalfeld, F. E., 164, **173**
Sacconi, L., 308, 310, 311, 312, 315, 319, **323**, **324**, **325**
Sadasivan, N., 305, **323**
Sadykh-Zade, S. I., 116, 119, **135**
Sakurai, H., 165, **173**
Samoilov, O. Ya., 19, **35**
Sandhu, S. S., 75, **85**
Sasaki, Y., 53, **64**
Savenkova, N. I., 90, **111**
Savoie, R., 332, 346, **360**
Schaeffer, G. W., 61, **64**
Schäffer, C. E., 216, **230**
Scherhaufer, A., 68, 79, **85**, **86**
Scheringer, C., 309, **324**
Schiff, H., 303, **323**
Schiff, H. I., 256, **261**
Schlemper, E. O., 7, **12**
Schlenk, W., 168, **174**
Schlesinger, H. I., 38, 48, 61, **63**, **64**, 156, **172**

Schmeisser, M., 151, 155, **170**, **172**, 336, 338, **360**
Schmidbauer, H., 60, 61, **64**
Schmidbauer, K., **360**
Schmidt, M., 38, 39, 53, 54, **63**
Schoening, F. K. L., 176, 212, **228**, **230**
Schomberg, G., 58, **64**
Schott, G., 151, **170**
Schott, P., 100, 104, 111, **113**
Schramm, G., 331, 332, **360**
Schrauzer, G. N., 300, **302**
Schriempf, J. T., 218, **230**
Schumb, W. C., 24, **36**, 151, 153, 158, 168, 169, **170**, **171**, **172**
Schunn, R., 286, 291, 293, **301**
Schunn, R. A., 15, 18, 19, 21, 22, 34, **35**, 76, **85**
Schwarz, R., 150, 151, 154, 155, 168, **170**, **171**, **172**
Schwartzmann, M., 151, **170**
Schwebke, G. L., 165, **173**
Sears, P. G., 256, **261**
Sedgwick, R. D., 164, **173**
Seitz, F., 263, **282**
Semlyen, J. A., 152, **171**
Senior, J. B., 343, 354, **361**
Sergeev, V. V., 153, **171**
Seuferling, F., 37, **63**
Shaposhnikova, A. D., 332, 347, **360**, **361**
Sharp, D. W. A., 340, **361**
Sharpe, A. G., xxiii, **xxix**, 74, **85**, 340, **361**
Shaw, B. L., 283, **300**
Sheat, S. V., 309, **324**
Sheka, I. A., 38, **63**
Shemanina, V. N., 92, 97, 109, **112**, **113**
Sheppard, N., xxvi, **xxx**, 108, 109, **113**, 162, 163, **173**
Sherwood, R. C., 215, 224, 226, 227, **230**, **231**
Shier, G. D., 4, 6, 9, **12**, **13**
Shindo, M., 10, **13**
Shreeve, J. M., 330, 331, 334, 335, 342, 351, 352, **359**, **361**, **362**
Shriver, D. F., 37, 38, 40, 41, 42, 53, 54, **62**, **63**
Siebel, H. P., 156, **172**
Siegel, B., 37, 60, **63**, **64**
Silverman, P. R., 77, **85**
Silverton, J. V., 19, 21, 26, 33, **35**
Simmons, G. L., 309, **324**
Simmons, R. F., 123, 125, 127, **136**, 164, **173**
Simons, J. H., 116, **135**

Simons, P. B., 9, **13**, 284, 294, 295, **300**
Simpson, C. C., 151, 152, 157, **170**, **171**, **172**
Simpson, J. H., 96, **112**
Simpson, W. B., 9, **13**
Singh, J., 75, **85**
Skripkin, V. V., 285, **301**
Sladky, F., 330, 358, **362**
Slater, J. C., 202, **229**
Sleet, R. J., 319, **325**
Smart, J. S., 177, **229**
Smith, A. J., 28, 29, **36**
Smith, G. B. L., 75, **85**
Smith, J. D., xxv, **xxx**
Smith, J. G., 294, 295, 296, **302**
Smythe, L. E., xxiv, **xxix**
Sokolov, B. A., 116, 119, **135**
Sokolov, D. M., 90, **111**
Sokolova, E. S., 332, 347, **360, 361**
Sokolskiĭ, G. A., 331, **360**
Somieski, C., 152, 156, **171**, **172**
Sommer, L. H., 118, **136**, 149, 161, 164, 166, **170**
Sommerfeld, A., 192, **229**
Spandau, H., **85**, 234, **260**
Spanier, E. J., 151, 166, **171**, **173**
Speier, J. L., 118, **135**
Speight, J. G., 133, 134, **136**
Speroni, G. P., 26, **36**, 315, **325**
Spiegler, K., 80, 81, **86**
Sreenathan, B. R., 238, 240, 241, **260**
Stafford, O. F., 234, **260**
Stall, B., 358, **362**
Stear, A. N., 143, 144, **147**
Stecher, O., 37, **63**
Steele, W. C., 164, **173**
Stefani, L., 323, **325**
Steinberg, H., 176, 214, **228**
Steininger, A., 66, **85**
Stephenson, I. L., 164, **173**
Stephenson, J. A., 287, 289, **302**
Stephenson, N. C., 216, 217, **230**
Sternbach, B., 159, **173**
Stewart, K., xxi, **xxviii**, 151, **170**
Stiebeler, P., 156, 158, **172**
Stock, A., 151, 152, 154, 156, 157, 168, **170, 171, 172**
Stockler, H. A., 7, **12**
Stokland, K., 168, **174**
Stolberg, U. G., 285, 287, 288, 291, 293, 294, 299, **301**
Stone, F. G. A., xxiv, **xxix**, 48, **64**, 149, 164, **170**, **173**, 284, 285, 289, 290, 293, 294, 295, 296, 297, 298, **300, 302**

Stone, P. J., 243, **261**
Storr, A., 37, 38, 40, 41, 42, 43, 44, 45, 46, 47, 48, 49, 50, 51, 52, 53, 54, 55, 56, 57, 58, 59, 60, 61, **62, 63, 64**
Strack, H., 152, **171**
Straub, D. K., 142, **147**
Straughan, B. P., 41, 42, **63**
Strausz, O. P., 151, **171**
Sujishi, S., 163, **173**
Sutcliffe, T., 92, **112**
Sutton, L. E., 161, 163, **173**
Svec, H. J., 164, **173**
Swaminathan, K., 309, 311, **324**
Swarts, F., 91, **112**
Sweemer, A. de, 80, **86**
Sweet, R. M., 286, 291, **301**

Tamao, K., 149, 165, 169, **170**
Tanenbaum, M., 38, **63**
Tannenberger, H., 75, **85**
Tarrant, P., 116, 119, **135**
Tattershall, B. W., xxvi, **xxx**
Taugher, G., 323, **325**
Taylor, H. S., xvii, **xxviii**
Taylor, R. C., 40, 41, **63**, 94, **112**, 284, 288, **301**
Teller, U., 283, **300**
Templeton, D. H., 35, **36**
Tesi, G., 97, **112**
Thiel, R., 150, **170**
Thielert, H., 320, **325**
Thomas, L. E., 168, **174**
Thompson, H. W., 90, **111**, 243, **261**
Thompson, J., xxiv, **xxix**
Thompson, J. A. J., 284, 294, 295, **300**
Thompson, J. C., 153, 154, 168, **171**
Thompson, R. C., 331, 343, 344, 356, **360, 361**
Timms, P. L., 151, 152, 153, **170, 171**
Tipping, A. E., 130, 131, **136**
Tissington, P., 132, 133, **136**
Tittle, B., xxvi, **xxx**
Tobe, M. L., 19, 31, **35**, **36**, 284, 288, **301**
Tobias, R. S., 3, **12**
Tomkinson, E., 37, **63**
Tomula, E. S., 80, 81, **86**
Traube, W., 339, 354, **360, 361**
Trautz, M., 339, **360**
Tromm, W., 152, **171**
Trotter, J., 6, **12**
Tsaë, B., 221, **231**
Tsai, J. H., 297, 299, **302**
Tucker, P. M., 284, 285, 290, 297, **300**
Tufail, R., xxvi, **xxx**

Tullock, C. W., 94, 95, **112**
Turco, A., 80, **86**
Turnbull, D., 280, **282**
Turner, J. A., 350, **361**
Turner, J. L., 166, **173**
Tyree, S. Y., 142, **147**

Ubbelohde, A. R., 271, **282**
Ugo, R., 284, 285, 294, 297, **301, 302,**
    322, **325**
Ulkii, D., 300, **302**
Unni, A. K. R., 256, **261**
Urenovitch, J., 155, 156, 157, 158, **172**
Urry, G., 155, 169, **172, 174**

Van Dyke, C. H., 151, 156, 157, 158,
    159, 160, 166, 167, **171, 172, 173**
Van Niekerk, J. N., 176, **228, 230**
Van Vleck, J. H., 176, 178, 181, 193,
    228, 229
Van Wazer, J. R., 168, **173**
Varma, R., 165, **173**
Vasil'eva, A. S., 109, **113**
Vasisht, S. K., 339, **361**
Vastine, F. D., 284, 288, 289, **300**
Vaughn, J. W., 253, **261**
Vekris, S. L., 337, **360**
Venanzi, L. M., 19, 31, **35,** 291, **302**
Venkatasetty, H. V., 78, **85**
Verhoek, F. H., 253, **261**
Videla, G. J., xxv, **xxx**
Vlcek, A. A., 287, **302**
Vogel, H. W., 80, **86**
Voigt, D., 96, **112**

Waddington, T. C., 33, **36**
Wade, K., xxv, **xxx**
Wadsley, A. D., 271, **282**
Wagner, R. I., 48, **64**
Walaschewski, E. G., xxiv, **xxix,** 96,
    112, 115, **135**
Walden, P., 244, **261**
Walker, A., 9, **13**
Wallbridge, M. G. H., 37, 38, 40, 41,
    42, 48, 53, 54, 59, **62, 63**
Walmsley, J. A., 142, **147**
Walton, R. A., 316, **325**
Ward, L. G. L., 157, 158, 162, 168,
    172, 173
Waser, J., 176, 214, **228**
Waterman, H., 176, 210, 211, 213,
    228, 229
Waters, T. N., 309, 316, 317, 318,
    324, 325
Watson, W. H., 176, 214, **228**

Watt, L. A., 93, **112**
Weber, H. P., 291, **302**
Wegleitner, K. H., 77, **85**
Weigold, H., 312, 316, **324**
Welch, A. J. E., xxii, **xxviii**
Welch, Z. D., 168, **174**
Welcman, N., xxv, xxvi, **xxx**
Wells, A. F., 10, **13**
Welo, L. A., 218, 220, 221, **230**
Wessel, G. J., 176, 214, **228**
West, B. O., 306, 307, 312, 313, 314,
    316, 320, 321, **323, 324, 325**
West, R., 165, **173**
Weyenberg, D. R., 155, **172**
White, D. G., 151, **170**
Whitley, A., 211, **229**
Wiberg, E., 37, 38, 39, 53, 54, 61, **63,
    64,** 152, **171**
Wilhoit, E. D., 256, **261**
Wilkins, C. J., xxii, **xxviii**
Wilkinson, G., 19, **35,** 91, **111,** 212,
    213, **230,** 283, 284, 285, 287, 288,
    289, 293, 294, **300, 301, 302**
Wilkinson, M. K., 194, **229**
Williams, E. J., 263, **282**
Williams, H. J., 216, **230**
Williams, R. J. P., 33, **36**
Willis, B. T. M., 265, 270, **282**
Willmarth, W. K., 271, **230**
Wilzbach, K. E., 156, **172**
Wing, R. M., 78, **85**
Winter, G., 214, 222, **230, 231**
Wintgen, R., 156, **172**
Witz, S., 163, **173**
Wolff, J., 358, **362**
Wollan, E. O., 194, **229**
Wood, J. F., xxii, **xxviii**
Wood, J. H., 202, **229**
Woodall, G. N. C., 151, **171**
Woodward, L. A., 21, **36**
Woodward, P., 290, 291, 292n., 293,
    302
Woolf, A. A., xxiii, **xxix,** 327, 333,
    336, 338, 339, 340, 344, 346, 353,
    355, 358, **359, 360, 361**
Wright, C. M., 15, 18, 19, 22, 26, 33,
    34, **35**
Wucher, J., 218, 221, **230, 231**
Wunderlich, F., 339, **360**
Wychera, E., 66, 67, 68, **85**

Yakubovich, A. Ya., 90, **111**
Yamada, S., 312, **324**
Yarovenko, N. N., 92, 97, 103, 109,
    112, 113

Yasuda, K., 7, **12**
Yoshioka, T., 167, **174**
Young, J. A., 92, **112**
Young, J. C., 116, 119, 125, 126, 132, **135**
Young, J. F., 284, 287, 288, 289, 293, 294, **301**
Yvon, J., 221 **231**,

Zackrisson, M., 65, 66, **84**
Zakharova, M. Ya., 286, **301**
Zalkin, A., 35, **36**
Zappel, A., 337, **360**
Zeidler, F., 156, **172**
Zellhoefer, G. F., 234, **260**
Ziman, J. M., 271, **282**
Zuckerman, J. J., xxvi, **xxx**

SUBJECT INDEX

acceptor molecules, reactions of donor solvents with, 67–8

acceptor solvents, 74

acetamide, complexes of Lewis acids and, 243–4; as solvent, 233–61

acetate ions, in paramagnetic clusters, 175, 176, 183, 211–12, 227

acetates of alkali metals, in amides, 237, 257

acetic acid, in amides, 248, 252, 254, 255, 256; as donor solvent, 74; in fluorosulphuric acid, 339–40, 344

acetic anhydride, as donor solvent, 69

acetone, cobalt complexes in, 80; as donor solvent, 67, 70

acetonitrile, cobalt complexes in, 78, 84; as donor solvent, 67, 69, 70, 71, 256;
fluorosulphates in, 340; reference acceptor in, 71

acetylacetone, in paramagnetic clusters, 187, 205, 215, 226, 227; as skeleton for
Schiff-base complexes, 307, 319

acetyl chloride, as donor solvent, 68, 69, 70, 75

acid-base neutralization reactions, in amides, 258–60; preparation of fluorosulphates by, 339–40

acid halides, as donor solvents, 74

acids, protonic, *see* protonic acids

actinide elements, coordination numbers of, 15

alanes, amine derivatives of, 44, 45, 47, 58

alcohols, carbonium ions from, 349–50

alkanes, reactions of dichlorocarbene with, 133

alkanes, halogeno-, compared with silane analogues, 168

alkoxides, in paramagnetic clusters, 223

alkyl mercurials, compared with fluoroalkyl mercurials, 137–8, 139

aluminium, fluorides of, 31; fluoroalkyl mercury compounds of, 143–4; mixed
hydride of gallium and, 54, 61

aluminium bromide, in amides, 238

aluminium chloride, in amides, 238, 259

amides as solvents, 233–61; acid-base neutralization in, 258–60; conductance
studies in, 244–57; formation of solvates with, 238–40; infrared spectra of
solvates of Lewis acids and, 240–4; protonation of, in acid solution, 348;
solvolytic reactions in, 257–8

amines, compounds of, with fluoroalkyl mercurials, 141; with polyfluoroalkyl
silicon compounds, 120–1; with silanyl compounds, 162; with transition-
metal fluorosulphates, 342; with trifluoromethylsulphenyl compounds, 104

ammonium ion, hydrogen bonding in, 33; in paramagnetic clusters, 176, 202,
204

amphoteric behaviour, of formamides, 260; of iodine trifluorosulphate, 352

aniline, in paramagnetic clusters, 213; in Schiff-base complexes, 306

anionic complexes, in donor solvents, 70–2, 77

anions, interaction of, with organometallic cations, 8, 10–12; superexchange via,
in paramagnetic clusters, 210

ansolvo-acids, in amides, 258

antiferromagnetic interaction of electron spins, 181, 184, 195, 197; examples of,
203, 210, 213, 214, 215, 218, 221, 226, 227–8

anti-Markownikov addition, of halogenogallanes to olefins, 60

antimony, coordination number of, 16; melting of crystals of, 275; organo-
compounds of, 10

antimony bromide, as acceptor in reactions with donor solvents, 67

[376]

antimony fluorosulphate, 356
antimony pentachloride, as acceptor in reactions with donor solvents, 67; in
    amides, 238–46 *passim*, 259
antimony pentafluoride, 344–5; compounds of chloryl fluoride and, 333
antimony trichloride, as acceptor in reactions with donor solvents, 67; in amides,
    238, 245, 246; as pseudo-tetrachloride, 76
antimony trifluoride, 93, 344
antiprism structure, 16, 17; square, 17, 25, 28, 32
arsenic, coordination number of, 33; mixed hydride of silicon and, 163; tri-
    fluoromethyl compounds of, 96
arsenic pentafluoride, compounds of chloryl fluoride and, 333
arsenic trichloride, as acceptor solvent, 74; in amides, 238, 245
arsenic trifluoride, 344; sulphur trioxide and, 353
arsines, compounds of gallium hydride and, 53–4
aryl groups, in Schiff-base complexes, 311, 312–13
autocomplexes, of cobalt salts in different solvents, 73–6, 78, 79, 80, 81, 83
auto-ionization, of amides, 242, 252, 257, 258
autoprotolysis, in amide solvents, 242, 244, 247
azide ion, cobalt compounds with, in different solvents, 79–83; as ligand with
    reference acceptor, 71
azides, perfluoro-, 89

barium, coordination number of, 17
barium fluorosulphate, 355
bases (nitrogen), in amides, 237, 248–52, 258
bent structure, 4, 5, 16, 24
benzene, as donor solvent, 75
benzoic acid, in amides, 251, 254, 256; in fluorosulphuric acid, 344, 346
benzonitrile, as donor solvent, 68, 69
benzoyl chloride, as donor solvent, 68, 69, 75
benzylamine, in amides, 248, 249–50
benzyl cyanide, as donor solvent, 69
beryllium, coordination number of, 16, 25; in paramagnetic clusters, 227;
    Schiff-base complexes with, 319
bipyramid structure, hexagonal, 17; pentagonal, 16, 32; trigonal, 3, 16, 20, 22,
    23, 26, 32, 42, 161, 293
bismuth, melting of crystals of, 275
bismuth chloride, in amides, 238
bismuth fluorosulphate, 346
boiling points, of fluorides and fluorosulphates, 327–37
bond angles, 290–1, 293
bond lengths, 163, 291, 292–3
bonds, chemical
    boron–gallium, 61
    carbon–carbon, 253; –chlorine, 117; –cobalt, 323; –iodine, 117; –metal, 2, 3,
        297; –nitrogen, in amide complexes, 242; –silicon, 121, 122
    covalent and ionic, 2, 4, 6, 11–12, 24, 31
    delta, 212
    gallium–phosphorus, 51
    germanium–manganese, 297; –platinum, 298; –silicon, 151
    Heitler–London scheme of, 193
    hydrogen, 33, 233, 238, 243, 260
    manganese–manganese, 287n.; –tin, 297
    metal–halogen, 286; –metal, 287, 296, 297; –solvent, 73
    molybdenum–molybdenum, multiple, 213, 214

bonds, chemical (*cont.*)
  Rundle's 3-centre-4-electron, 29
  silicon–hydrogen, 146, 157; –oxygen, 122; –phosphorus, 163
  silicon–silicon, 149, 156, 162; cleavage of, 167–9; physical properties associated with, 163–7; synthesis of, 150–5
  tungsten–tungsten, 214
boranes, amine derivatives of, 44, 45, 48, 58; deca-, 61, 62; phosphine derivatives of, 50, 53; poly-, 29; pyridine-, 46
borides, 34
boron, coordination numbers of, 16, 34; mixed halides of, 91; mixed hydrides of gallium and, 61–2
boron trichloride, reactions of, with fluoroalkyl mercurials, 144; with silicon ethers, 160–2
boron trifluoride, 144
boron tri-iodide, in amides, 238
bromide ion, cobalt compounds with, in different solvents, 79–83; as ligand with reference acceptor, 71
bromine, bridges of, in paramagnetic clusters, 176, 196
bromine fluorosulphates, 353–4
bromine trifluoride, as acceptor solvent, 74; as base, 344; disproportionation of, 351; in fluorosulphuric acid, 339, 340; sulphur trioxide in, 353–4
bromogallanes, 61
bromonium ions, 353
bromo-silanes, 157
bromotrifluoromethane, 116
n-butylamine, in amides, 248, 250
tert-butylamine, in Schiff-base complexes, 307

cadmium, halogenoperfluoroalkyl mercury compounds of, 141
cadmium chloride, in amides, 239–40
cadmium formamide, amphoteric behaviour of, 260
caesium, coordination number of, 17; in paramagnetic clusters, 214
caesium halides, structure of, 33
calcium fluorosulphate, 346, 355
carbamyl compounds, trifluoromethylsulphenyl-, 98, 99
carbenes, halogeno-, as intermediates in breakdown of polyfluoroalkyl silicon compounds, 126, 127, 128, 129, 132; reactions of, 132–4
carbon, theoretical coordination number for, 25; *see also* bonds
carbon tetrachloride, as donor solvent, 74
carbonium fluorosulphates, 348–50
carbonylation, of Schiff bases, 322
carbonyl compounds, reactions of dichlorocarbene with, 132
carbonyl group of amides, formation of complexes through oxygen of, 240, 241, 244
carbonyls, alkyl-metal, conversion of, to acyl-metal, 286; transition-metal complexes with, 283–98 *passim*
carboxylate anions, bridges of, in paramagnetic clusters, 210, 211–13, 218–19, 221; in triorganotin complexes, 8
carboxylates of mercury, in synthesis of fluoroalkyl mercurials, 143
catalysts, coordination compounds as, 298–9
catenation, in inorganic silicon compounds, 149–74
cations, organometallic, 1–13; free, 9, 10, 11
cerium, coordination number of, 17, 35
chloride ion, cobalt compounds with, in different solvents, 79–83; as ligand with reference acceptor, 71

chlorides, inorganic, in amides, 236, 238; ionization of, in different solvents, 73

chlorine, addition of, across carbon–sulphur bond, 105; bridges of, in organolead compounds, 9, and in paramagnetic clusters, 176, 196, 214; coordination number of, 16; formation constants of complexes containing, in different solvents, 72, 76; and metal–metal bonds, 297

chlorobenzenes, as donor solvents, 75

chlorofluoromethylsulphenyl compounds, 104–13

chlorofluorosulphates, 338

chloroform, as donor solvent, 75

chlorogallanes, 60

chloryl fluoride, 333

chromium, coordination numbers of, 18–19; Group IVB derivatives of, 284; in paramagnetic clusters, 175, 176, 183, 202–22 *passim*; Schiff-base complexes with, 314

chromium fluorosulphate, 346

clusters, *see* paramagnetic clusters

cobaloximes, as coordination analogues of vitamin B_{12}, 300

cobalt, bond between carbon and, 323; complexes of, in different solvents, 77–83, 84; coordination numbers of, 16, 17, 34, 35; ferromagnetic, 193; Group IVB derivatives of, 284, 285, 287, 293, 294, 295, 296, 297; in paramagnetic clusters, 190, 205, 206, 224, 227; Schiff-base complexes with, 308, 310, 311–12, 315, 319, 320, 321

cobalt bromide, autocomplex formation by, 73

cobalt carbonyl compounds, 289, 290, 322

cobalt chloride, in amides, 238, 239

cobalt fluorosulphate, 346

conductance, of amides, 234; equivalent, of solvates in amides, 247–8; specific, of solvates in amides, (Lewis acids) 240, 244–8, (nitrogen bases) 249–52, (protonic acids) 252–6, (quaternary ammonium salts) 256–7

coordination numbers of atoms or ions, conditions affecting, 24; equal to number of bonding orbitals, 27–8; nature of central atom and, 26–9; of organometallic ions, 2, 3; unusual, 15–36

copper, coordination numbers of, 15, 16, 28; halogenoperfluoroalkyl mercury compounds of, 141; in paramagnetic clusters, 175, 176, 177, 190, 211, 215, 218, 228; Schiff-base complexes with, 303, 305–7, 308, 309, 314, 316, 319, 320, 321

copper(II) chloride, in amides, 239–40

copper formamide, amphoteric behaviour of, 260

copper fluorosulphate, 346, 355

cross-linking by silicon–silicon bonds in polysilanes, 156

crystal-field stabilization energies, for various coordination numbers, 31–2

crystals, aggregation of point defects in, and model for melting of, 263–81; lattice energy of, 33, 65, 237; of organometallic ions, 3–4, 7, 11; structure of, and participation of orbitals in bonding, 29; variation in structure of, for same Schiff-base complex, 309; *see also* solid state

cube structure, 17, 25, 28

cuboctahedron structure, 17

cyanides, of alkali metals in amides, 236; cobalt compounds with, in different solvents, 77–83; in coordination compounds, 33; in paramagnetic clusters, 210

cyclohexylamine, in Schiff-base complexes, 306

cyclo-octadiene, in Group IVB complexes, 291

cyclopentadienyl, in Group IVB complexes, 291, 293; in paramagnetic clusters, 214

cyclopropanes, from reaction of dichlorocarbene and olefins, 134

cyclopropyl carbonium ions, 350

defluorination, of fluoroalkyl mercurials, 144, 145, 146

diamagnetism, 176, 194, 212, 216, 291, 352; of some Schiff-base complexes, 308, 309, 310–311, 316

diazoaminobenzene ion, in paramagnetic clusters, 213

diazonium fluorosulphates, 341

dichloroacetic acid, in amides, 248, 252, 253, 255, 256

dichlorocarbene, reactions of, 132–4

1,2-dichloroethane, as donor solvent, 69; as inert solvent, 66

dihalogenogallanes, 60–1

dielectric constants, of amide solvents, 237, 238, 256; and autocomplex formation, 73–4; as criterion of solvent, 74, 76, 253; of donor solvents, 69

diethyl ether, complex of lithium tetrahydridogallate and, 38; as donor solvent, 67, 69, 74

diethylether gallane, 53–4

difluorocyanamide, 93

difluoroazirine, 93

1,1-difluoroethyltrichlorosilane, rearrangement of, during thermal breakdown, 130–1

2,2-difluoroethyltrifluorosilane, kinetics of thermal breakdown of, 124–5

digermanyl compounds, 150, 166

dimers, of dihalogenogallanes, 60–1; of fluorosulphuric acid, 332; of NF radicals, 90; of trifluoromethylsulphenyl isocyanate, 98

NN-dimethylacetamide, cobalt complexes in, 82, 83; as donor solvent, 67, 69

dimethylamine gallane, infrared data for, 45

dimethylarsenic chloride, reaction of, with hexafluorodisilane, 169

NN-dimethyl formamide, complexes of Lewis acids and, 241–2; as donor solvent, 67, 69; as ligand with reference acceptor, 71; as solvent, 233–61

dimethylphosphine gallane, infrared data for, 51

dimethylsulphide gallane, 54

dimethyl sulphoxide, cobalt complexes in, 83, 85; as donor solvent, 67, 68, 69, 73, 74, 75; as ligand with reference acceptor, 71; as 'selective' solvent, 72

diphenyl phosphonic chloride, as donor solvent, 69

dipolar coupling, in paramagnetic clusters, 210

dipole moments, of amides, 234; of solvates in amides, 240, 244

direct (Heisenberg or potential) exchange, in paramagnetic clusters, 197

disilanes, see under silanes

disilanyl compounds, 157, 158, 162; as analogues of ethyl compounds, 150

disilyl compounds, 150

dispersion attractions, and coordination numbers, 33

disproportionation reactions, of halogen fluorides, 351; of silanes, 154, 155; of trifluorosulphates, 356

dissociation constants, of vanadyl acetoacetonate with different ligands, 71

dodecahedron structure, 17, 28

donor number, donor solvents, donor strength, see under solvents

double bonds, additions across, 105, 116, 118–20, 351, 352; between fluorine and sulphur, 358–9

electron diffraction, in gas phase, 22–3; time scale for, 20

electronegativity, of aluminium/gallium/boron sequence, 44; of anion, and coupling in paramagnetic clusters, 196; of fluoro- and perfluoroalkyl mercurials, 138–9; of ligands, and coordination numbers, 33

electroneutrality principle, 26–7

electron-pair repulsions, and stereochemistry, 29

electrons, delocalized, 18; non-bonding valence, 31; unshared pair of, considered as ligand, 76

electron spin resonance, 177, 346; time scale for, 20
electron spins in paramagnetic clusters, coupling of, 176, 177–80, 191; interaction of, 176, 181–91 (*see also* ferromagnetic *and* antiferromagnetic interactions); levels of, correlation diagrams for, 183, 187, 191; polarization of, 197
electrostatic effects, on coordination numbers, 24, 25, 30–3
energetics, of interconversion of systems with unusual coordination numbers, 18
enthalpies, effect of physical state on, 31
equilibrium constants, between donor solvents and acceptors, 68
ethers, silicon, 158–9; reactions of, with boron trichloride, 160–2
ethyl acetate, as donor solvent, 67
ethylacetoacetate, as donor solvent, 67
ethyl alcohol, cobalt complexes in, 81
ethylene carbonate, as donor solvent, 68, 69
ethylene-diamine, halogenoperfluoroalkyl mercury compound with, 140–1; Schiff-base complexes with, 305, 315
ethylene sulphite, cobalt complexes in, 79–80; as donor solvent, 68, 69, 76
ethyl fluorosulphate, 341
exchange integral, for metal–metal interaction in paramagnetic clusters, 181, 191, 192–3, 196, 200

ferri- and ferro-cyanides, in amides, 256, 257
ferromagnetic interaction of electron spins, 181, 184, 196, 197; examples of, 204, 205, 224, 227
ferromagnetism, theory of, 193
ferrous oxide, point defects in, 265
fluoride anion, 194–5
fluorides, comparison of, with fluorosulphates, 327–37; reactions of, with sulphur trioxide, 338; of triorganotin, 6–7; variety of structure of solid complexes of, 33; volatile, prepared with fluorosulphuric acid, 358
fluorine azide, 90, 91
fluorine cyanate, 90
fluorine cyanide, 90, 91
fluoroalkyl mercury compounds, 137–47; coordination compounds of, 139–43; as synthetic intermediates, 143–7
fluoroalkyl and perfluoroalkyl groups, electronegativities of, 138–9
(poly)fluoroalkyl silicon compounds, 115–36; nucleophilic attack on, 120–1; polymers of, 121–3; synthesis of, 116–20; thermal breakdown of, 123–32
fluoroborates, resemblances of fluorosulphates to, 357–8
fluorocarbonyl cyanide, 95
(poly)fluoro-olefins, addition of silicon–hydrogen across double bond of, 116, 118–20; from breakdown of fluoroalkyl silicon compounds, 121, 123, 126; from reactions of fluoroalkyl mercury compounds, 144, 145, 146
fluorophosphoryl pseudohalides, 93–4
fluoropolysilanes, organo-substituted, 153
fluoropseudohalides, 87–103
fluoroselenic acid, 357
fluoroselenium fluorosulphate, 336–7
fluorosilyl isocyanates, 93
fluorosulphates, carbonium, 348–50; comparison of, with fluorides, 327–37; di- and mono-, comparison of, 334; halogeno-, 351–4; hydrolytic stability of, 354–5; of metals, 357–9; perfluoro-, 355; preparation of, 337–42; source of radicals of, 340–2, 351; thermal stability of, 355–7
fluorosulphuric acid, in amides, 238, 248, 251, 256; anhydride of, 355; comparison of, with hydrofluoric acid, 332, 343, and sulphuric acid, 343, 347; displacement reactions in, 338, 339; reactions of, 342

fluorothioformyl cyanide, 95
formamide, complexes of Lewis acids and, 242–3; as solvent, 233–61
formate ions, in paramagnetic clusters, 210, 212, 213
formation constants, of chloro-complexes in different solvents, 72, 76; of compounds of reference acceptor with different solvents, 66–7
formic acid, in amides, 256
four-centre reaction, 120
four-centre transition states, 125, 126

gadolinium, coordination number of, 16; ferromagnetic, 193; melting of crystals of, 275
gallanes, amine derivatives of, 40–7; arsine derivatives of, 53; deuterium-substituted, 40, 43, 45, 47, 49, 55, 57; halogeno-, 55–61; oxygen and sulphur derivatives of, 53–4; phosphine derivatives of, 47–53
gallates, tetrahydrido-, 38–9
gallium, organo-compounds of, 4; mixed hydrides of, 61–2
gallium hydride, 37–8, 59–60; see also gallanes
gallium triazide, 54
gallium tribromide, 38
gallium trichloride, 38
gas phase, structural analysis of compounds in, 22
geometry, idealized, of systems with unusual coordination numbers, 16–17
germanium, mixed hydrides of, 151, 152; organo-compounds of, 6, 7; transition-metal derivatives of, 283–300 passim
germanyl halides, 166
germyl compounds, 150
glycine, in Schiff-base complexes, 305–6
gold trifluoride, 344
Grignard compounds, per- or polyfluoroalkyl, 116, 117–18; reactions of, with Group IVB complexes, 297

hafnium, coordination numbers of, 25, 34
halide solvents, 74
halides, anionic complexes containing, 71; cobalt compounds with, in different solvents, 77–83; fluoroalkyl mercury compounds with, 140; inorganic, in amides, 236; ionization of, and donor number of solvents, 73; of organo-metallic compounds, 2, 4
halogenocarbenes, 132
halogenofluorosulphates, 351–4
halogenogallanes, 55–61
halogenoperfluoroalkyl mercurials, 140–1
halogenosilanes, 153, 155, 156–8, 166, 168–9
halogens, bridges of, in organometallic compounds, 4, 6, 7, 9, 11, and in paramagnetic clusters, 176, 196; exchange of, in thermal breakdown of poly-fluoroalkylhalogenosilanes, 129–31; latent heats of vaporization plotted against melting temperatures of, 280; mixed fluorosulphates of, 351
heptafluoroiodopropane, 117
hexafluoroethane, insertion of sulphur trioxide into, 332
hexamethylphosphoramide, basicity of P=O group in, 70; as donor solvent, 69, 70; as ligand with reference acceptor, 71
'hydration numbers', 19
hydrochloric acid, in amides, 248, 254
hydrogen, reaction of Group IVB complexes with, 298
hydrogen bonds, 33, 233, 238, 243, 260
hydrogen bromide, in amides, 238

hydrogen chloride, in amides, 238
hydrogen fluoride, compared with fluorosulphuric acid, 332, 343; in fluoro-
    sulphuric acid, 339, 344, 346; as solvent, 233; system of sulphur trioxide
    and, 342–7
hydrogenation, promoted by transition metal ions and complexes, 299
hydrolysis, of fluorosulphates, 354–5; solvation of halides in water as, 73;
    stability of polysiloxanes towards, 122
hydrosilation, of fluoro-olefins, 116, 118–20
hydroxonium ion, 354, 355

icosahedron structure, 17, 28
imines, reactions of dichlorocarbene with, 133
indium, coordination number of, 21; organo-compounds of, 4
indium fluorosulphate, 346
infrared spectroscopy, of amides and solvates in amides, 238, 240–4; of carbon
    and silicon ethers, 159; of chlorofluorofloromethylsulphenyl compounds,
    106; of fluorosulphates, 358–9; of gallanes, 40–60; of Group IVB deriva-
    tives, 294; of organometallic compounds, 3–11 passim; of perfluoroalkyl
    mercury compounds, 141–2; of ruthenium complexes, 23; time scale for, 20
insertion reactions, 286–7, 337
iodide ion, cobalt compounds with, in different solvents, 79–83; as ligand with
    reference acceptor, 71
iodides, inorganic, in amides, 236; ionization of, in different solvents, 73
iodine, coordination number of, 16, 29; as acceptor in reactions with donor
    solvents, 67
iodine fluorosulphates, 351–3, 356
iodine heptafluoride, 20, 22
iodine pentafluoride, 333
iodine trichloride, as pseudopentachloride, 76
iodo-complexes, conditions for preparation of, 76
iodonium ions, 352
iodyl fluorosulphate, 353
iridium, Group IVB derivatives of, 284, 287, 288, 291, 294; Schiff-base com-
    plexes with, 322
iridium pentafluoride, 344
iron, coordination numbers of, 16, 22, 34, 35; in paramagnetic clusters, 217,
    219–21, 221–3; Schiff-base complexes with, 318–19, 320
iron carbonyls, 22, 284, 290, 322
iron(III) chloride, in amides, 238, 246
iron(II) oxide, point defects in, 265, 266, 270
isobutyric acid, in amides, 248, 255
isobutyronitrile, as donor solvent, 69
isocyanates, perfluorinated, 92, 93, 94, 98, 99
isomerization, of olefins by platinum–tin complexes, 299; as possible cause of
    polymorphism, 21
isomers, time scale for determination of molecular structure by separation of, 20

β-ketoamines, in Schiff-base complexes, 310, 311, 313

lanthanum, coordination numbers of, 17, 26, 29, 34, 35
lanthanum trifluoride, 22
lead, coordination number of, 17; organo-compounds of, 3, 9–10; transition-
    metal derivatives of, 283–300 passim
lead fluorosulphate, 346
Lewis acids, in amides, 237, 239–48, 258; compounds of chloryl fluoride with, 333

ligands, cobalt complexes with, in different solvents, 77–83; effects of, on coordination numbers, 24–6, 27, 33–5; exchange reactions of, 65–70; number of, and number of bonding orbitals, 29; repulsion between, and coordination numbers, 31, 33
linear structure, 16, 24, 190, 191, 224; of organometallic groups, 3, 6, 7, 9
liquid crystals, 281
liquid state, determination of coordination properties in, 21, 23–4
lithium, per- and polyfluoroalkyl compounds of, 116, 117
lithium tetrahydrido compounds, 38–9
lutetium, coordination numbers of, 25

magnesium fluorosulphate, 355
magnesium halides, in amides, 238, 239–40
manganese, coordination number of, 16, 23; Group IVB derivatives of, 284, 285, 287, 290, 291, 296, 297–8; in paramagnetic clusters, 201; Schiff-base complexes with, 310, 320
manganese carbonyls, 287, 297
manganese(II) chloride, in amides, 239–40
manganese fluorosulphate, 346
melting, defect aggregation mechanism for, 271–81
melting points, of fluorides and fluorosulphates, 327–37; plotted against latent heats of vaporization, 279, 280
mercury, coordination numbers of, 16, 19, 25, 28; dimethyl-, 20; fluoroalkyl compounds of, 115, 137–47; pentafluorophenyl compounds of, 143
metals, borides and silicides of, 34; carbonyls of, 287; fluorosulphates of, 357–9; in fluorosulphuric acid, 346; interaction between, in paramagnetic clusters, 176–231; organo-cations of, 4–10; Schiff bases and, 319–21; transition, see transition metals
methoxide, sodium, in amides, 259
N-methyl acetamide, as donor solvent, 256
methyl acetate, as donor solvent, 67
methyl benzenes, protonation of, 348–9
methylene chains, in Schiff-base complexes, 316–17
methylene chloride, reference acceptor in, 71
microdomains, in crystals, 270
microwave spectroscopy, time scale for, 20
molecular structure, time scales for techniques for determining, 20
molybdenum, coordination numbers of, 17, 18, 19, 20, 21; Group IVB derivatives of, 285, 286, 287; in paramagnetic clusters, 213, 214
monochloroacetic acid, in amides, 238, 248, 252, 253, 255, 256
Mössbauer resonance, 7; time scale for, 20

neodymium, coordination number of, 17
(neophyl)₃ tin fluoride, 6–7
neutron diffraction studies, on defects in crystals, 270; time scale for, 20
nickel, coordination numbers of, 17, 28; ferromagnetic, 193; Group IVB derivatives of, 285, 294; halogenoperfluoroalkyl mercury compounds of, 141; in paramagnetic clusters, 187–90, 193, 199, 204–5, 205, 207–8, 224–6; in perovskite crystal, 194–5; Schiff-base complexes with, 305, 308, 309, 310, 315, 316, 319, 320
nickel chloride, in amides, 238, 239
nickel fluorosulphate, 355
niobium, coordination numbers of, 16, 21, 34
niobium fluorosulphate, 346
nitrate ion, as ligand, 34–5; in complexes with organometallic compounds, 8, 10

nitrates of alkali metals, in amides, 236
nitrites, reactions of dichlorocarbene with, 133
nitroalkanes, as solvents, 75
nitro-aromatics, in fluorosulphuric acid, 344, 346
nitrobenzene, as donor solvent, 68, 69, 75
nitrogen, coordination number of, 16
nitromethane, cobalt complexes in, 77–8, 84; dielectric constant of, 77; as donor solvent, 68, 69, 71, 76, 256; as 'levelling' solvent, 72
nitrosyl chloride, as solvent and chloride-ion donor, 74–5
nuclear magnetic resonance spectroscopy, of amides, 238, 242; of chloro-fluoromethylsulphenyl compounds, 108; of fluorosulphates, 336, 337, 347, 353; of gallium hydride and derivatives, 48; of silicon compounds, 131, 159, 163, 166–7; use of, restricted by stereochemical non-rigidity, 23
nucleophilic attack, on silicon in polyfluoroalkyl compounds, 120–1, 125, 129

octahedron structure, 3, 21, 22, 32, 79, 80, 81, 202, 227
olefins, addition of dichlorocarbene to, 134; catalysis of double-bond migration in, 299; in synthesis of per- and polyfluoroalkyl silicon compounds, 116, 118–20
onium complexes, in donor solvents, 73
optical isomers, stereochemical non-rigidity and, 20–1
orbitals, bonding, coordination number and, 27–8, 29; ionization energies of, 27; overlap of, 197–9, 350; in paramagnetic clusters, 179, 194, 196, 200–9
organometallic cations, 1–13
organometallic compounds, reactions of carbenes and, 134
oxidation, state of, and coordination numbers, 31
oxides, anions of, as bridges in paramagnetic clusters, 216–17
oxyfluorides, 357

palladium, coordination number of, 28; Group IVB derivatives of, 285, 287, 288, 295; Schiff-base complexes with, 308
paramagnetic clusters, metal–metal interaction in, 175–231; binuclear, 182–5; dimers, 211–16; energy-level schemes for, 220, 222, 223, 225; magnetic dipolar coupling in, 210; magnetic moment v. temperature in, 189, 225; magnetic susceptibility v. temperature in, 176–7, 185, 211, 216, 218, 221, 222; spin–spin coupling and exchange effect in, 176, 177–80, 191; spin–spin interaction in, 176, 181–91; superexchange in, 176, 193, 202–9, 210, 226; tetramers, 227–8; tetranuclear, 190–1; trimers, 218–26; trinuclear, 186–90
paramagnetism, of some Schiff-base complexes, 308, 309, 310–11
pentafluoroiodoethane, 115
pentafluorophenyl mercurials, 143
pentafluorophenyl pseudohalides, 92–3, 97
pentafluorophenyl thallium halides, 5
perchlorates, resemblances of fluorosulphates to, 357–8
perchloric acid, in fluorosulphuric acid, 344
perchloryl fluoride, 357
perfluoroalkyl silicon compounds, see fluoroalkyl-
perfluoro cyanides and cyanates, 89
perfluoropseudohalides, 87–103
perovskite, superexchange in, 194–6
peroxydisulphuryl difluoride, as source of fluorosulphate radicals, 340–1, 351; reactions of, 342
perrhenates, resemblances of fluorosulphates to, 357–8
phenol, as acceptor in reactions with donor solvents, 67; in amides, 255
phenonium ions, 350

phenyl phosphonic dichloride, as donor solvent, 67, 69, 75
phenyl phosphonic difluoride, as donor solvent, 67, 69
phosphines, compounds of gallium hydride and, 47–53; compounds of transition
    metals and, 293, 297
phosphorus, coordination numbers of, 16, 19, 20; fluorinated pseudohalides of,
    93; fluoroalkyl compounds of, 96, 97, 146; silane compounds of, 163
phosphorus oxychloride, basicity of P=O group in, 68; as donor solvent, 66,
    67, 68, 75; reaction of fluoroalkyl mercurials with, 145
phosphorus oxy-trifluoride, 356
phosphorus pentafluoride, 20, 22
phosphorus trichloride, reaction of, with fluoroalkyl mercurials, 145
phosphorus trifluoride, 145
phosphoryl compounds, fluoro- and fluorochloro-, 95
picolines, in amides, 256
picric acid, in amides, 256
piperidine, in amides, 248, 249–50
pivalate anions, in paramagnetic clusters, 227
planar structure, 16; of organometallic groups, 3, 7, 9; of Schiff-base complexes,
    308, 310–11, 313; of triphenylamine group, 53
platinum, coordination numbers of, 25, 28, 293; Group IVB derivatives of, 284,
    287, 288, 291, 293, 294, 295, 298, 299
platinum hydrides, anionic, 293
platinum tetrafluoride, 344
point defects in solid state, 263; aggregation of, as model of melting, 271–81;
    interaction of, 263–6; premonitory action of, 270–1; statistics of, 266;
    types of, 264, 265, 272
polarizability, of ligands, and coordination numbers, 31, 33
polyfluoro-, see fluoro-
polymers, of gallium hydride, 59; of halogenosilanes, 151; of polyfluoroalkyl
    silicon compounds, 121–3, 125; of silicon difluoride, 153; of thiocarbonyl
    fluoride, 105; of triorganotin compounds, 6, 9
potassium, in paramagnetic clusters, 214
potassium fluorosulphate, 346, 355
prism structure, 16, 17
propane-1,2-diolcarbonate, cobalt complexes in, 78–7, 84; as donor solvent, 67,
    69, 75
propionic acid, in amides, 248, 254, 255
propionitrile, as donor solvent, 69
protonation, of amides in acid solution, 348; in fluorosulphuric acid, 344; of
    methyl benzenes, 348
protonic acids, in amides, 237, 248, 252–6
pseudohalides, cobalt complexes with, 77–83; perfluoro-, 87–103
pseudohalogens, 88, 351
pseudo-interhalogens, 351
pyramid structure, 16; square, 16, 22, 32
pyridine, in amides, 248, 249–50; attached to Schiff-base complexes, 312, 313;
    as donor solvent, 67, 69, 71, 73, 74; as ligand with reference acceptor,
    71; organometallic complexes with, 4, 6, 10; in paramagnetic clusters, 176,
    213
pyridine-borane, 46
pyridine-gallane, 46
pyridine N-oxide, perfluoro mercury compound of, 141, 142
pyrolysis, of fluoroalkyl silicon compounds, 123–32
pyrosulphuric acid, in amides, 248, 254, 256
pyrosulphuryl fluoride, 355

quaternary ammonium salts, in amides, 237, 256–7
quinoline, in amides, 248, 249–50

radical-chain mechanisms, 119, 123
radical reactions, preparation of fluorosulphates by, 340–1
radius-ratio rule, for disposition of an ionic aggregate, 24–5
Raman and infrared data, for fluorosulphuric acid, 346, 347; for gallane compounds, 41; for systems in solution, 23; time scale for, 20
randomization, in disilanyl halides, 157–8
rare gases, latent heats of vaporization plotted against melting temperatures of, 280
reactivity, chemical, of Group IVB transition metal complexes, 296–9
redox reactions, in fluorosulphuric acid, 346
rhenium, coordination number of, 16, 17; Group IVB derivatives of, 284, 286; in paramagnetic clusters, 213
rhenium carbonyl, 286, 296
rhenium hydride, 20
rhodium, coordination number of, 34; Group IVB derivatives of, 284, 287, 288, 289, 294, 296; in paramagnetic clusters, 213; Schiff-base complexes with, 322
rhodium carbonyl, 294
rubbers, silicone, 122
ruthenium, coordination number of, 16, 23; Group IVB derivatives of, 284, 287, 288, 294; infrared spectroscopy of complexes of, 23; in paramagnetic clusters, 216–17
ruthenium carbonyl, 294

salicylaldehyde, in Schiff-base complexes, 303, 305, 307
salicylaldimines, reduction of, 321; in Schliff-base complexes, 303, 307, 309, 310, 313, 314, 317
salicylic acid, in amides, 251, 256
samarium, coordination number of, 16
Schiff bases, coordination compounds of, 303–25; in paramagnetic clusters, 210, 215–16
selenium, coordination number of, 16
selenium oxychloride, as donor solvent, 69, 75
selenium oxyfluoride, 357
selenocyanates, 89, 103
selenyl compounds, trifluoromethyl-, 92, 96, 97
silanes, alkylpolychlorodi-, 155; bromo- and bromodi-, 157; difluoroethyltrichloro-, 130–1; difluoroethyltrifluoro-, 124–5; hexahalogenodi-, 151, 155, 168, 169; hexaphenyldi-, 168; organo-substituted fluoropoly-, 153; perhalogenopoly-, 153, 168; poly-, 121–3, 155–63; synthesis of poly-, 150–2; tetrafluoroethyltrihalogeno-, 127; tetraphenyl, tetravinyl, 290; tribromo-, 151
silenes, 153, 154
silicides, 34, 150, 152, 156
silicon, catenation in inorganic compounds of, 149–74; coordination numbers of, 16, 23–4, 25; hydrides of, see silanes; mixed hydrides of, 151, 152, 163; perfluoro-, isocyanates of, 93; polyfluoroalkyl compounds of, 115–36; reaction of per- and polyfluoroalkyl halides with, 116–17; transition-metal derivatives of, 283–300 passim; organo-compounds of, 6
silicon chlorides, 150–1
silicon difluoride, 153
silicon tetrabromide, in amides, 238

silicon tetrafluoride, 117, 153; in amides, 238
silicone rubbers, 122
siloxanes, polyfluoroalkylpoly-, 121–3, 124
siloxene, 156
silsesquioxanes, 121–3
silylamines, 162–3
silyl compounds, 150; trifluoro, 92, 93
silver fluorosulphate, 346, 355
silver tetrahydridogallate, 39
sodium fluorosulphate, 346, 355
solid state, aggregation of point defects in, 263–82; organometallic ions in, 1–2, 3–4, 8; systems with unusual coordination numbers in, 18, 21; see also crystals
solubilities, of inorganic compounds in amides, 234, 236–8; of sulphates in sulphuric acid and fluorosulphates in fluorosulphuric acid, 358
solutions, organometallic ions in, 1, 2–3; coordination numbers in, 19
solvation, in amide solvents, 237, 238–40; coordination numbers in, 19, 23; followed by ionization, 73; of organometallic ions, 1, 3, 6, 9, 19; of solute by donor solvent, 65; in water, as hydrolysis, 73
solvents, amides as, 233–61; donor, coordination chemistry in, 65–86; donor numbers of, 66–74; 'levelling' and 'selective', 72; relative donor strength of, 65–6
solvo-acids, solvo-bases, in amides, 258
solvolysis, in amides, 257–8, 259; in fluorosulphuric acid, 356, 357; in strong acids, 233; in trimethylphosphate, 81
specific heat, up to melting point, 272, 281
specific volume, up to melting point, 281
spectroscopy, see infrared, microwave, Raman, ultra-violet/visible, and vibrational/rotational spectroscopies
stability constants, for Schiff-base complexes, 319
stannyl compounds, 150
stereochemical non-rigidity, 19–20, 23
steric factors, and coordination numbers, 15, 24–6, 34; in fluoroalkyl silicon compounds, 129; and optical isomers, 21; in Schiff-base complexes, 307, 311, 313; and solvent coordination, 74
strontium fluorosulphate, 355
sulphates, diorganothallium, 5
sulphenyl compounds, chlorofluoromethyl-, 102–3, 105, 108–10; trifluoromethyl, 97–9, 104–5, 106–7
sulpholane (tetramethylene sulphone), as donor solvent, 68, 69, 76; perfluoroalkyl mercury compound of, 141, 142
sulphur, coordination number of, 16
sulphur fluorides, 94
sulphuric acid, in amides, 248, 251, 254, 256; compared with fluorosulphuric acid, 343, 347; as solvent, 233
sulphur trioxide, in amides, 238, 239, 240, 241, 242, 243, 246; in preparation of fluorosulphates, 337–8; system of hydrogen fluoride and, 342–7
sulphuryl chloride, as donor solvent, 69, 75
(poly)sulphuryl fluorides, 347
supercooling, 280
superexchange (kinetic exchange), in paramagnetic clusters, 193–9, 202–9, 210

tantalum, coordination numbers of, 17, 21, 34
tantalum fluorosulphate, 346
tantalum pentafluoride, 344

tellurium chloride, in amides, 238
tellurium fluorosulphates, 357
temperature, and decomposition of fluorosulphates, 355–7; and extent of disorder in crystals, 278; and magnetic moment, 189, 225; and magnetic susceptibility, 176–7, 185, 211, 216, 218, 221, 222
tensiometric titrations, 41–2, 47
terbium, coordination number of, 17
terpyridyldichlorozinc, basic geometry of, 22
tetrafluoroethylene, photochemical reaction of trichlorosilane and, 119–20
1,1,2,2-tetrafluoroethyltrichloro- and trifluoro-silanes, thermal decomposition of, 127–8
tetrahedron structure, 77, 190, 227; proposed for triorganotin compounds, 2, 7; of Schiff-base complexes with cobalt and zinc, in equilibrium with planar structure, 310–11, 313; truncated, 17
tetrahydridogallates, 38–9
tetrahydrofuran, as donor solvent, 69; as solvent for fluoroalkyl aluminium compound, 144
tetramethylene sulphone, see sulpholane
tetrapseudohalogen halites, 351
thallium, organo-compounds of, 2–3, 4–5
thallium fluorosulphate, 346
thallium tetrahydridogallate, 39
thioamides, protonation in, 348
thiocarbonyl fluoride, 105
thiocyanate ion, cobalt compounds with, in different solvents, 77–83; as ligand with reference acceptor, 71
thiocyanates, inorganic, in amides, 236; perfluoro-, 89
thiol, trifluoromethyl-, 97, 98
thionyl chloride, as donor solvent, 69, 75
thorium, coordination numbers of, 25
tin, coordination numbers of, 25, 28; organo-compounds of, 2–3, 6–9, 67; tetravinyl-, 289; transition-metal derivatives of, 283–300 passim
tin(III) chloride, trans-activating ability of, 299
tin(IV) chloride and bromide, in amides, 238, 239, 240, 241, 244, 245, 246, 259
tin fluorosulphates, 346, 356
titanium, coordination numbers of, 16, 17, 28, 45; in paramagnetic clusters, 214
titanium(IV) chloride, in amides, 238, 240, 245, 246
titanium fluorosulphate, 356
titanium oxide, point defects in, 270
p-toluenesulphonic acid, in amides, 248, 253, 254
trans configuration, in Schiff-base complexes, 308, 313
transition metals, Group IVB derivatives of, 282–302; halogenoperfluoroalkyl mercury compounds of, 140–2; penta-coordinate complexes of, 22
tributyl phosphate, basicity of P=O group in, 69–70; as donor solvent, 67, 69, 75
trichloroacetic acid, in amides, 248, 252–3, 256
trichlorosilane, photochemical reaction of tetrafluoroethylene and, 119–20
triethylamine, in amides, 248, 250
triethylphosphine gallane, infrared data for, 52
trifluorocyanuric acid, 90
trifluoroethylene, 133
trifluoroethylidene, 133
trifluoroethyl trichlorosilane, thermal decomposition of, 127–8
trifluoroiodomethane, 115, 137
trifluoroethyl mercurials, compared with methyl analogues, 138
trifluoromethyl pseudohalides, 91–2

trifluoromethyl sulphuric acid, 332
trifluoropropyl silicone rubbers, 122
trimethylamine gallanes, 40–2; infrared data for, 41
trimethylamine halogeno gallanes, infrared data for, 56–8
trimethyl phosphate, basicity of P=O group in, 68; cobalt complexes in, 81–2, 84; as donor solvent, 67, 68, 69, 73, 74
trimethylphosphine gallane, infrared data for, 49
trimethyltin chloride, as acceptor in reactions with donor solvents, 67
triphenylchloromethane, as donor solvent, 72, 75
triphenylphosphine, organometallic complexes with, 4
triphenylphosphine gallane, infrared data for, 52–3
triphenylphosphorus oxide, as ligand with reference acceptor, 71
tungsten, coordination numbers of, 16, 17, 18, 35; Group IVB derivatives of, 284, 285; in paramagnetic clusters, 176, 214

ultra-violet/visible spectroscopy, time scale for, 20
uranium, coordination numbers of, 15
uranium fluorosulphate, 346
uranium oxide, point defects in, 265–6, 270
urea, fluorinated derivatives of, 98–100

vanadium, coordination numbers of, 21, 25; in paramagnetic clusters, 215–16
vanadium oxide, point defects in, 270
vanadyl acetylacetonate, as reference acceptor in reactions with donor solvents, 70–1
vaporization, latent heats of, plotted against melting temperatures, 279, 280; molar entropy of (Trouton constant), in fluorides and fluorosulphates, 329–37
vibrational/rotational spectroscopy, for gas phase, 22
vinyl fluoride, 124–5, 131
viscosity, of amide solvents, 247, 255

water, amide solvents compared with, 235; cobalt complexes in, 80; as donor solvent, 67, 74; melting of crystals of, 275; reaction of perfluorinated isocyanates with, 92

X-ray diffraction studies, determination of coordination numbers by, 21, 22; of Group IVB complexes, 290, 291, 294n.; up to melting point, 281; of organometallic compounds, 3, 7, 11; of paramagnetic clusters, 188; of Schiff-base complexes, 307; time scale for, 20
xenon, coordination number of, 16, 29
xenon difluoride, 357

zinc, coordination numbers of, 25; in paramagnetic clusters, 227; reaction of, with fluoroalkyl mercurials, 145; Schiff-base complexes with, 308, 310
zinc chloride, in amides, 239–40
zirconium, coordination numbers of, 16, 17, 25, 34
zirconium chloride, in amides, 238, 239, 246, 259